The Eocene oil shales of M

Messel

Messel

An insight into the history of life and of the Earth

Edited by

Stephan Schaal
Head of the Messel Section

and

Willi Ziegler
Director

*Senckenberg Society for Nature Research,
Frankfurt am Main, Germany*

Translated by Monika Shaffer-Fehre

CLARENDON PRESS · OXFORD
1992

Oxford University Press, Walton Street, Oxford OX2 6DP

Oxford New York Toronto
Delhi Bombay Calcutta Madras Karachi
Petaling Jaya Singapore Hong Kong Tokyo
Nairobi Dar es Salaam Cape Town
Melbourne Auckland

and associated companies in
Berlin Ibadan

Oxford is a trade mark of Oxford University Press

Published in the United States
by Oxford University Press, New York

First published (in German) by Waldemar Kramer, Frankfurt am Main 31988
German edition © Senckenbergischen Naturforschenden Gesellschaft, 1988
This translation © Oxford University Press, 1992

A catalogue record for this book is available from the British Library

Library of Congress Cataloging in Publication Data
(Data available)

ISBN 0 19 854654 8

Typeset by Joshua Associates Ltd, Oxford
Printed in Singapore by
Times Offset Ltd

Authors

Michael Ackermann
Jens-Lorenz Franzen
Jörg Habersetzer
Thomas Keller
Wighart von Koenigswald
Herbert Lutz
Norbert Micklich

Dieter Stefan Peters
Gotthard Richter
Stephan Schaal
Friedemann Schaarschmidt
Rainer Springhorn
Gerhard Storch
Michael Wuttke

Acknowledgements

Line drawings: K. Albrecht, G. Eder, J. L. Franzen, A. Helfricht, M. Hinkel, E. Junqueira, T. Keller, I. Lehnen, M. Micklich, M. Möller, H. Schäfer, M. Wuttke

Photographs: J. F. Bornhardt, E. Haupt, W. v. Koenigswald, G. Krebs, W. Kumpf, T. Martin, N. Micklich, G. Richter, S. Schaal, F. Schaarschmidt, M. Starck, G. Storch, H. Thiele-Pfeiffer, N. Wolz

Radiographs: J. Habersetzer, W. Stürmer, S. Tuengerthal

To all who helped in the recovery of and research into the invaluable fossils of Messel:

in particular, to the Federal Minister for Research and Technology; the Minister for Science and Art, Hessen; the Volkswagenwerk Foundation; and the Deutsche Forschungsgemeinschaft, whose financial support made the excavations possible and supported the research;

to the politicians who, in the conflict between science and refuse disposal decided in favour of science, as a testament for others;

to all who are interested, as a document setting out the excavation results so far; and as an appeal for the maintenance of access for research to the immeasurable and unique scientific treasure which still remains, unrecovered, in Messel.

Preface to the English edition

In the last few years Messel in special measure has experienced very fast-moving developments. On the one hand, the steady flow of new discoveries has been uninterrupted, so that new palaeontological results have been published continually. On the other hand, the future of the Messel pit has been at risk in a series of controversial decisions on waste disposal policies. This book, written between 1987 and July 1988, should not, however, be considered a final report. The literature listed in the references for this preface to the English edition reports on progress during the period 1988–90. It was, however, not possible to revise all chapters. In the meantime, the second edition of the original German version has been published.

Political developments reached a turning point on 13 February 1990. The Ministers for the Environment and for Science finally announced that this world-famous fossil discovery site would not be used for waste disposal. At the same time, however, concern about the stability of the walls of the mine came to the fore, and a plan emerged to put huge amounts of slag, which has come to be considered an economically valuable substance, on the oil shale containing the fossils as a stabilization measure. 'The stabilization measures will be kept to the absolute minimum, and the interests of science will be taken into account in a comprehensive way . . .' said the politicians. Thus the discussion still continues and, as there is often a wide gap between theory and practice, future events concerning the Messel pit must be monitored with the utmost care by palaeontologists. Finally and fortunately the Government of the State of Hesse acquired the Messel pit in order to allow long-term scientific utilization.

The scientific discussion on the fossil site and the new excavation results is a much more cheerful subject. In April 1987, an international symposium on research into the Messel fossil site was held at the Senckenberg Museum in Frankfurt, Germany and, in 1988, a symposium volume was published by J. L. Franzen and W. Michaelis, the organizers of the symposium. This volume contains the most recent geological, geochemical, and palaeontological data. With a view to achieving an analysis and understanding of the sedimentation model and the biosphere of Messel as it existed approximately 50 million years ago, dozens of geological profiles have provided important information about the stratigraphic classification of the excavation sites (Möller, in preparation). In work concluded in 1989, Goth (1990) demonstrated that the high organic content of the Messel oil shale can be mainly traced back to the resistant cell walls of the monocellular green alga *Tetraedron minimum* (Chlorophyceae). They grew in large numbers in annual blooms so that, as a general rule, the oil shale is finely laminated and even consists in part of pure *Tetraedron* layers. The formation and preservation of the finely laminated sediments indicates the existence at that time of a relatively deep meromictic lake whose solids were being continously deposited in layers. Other algae like *Botryococcus* and diatoms were also present but did not play a significant role. Wilde (1989) conducted comprehensive research into the taxonomy of the often very well preserved leaves. He describes 65 species which were subjected to cuticular analysis, indicating that, in terms of leaves with cuticular structure, the flora of Messel is one of the most richly varied of the early Tertiary period.

Systematic and palaeoecological research on insects has continued, and the analysis of fossilized ants concluded (Lutz 1990). Our excavation team made some spectacular finds in the summer of 1989. We have been able to find proof of probably autochthonous dwellers of the parts of the lake nearest to the bank, i.e. water-skaters and waterboatmen; we also found taxa of water-fleas (crustaceans: order Cladocera; fossilized ephippia have been discovered).

Birds are found regularly in the course of excavations at Messel. With great excitement we are investigating the find of a small bird, which in some features is reminiscent of sparrows (order Passeriformes), in others of small Coraciiformes. It has many features in common with the still extant Todi birds of the Caribbean islands; what makes the bird particularly interesting is that the origin of the Passeriformes has so far remained obscure. The only known owl (Fig. 207), a new species, has been described by Peters (in press). On the basis of the combination of features present this owl has been classified as belonging to a separate family.

The bat shown in Fig. 261 has also been named, as *Archaeonycteris pollex*. In 1988 Storch and Habersetzer described this bat, in which features of the teeth differ from the slightly smaller species, *A. trigonodon*. Radiological examinations of the inner ears of Messelian and modern bats have continued. New bat finds at Messel have also made it possible to establish bio-acoustical differentiation within this bat community. *Archaeonycteris* shows less specialization than the representatives of the other two bat families found at Messel. The morphology of the inner ear of the Messelian bats is very similar to that of modern bats which have active/passive acoustic orientation, as well as that of flying foxes which as a rule do not use echolocation (Habersetzer and Storch, 1990).

To the only skeleton find of the ant-eater *Eurotamandua* from Messel (Fig. 305), Storch and Haubold (1989) were able to add humerus finds from the Lower Coal areas of the Geiseltal (Geisel Valley), of approximately the same age as Messel, and to describe further characteristics of this group of mammals, normally only found in South America.

MacPhee *et al.* (1988) have given a detailed description of the structure of the base of the skull and the auditory region of the Messel scaly-tail hedgehog *Pholidocercus* (Fig. 245ff); comparisons made with early Tertiary hedgehog relatives and other primitive groups were primarily directed toward the question of the origin of primates.

Further observations and discussions on the primate hand found in Messel have been provided by Franzen (in press); the apatemyide from Messel has been described as *Heterohyus* sp., and evaluated functionally with the help of three skeletons (Koenigswald 1990).

The first specimen of *Hallensia matthesi* has been found at Messel, and recognized as an odd-toed ungulate (Franzen 1990). The extremely well preserved soft tissue contours imply that the food remains that have been recovered in abundance are not stomach but intestinal contents. In the chapter on horses and other odd-toed ungulates, Franzen mentions an extinct group of odd-toed ungulates, the lophodonts, of which—although they were so common in the Eocene—only one tooth had been recovered at Messel by 1988. The 1988 Senckenberg excavation yielded the first specimen of *Lophiodon* as large as an Alsatian dog, a young animal with soft tissue contours, with the contents of the digestive tract preserved.

A further spectacular discovery was made in 1989. For the first time, a complete skeleton of *Kopidodon macrognathus* has been recovered, with the remains of intestinal contents and with skin markings. This tree-dwelling mammal is described in this volume in the chapter by W. v. Koenigswald entitled 'The arboreal *Kopidodon*, a relative of primitive hoofed animals'. The find is being prepared in the Messel laboratory.

A comprehensive analysis of coprolites was concluded in 1988 by Schmitz (1991). In the course of his investigations he found phosphatized globular formations and stick-shaped individual bodies identified as bacteria.

The palaeozoogeographical puzzle of the origin of Messelian mammals (Fig. 404) has been taken up again by Storch (1989*a*, *b*, in press *a*, *b*); in particular, the often underestimated role of West Gondwana (Africa and South America) has been emphasized and substantiated by new evidence and finds.

Senckenberg   W.Z.
January 1991

Preface to the German edition

The book offered here by the Senckenberg National History Museum and Research Institute, treating as it does the fossils and the oilshale of the Messel pit, fulfils a long-held ambition. It also expresses the wish of the Senckenberg team to share the results of their research with the public in this form, as well as Messel in the form of exhibitions and individual publications. Two books have already been written about the Messel pit (Heil *et al*. 1987; Behnke *et al*. 1986). This fact, however, by no means makes this Senckenberg book redundant as this publication is more inclusive and in many details supplements those earlier publications. Yet, this book is not a monograph that treats all of the Messel data in an integrated manner. For such a work, which might be misunderstood to be a concluding report, it is far too early. Rather, we have tried to give a feeling for the dynamics of the research at Messel. In this collective volume many authors have their say; not just members of the Senckenberg team, but also other researchers who contributed results at the Messel fossil site. Topicality and first-hand information make this book special. Overlaps cannot be avoided in reports coloured by personal experience and enthusiasm. By the same token, gaps in our knowledge become apparent which foreshadow the enormous tasks still lying before the researchers at Messel.

In the different chapters, personal views have deliberately not been eliminated but, where possible, we have attempted a presentation which is accessible to the layman without reducing scientific information more than is necessary. The understanding of this information is assisted by the lavish illustrations, with colour and black and white photographs and radiographs as well as supplementary drawings.

The majority of the fossils illustrated was recovered by the Senckenberg excavation team. The plant fossils derive from the excavations of the Senckenberg Research Institute and other organizations, for example, the Hessischen Landesmuseum, Darmstadt, the Landessammlungen für Naturkunde, Karlsruhe, and the Institut Royal des Sciences Naturelles de Belgique, Brussels, who graciously donated some of their finds to the Senckenberg Collection. These museums as well as the Naturhistorische Museum of Mainz and private collectors (a complete list is given in the 'Details of illustrations' at the end of this book) have made their finds from Messel freely available for scientific research and publication. The firm of YTONG Rhein-Main GmbH donated all fossil finds, recovered by us to the Senckenberg Museum. Our special thanks go to all of them.

We should also like to thank here all who helped to produce this book on Messel, aimed equally at the layman and the specialist. We thank the authors for their articles, and Mrs E. Junqueira, who was responsible for the layout and was involved with the graphics. Others also took part in this work: H. Schäfer, G. Eder, M. Möller, A. Helfricht, K. Albrecht, and, in part, the authors themselves. All the typing was carried out by Mrs E. Richter: the majority of the photographs (almost all of the colour photographs) were taken by our photographer, E. Haupt. Particular mention must here go to the conservators and technical assistants of the Senckenberg Museum and of the Senckenberg field station in Messel, who were involved with the recovery and the exquisite preparation of the fragile fossils. They are: A. Ackermann, I. Dröhmer, U. Hänig, A. Hentschel, G. Krebs, G. Kreiskott, M. Müller, K. Schmidt, C. Wagner, and S. Weigelt. Finally, we thank the Senckenberg Society for Nature Research for generously providing the means by which this book was made possible.

The unique fossil assemblage at the Messel pit places it among the most famous fossil sites on Earth. There is, however, another reason why the fame of this pit has spread beyond the German border. To the chagrin of all who deem the Messel site to be a unique geological inheritance, many politicians, because of the worries about rubbish in our consumer society, wish to turn this former open-cast mine into a rubbish dump. A final decision has not yet been taken and the outcome still seems uncertain. Because of this we hope that our book will contribute to broadening the distribution of knowledge of the exciting findings of the fauna and flora of approximately 50 million years ago. The reader may also gain the impression that only a small part of the scientific treasure in Messel has yet been discovered. Much remains to be done in order to reveal this heritage and to solve the mystery of these traces of a long-vanished world. Future generations of scientists will have to continue this work using new methods as they become known. But we are optimistic enough to believe that the final deciphering of the secrets of the pit will, one day, be possible. For all these reasons it must not disappear under rubbish!

Frankfurt am Main W.Z.
1988 S.S.

Fig. 1: The Messel pit,
autumn 1987.

Contents

1

Curriculum vitae of the Messel pit

Curriculum vitae of the Messel pit

STEPHAN SCHAAL

4 February 1859

Chance finds of limonite lead to permission being granted for a limonite open-cast mine close to the Messel railway station.

1875

Several very promising indications lead the joint-stock company Eisenhüttenwerk Michelstadt to dig for brown coal in the vicinity of Messel. (The rock is termed 'brown coal', but differs from the usual brown coal by its slaty appearance. It is a type of oil shale.) The successful search is followed by permission being granted for Mines I and II.

30 December 1875

First fossil find at the Messel pit. This consists of the remains of a crocodile and was found by R. Ludwig.

1883

Installation of a brown-coal tar distillation plant.

24 July 1884

Founding of the Messel mining company. Operation of a petroleum and paraffin factory. Under the guidance of the entrepreneur Dr Spiegel, distillation (under hermetic seal) of the so-called brown coal into mineral oil and paraffin takes place.

1898

First survey of the Messel site by E. Wittich.

1884–1909

Successful drillings lead to the acquisition of further mines: Maria mine near Offenthal and Zimmern, Eugen, and Max as well as the Prinz von Hessen mine.

1912

Dr Spiegel, the representative of the Messel mining company, increasingly uses his influence to have the fossil finds transferred to the Archduke's Regional Museum in Darmstadt. A contract is drawn up between the museum and the mine owner to cover the salvage of finds.

1919–22

Further survey reports concerning the Messel fossil site, by O. Haupt (fauna), H. Engelhardt (flora), and H. L. F. Harrassowitz (reptiles).

1923

Incorporation of the Messel mining company into the Hugo Stinnes-Riebeck mining and petroleum plant AG (joint-stock company) in Halle/Saale. The Messel mining company operate under Riebeck management until 1945.

5 July 1945

The Messel pit and factory are seized by the American Military Government.

1951

Building of a plant for manufacturing gas-concrete building blocks by the YTONG group. The refuse from the distillation of the bituminous shale is processed and used in block manufacture.

24 June 1954

Founding date of the paraffin and petroleum factory, Messel GmbH (a limited liability company).

19 October 1954

New contract between the State of Hessen and the paraffin and petroleum factory, Messel GmbH. On the basis of Prussian legislation concerning petroleum, the Hessian Board of Mines makes the Messel shale subject to a law on petroleum and a new contract is concluded. All workings undertaken at the open-cast mine of the Messel pit are now subject to the supervision of the Board of Mines.

1955 and 1969

Further survey reports on the Messel fossil site by H. Tobien (mammal palaeontology).

9 November 1959

Merger with the Swedish YTONG group to form YTONG–Messel GmbH.

1961

Professor Kühne, Berlin, for the first time successfully transfers Messel fossils to a bed of artificial resin. This so-called transfer method makes possible the permanent conservation of the Messel fossils and represents a decisive step forward in the investigation of the fossil site.

January 1962

The YTONG manufacture of building stone is changed from a process using distillation residues to one using sand.

3 March 1962

YTONG–Messel GmbH largely ceases distillation of bituminous shale.

1966

First methodical excavations for fossils in the Messel pit, carried out by the Department of Geology, Palaeontology, and Mineralogy of the Hessen Regional Museum.

1970

A chronicle about the open-cast Messel pit is published by BEEGER.

1971

Private collectors begin the search for fossils and use the transfer method for preparing them. The collections contain spectacular finds including, for example, the first skeleton of an early horse, a tapir skeleton, and the skeleton of an ant-eater. They help to emphasize the importance of the fossil site.

13 July 1971

The new Hessen waste-disposal law forms the legal basis for the establishment and use of a planned refuse repository.

Up to the end of 1971

Limited quarrying of rock for the production of electricity. During the period of quarrying, approximately 20 million tonnes of rock were quarried and from these approximately 1 million tonnes of oil were obtained. A decision has to be made as to whether the site should be recultivated or whether open-cast mining should be recommenced. The governing bodies of the state conceive the idea of filling the Messel pit with refuse.

1973–1974

Private individuals, citizens' initiatives, scientists, scientific institutions, and politicians raise objections against the planned refuse dump.

1974

The frequency of fossil finds and the high prices paid for rare finds leads to invasion of the pit by private excavators and fossil merchants.

1 January 1975

On the advice of the mine owner, safety and liability considerations lead to the closure of the pit to the public by the Board of Mines.

March 1975

The Hessen Regional Museum, Darmstadt, begins to intensify scientific excavations.

June 1975

Under the threat of the refuse dump, the Senckenberg Natural History Museum and Research Museum is granted permission to join the Hessen Regional Museum in excavating and salvaging finds. Following this, other institutions gradually obtain permission to excavate: Natural History Museum, Dortmund (1975); University of Hamburg, Institute for Geology and Palaeontology (1975); Johannes Gutenberg University, Mainz (1976); Regional Natural History Collection, Karlsruhe (1979); Royal Institute of Natural Sciences, Belgium, Brussels (1983); University of Tübingen, Institute for Geology and Palaeontology (1987).

November 1975

The 'Zweckverband Abfallbeseitigung Grube Messel' (Rubbish Disposal Committee: Messel Pit) purchases the land containing the Messel pit from the YTONG group.

1976–1977

A multitude of scientific research projects concerning the fossils are initiated. By this means the importance of the Messel pit is brought to the attention of the scientific world and the public.

27 April 1977 to 8 January 1978

The Senckenberg Natural History Museum, together with a group of amateurs, organizes a special exhibition on the topic: Messel fossil site. From this the present Messel exhibition has emerged.

1979

The Hessen Regional Museum in Darmstadt opens a permanent exhibition on the theme of 'the Messel fossil site'.

30 December 1981

A planning decision in favour of the central refuse deposition site. Complaints by private individuals are registered with the administrative courts of law against this decision and the order to proceed immediately. They had little success.

30 April 1983

New form of statute concerning the Rubbish disposal Committee: Messel pit. It is henceforth known as: Zweckverband Abfallverwertung Südhessen (ZAS) (Refuse Utilization Committee: South Hessen).

4 June 1984

A treaty was signed between the SPD and the Green Party: agreeing about the maintenance of the Messel pit as a cultural monument.

12 December 1984

Split between the SPD and the Green Party.

1985

Work in connection with the installation of a refuse disposal facility is increased.

July 1985

The Hessian Minister for Science and Art, Dr V. Rüdiger, orders an expert report to elucidate questions that might arise for palaeontological research were the pit to be kept open.

8–11 April 1987

International Messel Symposium with over 100 participants from 12 countries at the Senckenberg Natural History Museum, Frankfurt. Superb research results emphasize the importance of the Messel fossil site.

21 April 1987

Following the State elections on the 5th of April and a change in the proportions of the majority, a green light is given to the completion of the pit for refuse disposal. The parties, CDU, FDP, and SPD, vote, at the meeting of the Refuse Utilization Committee (ZAS), for the completion of the pit as a refuse dump. Only the Green Party supports the permanent maintenance of the pit for the purpose of research.

16 December 1987

On the basis of shortcomings in the planning decisions, the law court in Kassel cancels the immediate start of the rubbish disposal project at the Messel pit, in response to an urgent application from the municipality of Messel.

13 February 1990

The Hessian Ministers for Environment and for Science and Art announce that the Messel pit will not be used as a dump site.

1991

The Federal Government and the Hessian Government approach the Senckenberg Natural History Museum to devise a plan for future investigations and management of the pit.

14 June 1991

The Hessian Government purchases the Messel pit after long and intensive discussions to guarantee long-term scientific use.

6–9 November 1991

Second International Messel Symposium with over 100 participants from 21 countries at the Hessian Regional Museum, Darmstadt. The research results again highlight the significance of the Messel fossil site.

2

Europe in the Eocene: Messel in time and space

In 1911, the first finds of primitive horses enabled the former curator of the Archduke's Regional Museum, Darmstadt, Oskar Haupt (1878–1939), to place the Messel community in the Eocene. Today, once again, the grade of evolutionary advancement of the fossil horses of Messel has led to a more precise date, by comparing these horses with the famous fauna of the Geiseltal near Halle/Saale. On the basis of this evidence, the Messel site must be assigned to the beginning of the Middle Eocene.

Europe in the Eocene: Messel in time and space

JENS LORENZ FRANZEN

The Middle Eocene of terrestrial European stratigraphy should strictly be called the Geiseltalian rather than the Lutetian (Franzen and Haubold 1985, 1986*a*, *b*). This is because the term 'Lutetian' was established from the chronology of marine sediments, to which the terrestrial mammals can only erratically be related. The type stratum of the Lutetian (after 'Lutetia', the Celtic/Roman name for Paris) is the coarse-grained limestone of Paris (Calcaire Grossier). This, however, is a marine deposit which in its middle and lower parts contains almost no mammalian fossils. On the other hand, the stratigraphy of the Geiseltal is composed mainly of brown-coal seams which are separated by sandy to argillaceous intercalations. This terrestrial sedimentary sequence is very well known through 300 years of mining activity and, in particular, through the excavations of the Institute of Geology and Palaeontology of the Martin Luther University of Halle–Wittenberg (Krumbiegel *et al.* 1983). The sediments were formed in a swampy environment which was fed by rivers from the limestone plateau (Muschelkalk) on its southern border. Because the humic acids were neutralized by the calcium-rich waters, alongside a luxuriant flora, an extraordinarily numerous and species-rich fauna was preserved. In its entirety the stratigraphical sequence of the Geiseltal covers the whole of the Middle Eocene and extends even beyond into the early Upper Eocene. It provides therefore an ideal locality for assessing the stratigraphy, based on land vertebrates and plants (Krutzsch 1976; Franzen and Haubold 1985, 1986*a*, *b*, 1987).

The mammals of Messel correspond in their stage of evolutionary development with the oldest fauna of the Geiseltal. This fauna from the Lower Coal is evolved just a little further

than that of the uppermost Lower Eocene which is known from the Paris Basin (sands with *Teredina* and *Unio*), and from southern France (Mas de Gimel). For this reason Messel must be classified with the lower Middle Eocene or, respectively, with the lower Geiseltalian. Apart from the primitive horses, the evolutionary advancement of the artiodactyl *Messelobunodon* (Franzen 1981; Franzen and Krumbiegel 1980), the rodent *Ailuravus* (Tobien 1968*a*), and the primate *Europolemur* (Franzen 1987) support this classification.

The fauna from the freshwater limestones of Argenton-sur-Creuse in central France (Département de l'Indre) and the vertebrate fossils from the Issel site in southern France (Département de l'Aude) are more recent. Evidence pointing to this includes the evolutionary advancement of the archaic horses. The fauna of the freshwater marls and limestones of Buchsweiler (Bouxwiller) in Alsace can be placed at the end of the Middle Eocene. In its turn, Buchsweiler corresponds with the faunas from the Calcaire Grossier Supérieur of Paris and from the upper Middle Coal of the Geiseltal. This is succeeded at the start of the Upper Eocene by the species-rich mammal fauna from the karst fissure fillings near Egerkingen in the Swiss Jura (Hartenberger 1970).

Where is the Eocene situated on the geological time scale? It belongs to the Lower Tertiary. With the aid of absolute dating, by studying the degree of disintegration of radioactive elements in volcanic rocks, it has been determined that the Eocene lasted from approximately 54 to 35 million years before the present (Savage and Russell 1983). The Eocene followed the earliest period of the Tertiary, the Palaeocene, and was succeeded by the Oligocene (Fig. 2).

The term 'Eocene' was coined by the English geologist Charles Lyell (1797–1875), who during the years 1831–3 was professor of geology in London. Lyell subdivided the Tertiary, the second youngest period of the Earth's history, on the basis of the evolutionary advancement of the fossil invertebrate faunas of the Mediterranean. He named the Eocene after *Eos*, the Greek goddess of dawn. It was

Fig. 2: Chronological table of the Earth's history. The Tertiary, the Eocene, and the position of Messel are shown in detail.

ERA	Ma	PERIOD	TERTIARY	EOCENE		
				MARINE AGE	MAMMAL AGE	ABSOLUTE AGE
CAINOZOIC ERA	65	Tertiary	1.8 Pleistocene / 5 Pliocene	Latdorfian	Headonian	
MESOZOIC ERA	135	Cretaceous	Miocene	Ludian		39
	190	Jurassic		Bartonian	Robiacian	
	230	Triassic	24			
	280	Permian	Oligocene			44
PALAEOZOIC ERA	350	Carboniferous	35	Lutetian	Geiseltalian	
	405	Devonian				
	435	Silurian	Eocene			
	500	Ordovician			Messel	49
	570	Cambrian	54	Grauvian		51
PRECAMBRIAN ERAS	3 Billion Years	Algonkian	Palaeocene / 65	Ypresian	Sparnacian	

the epoch during which there occurred, in the sediments of the Mediterranean, the first precursors of our present mollusc fauna (3.5 per cent of species still extant). In the Miocene (Gr. *meion*: less) these increased to 17 per cent of species still extant, and in the Pliocene (Gr. *pleion*: more) with a further increase of 30–50 per cent, still more. In the middle of the last century, after examination of Tertiary

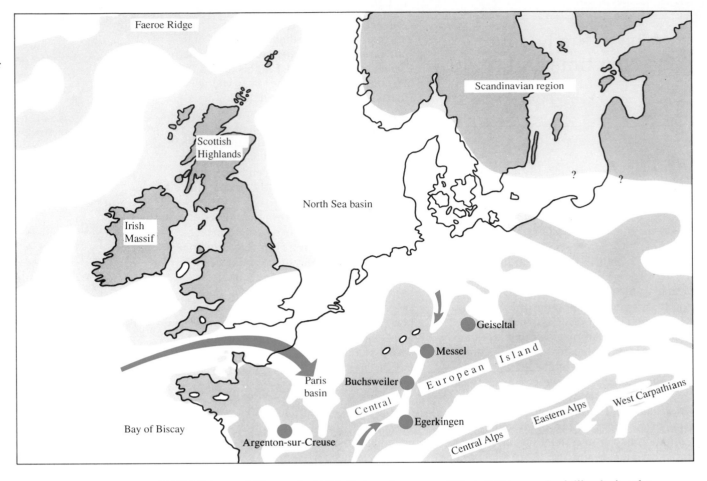

Fig. 3: Europe in the Eocene. The distribution of land (grey) and sea (white) at that time is shown superimposed over the outline of the continent as it is today (southern Europe is not shown). At that time, Europe was an island which lay approximately 10 degrees of latitude further south than today. Messel, the Geiseltal, and some other fossil sites are marked.

Labels on map: Faeroe Ridge · Scandinavian region · Scottish Highlands · North Sea basin · Irish Massif · Geiseltal · Messel · Buchsweiler · Central European Island · Paris basin · Bay of Biscay · Egerkingen · Argenton-sur-Creuse · Central Alps · Eastern Alps · West Carpathians

sediments from the North Sea, it was deemed necessary to insert a new epoch, the Oligocene (Gr. *oligos*: little, modest), between the Eocene and the Miocene. Towards the end of the nineteenth century a further epoch, the Palaeocene, or old Eocene (Gr. *palaios*: old), was inserted between the Cretaceous and the Eocene.

The middle Eocene started approximately 49 million years ago. This chronological statement is based on the absolute dating of 48.7 million years that was obtained for glauconitic sands at the base of the Calcaire Grossier of the Paris Basin. It correlates well with the dating of volcanic activity in the environment of the Eocene Lake Messel of approximately 49 million years and the occurrence of numerous volcanic minerals and occasional tufas in its sediments (Lippolt *et al.*

1975; Weber and Zimmerle 1985; Kubanek *et al.* 1988).

Forty-nine million years is a time span that is almost impossible to visualize. Yardsticks are necessary to understand properly such geological dimensions. Such a yardstick might be the start of our western chronology, the birth of Christ. This is a moment that lies far back in history. That age, however, of which the Messel pit offers a glimpse, lies 25 000 times further back in time. Another comparison can be made by anyone who has crossed the Alps during a journey to the south. When the plants and animals lived in the Eocene Lake Messel and its surroundings, the Alps did not yet exist as a mountain range. It is only since that time that millimetre by millimetre, the mountains have pushed up towards the sky.

What did Europe look like during the Eocene? The contours of the continent that are familiar to us today could hardly be recognized (Fig. 3). Europe continued to be flooded by shallow epicontinental seas and their side branches. Not only did the Alps not yet exist, but neither did the Pyrenees, nor the mountains and gorges of the Balkans. All the Alpine mountain chains that frame the Mediterranean today originated as compression zones caused by a collision between the continents of Europe and Africa (Barron *et al.* 1981; Panza *et al.* 1985). The African continent edged inexorably northward, narrowing the Mediterranean and triggering numerous major earthquakes. This process still continues today.

During the Eocene the entire Central European Island lay about 10 degrees further south than at present (Smith *et al.* 1982). The area of the Eocene Lake Messel therefore lay a little to the south of the present position of Naples. The Mediterranean was far larger then and covered the area of present-day Turkey, Iraq, and Iran with a wide passage linking it to the Indian Ocean (Fig. 404). This situation may explain, at least in part, the entirely different climatic conditions (see Chapter 27, 'The Messel fauna and flora: a biogeographical puzzle'). The Middle Eocene fish fauna of Monte Bolca in northern Italy, with its exotic character, gives us an idea of the tropical–subtropical conditions that existed at that time on the southern coast of Europe.

The Iberian Peninsula, as well as Sardinia and Corsica, were small fragmentary plates between the continents of Africa and Europe, with which they were connected only occasionally (Fig. 404). Italy and the greater part of the Balkans were covered by the sea. To the east the Russian Platform was also still flooded by the sea and separated from the Asian land-mass by the Turgai Strait which then connected the Arctic with the Indian Ocean.

To the west the North Atlantic was in an early phase of its evolution (Owen 1983; Smith *et al.* 1982). At least at the start of the Eocene, Europe and North America were still attached to one another at the latitude of Spitsbergen. Apparently there was another land bridge further south, in the region of today's Faroe Islands, Shetland Islands, and Iceland, which led, via Greenland and Ellesmere Island, right into western North America. It is not surprising, therefore, that approximately 50 per cent of all genera of mammals known to exist at the beginning of the Eocene appeared on both sides of the North Atlantic, in Europe and in North America at the same time (Russell 1968; Savage and Russell 1983). On the other hand, this land bridge proved an insurmountable barrier for the marine molluscs of the Atlantic to the south and the Scandic Ocean to the north, which led to separate development of these faunas (Strauch 1970). In the course of the Lower Eocene, however, the Atlantic expanded, so that the land bridges were inundated.

In the area of the Middle European Island, at the centre of which lay the Eocene Lake Messel, the influence of the sea became increasingly noticeable (Fig. 3). From the west, over the area now occupied by the Channel and southern England, the sea advanced into the wide bowl of the Paris Basin. There it formed a lagoon where the Calcaire Grossier of Paris was deposited during the Middle Eocene and where later, during an evaporation phase, the famous gypsums of Montmartre were deposited. From quarries into these deposits, the founder of vertebrate palaeontology, Georges Cuvier (1769–1832), described, at the start of the nineteenth century, one of the first and oldest mammalian faunas in Europe. (Belgium, the Netherlands, and north Germany were also still flooded by the North Sea at that time.)

Towards the close of the Eocene, another marine strait extended from the Mediterranean into the intermittently sinking rift of the Rhine valley, only to unite in the Oligocene with a channel that reached southward from the North Sea. As a result, the Middle European Island was separated into at least two faunistic provinces (Franzen 1968; Schmidt-Kittler and Vianey-Liaud 1975).

Volcanism was not inconsiderable in Europe during the Eocene. It affected not only the immediate vicinity of Lake Messel, but also the entire reach of the Upper Rhine rift valley (Lippolt *et al.* 1975). The deep layers of ash of the Moler series, whose origin derived from the region of the present-day Skagerrak Channel, were already deposited in northern Jutland at the start of the Eocene. In the Eifel the first maars formed more or less contemporaneously with Messel (Negendank *et al.* 1982).

Overall, the Eocene was a time of transition between the peaceful and equable conditions of the Mesozoic and the increasingly restless Cainozoic era during which the geological face of Europe changed fundamentally.

The transitional character of the Eocene is also expressed by the composition of the European flora and fauna. This is particularly relevant for the mammals. On the one hand, there were ant-eaters and pangolins ('scaly ant-eaters') at Messel, survivors of a fauna that had probably already evolved during the close of the Mesozoic. These forms exist as 'living fossils', still almost unchanged today, in the favourable climatic conditions of the tropics and subtropics. For other mammals, however, the Eocene proved to be a time of almost explosive evolution. In their evolutionary development numerous orders of modern mammals hail back to the Eocene. This is true, for example, for the primates, the proboscidians, the carnivores, the odd-toed hoofed ungulates, the even-toed hoofed ungulates, and the rodents. Their representatives at the time of the Eocene, however, still had quite a primitive structure.

3

The genesis of the Messel oil shale

Between Frankfurt-am-Main and Darmstadt in the northern foothills of the Odenwald, there exist deposits from an Eocene lake region, the Messel formation. Here during the Palaeocene and the Eocene, pressures developed over a wide area of the basement rocks, which were released through the formation of tectonic rifts. The basement consists of igneous rocks, which are overlain in the foothills of the Odenwald by 260-million-year-old sediments of the Lower Permian (Rotliegende). In a few places the basement sank in several stages, almost in the way that a lift goes down. In these tectonic rifts clays and sandstones were gradually deposited.

The genesis of the Messel oil shale

STEPHAN SCHAAL

Sediments of the Messel formation have been traced in the Messel and Prince of Hessen mines, in the open-cast mines at Eppertshausen and Offenthal, in boreholes at Gundershausen and Dieburg, and also in the rift valley ('graben') of the Rhine at Stockstadt. It is thanks to the rift formations that the lake sediments have remained protected up to the present day. The other deposits from the primeval forest and lakes succumbed to later processes of erosion. The graben and the oil shale of Messel, to which this summary is limited, are of particular scientific interest owing to the preservation of unusually complete vertebrate skeletons. The graben is approximately 1 km north–south, and approximately 700 m east–west.

The Messel formation (Fig. 4) consists at its base of sediments like gravel and sand, flaky claystone, commonly also referred to as 'oil shale', and, in the upper part, is composed of sands and clays (Matthess 1966; Weber and Hofmann 1982; Weber 1988). Because of the profile of the lake bottom, watery muds and sands periodically could start to move down the slope in the border region of the graben. The sediments of those slumps and numerous tear marks (small fissures in the sediment which one can find today), indicate that the graben has deepened repeatedly. In the Messel mine the sediments of the lake reach a thickness of 190 m.

At the time of deposition, the constituents of the decaying mud on the lake bottom were packed closely together. Pores filled with water made up a large part of the entire volume. By compression, that is, through pressure from the layers forming above, the density of the decaying mud was so increased that its porosity was decreased and solid rock was formed.

Even today, however, the Messel oil shale has a water content of 40 per cent.

Subdividing the oil shale

For an evaluation of the results of the excavation, it is necessary to compare the excavation sites with each other. This task is made very difficult because the shale has a uniform, almost black hue (Fig. 6). During the investigation of the Messel formation, layers have been found that, in composition and colour, differ distinctly from the oil shale. They are termed key horizon layers, and with their help the discovery sites can be precisely categorized.

Two prominent key horizons are introduced here. The key horizon beta (Franzen *et al.*

Fig. 4: Geological sketch showing the position of the Messel pit and its surroundings.

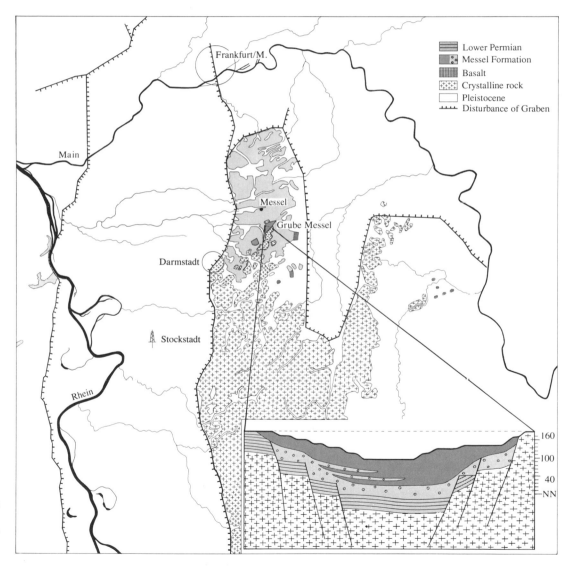

Legend:
- Lower Permian
- Messel Formation
- Basalt
- Crystalline rock
- Pleistocene
- Disturbance of Graben

Frankfurt/M.
Main
Messel
Grube Messel
Darmstadt
Stockstadt
Rhein

160
100
40
NN

1982) is a double horizon of micaceous shale, which indicates a twice-renewed input of sediment. While being deposited these formed a series of layers which contain grains of different sizes (Fig. 7). At first the larger grains sedimented out, then the fine-grained clays. Sedimentologists call this form of deposition 'graded bedding'. As this horizon was the second key horizon to be discovered, it was called beta.

The second key horizon to be discussed here is a layer containing phosphate which consists predominantly of the mineral montgomeryite (Schaal *et al.* 1987). It is therefore called key horizon 'M'. In the field the phosphate minerals (Fig. 5) are recognized by their light colour and greater hardness. While the shale can be sliced like hard chocolate, the phosphate minerals, neomesselite and montgomeryite, can only be scratched. The source material for this layer may have been a type of phosphate mud.

In addition to this temporal sequence of formation of the individual layers, the oil shale can vary in its lithology and fossil content in both vertical and horizontal directions. This is

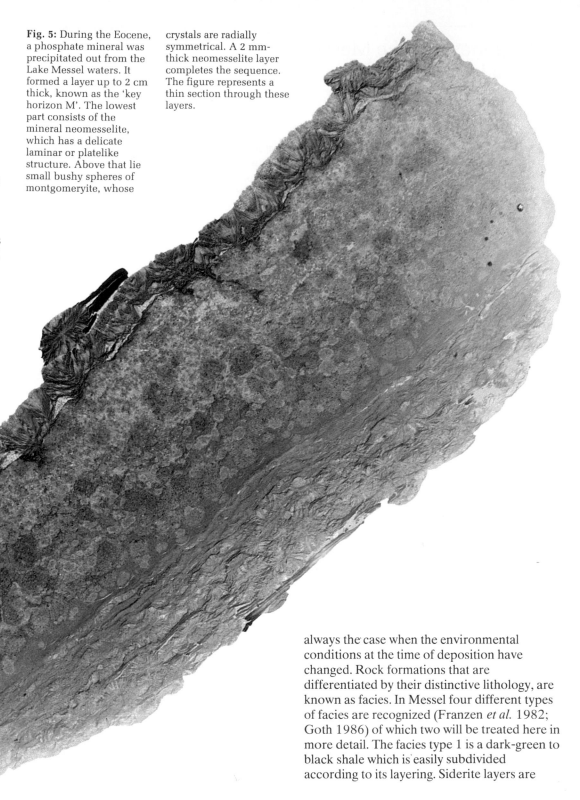

Fig. 5: During the Eocene, a phosphate mineral was precipitated out from the Lake Messel waters. It formed a layer up to 2 cm thick, known as the 'key horizon M'. The lowest part consists of the mineral neomesselite, which has a delicate laminar or platelike structure. Above that lie small bushy spheres of montgomeryite, whose crystals are radially symmetrical. A 2 mm-thick neomesselite layer completes the sequence. The figure represents a thin section through these layers.

always the case when the environmental conditions at the time of deposition have changed. Rock formations that are differentiated by their distinctive lithology, are known as facies. In Messel four different types of facies are recognized (Franzen *et al.* 1982; Goth 1986) of which two will be treated here in more detail. The facies type 1 is a dark-green to black shale which is easily subdivided according to its layering. Siderite layers are characteristic for type 1. In the facies of type 3 the shale is, by contrast, browner in colour; it can also be subdivided according to the layer sequence. In the facies of type 3 the algae have accumulated in such numbers that they can occur in layers up to 1 mm thick. This facies is also well known for its good preservation of the soft parts of vertebrates.

The stored energy of the sun

The content of almost 15 per cent kerogen has transformed the Messel oil shale into a valuable resource and, from the 20 million tonnes of mined stone, approximately 1 million tonnes of oil have been produced by means of carbonization. The high kerogen content is the result of a massive production of algae encouraged by a well-balanced ecosystem of sunlight, temperature, and nutrient availability.

However, even the most luxuriant tropical rain forest with its lakes and rivers leaves, under normal circumstances, no trace of kerogen. Decaying flora and fauna would be degraded by bacteria and would decompose. During this process, a transformation of organic particles into the gas CO_2 and into water takes place. Putrefaction is, in the end, nothing but oxidation of the chemical elements carbon and hydrogen. Oxidation could not, however, take place in the anoxic environment at the bottom of the lake, and anaerobic micro-organisms through their metabolic activities made hydrocarbon formation possible. Thus the energy from thousands of years of sunlight is ultimately stored here.

The oil shales of Messel: a geological mystery

From the commencement of the scientific observations at Messel the researchers were particularly concerned about the question of the genesis of the shale. The following points needed to be clarified: (1) the reason for the formation of the graben; (2) the stratification of the oil shale; (3) the abundance of the flora and fauna and its composition; (4) the state of preservation of the fossils.

The researchers agree at least on the last

Fig. 6: The oil shale of the middle Messel Formation. This almost black rock sometimes contains light-coloured siderite layers. The brown coloration of the upper (laminated) layers, which has disappeared from the lower part, is due to the precipitation of iron-containing minerals.

Fig. 7: The 'key horizon beta' is characterized by graded bedding containing mica. Such layers may form by means of a single input of sediment into an area of standing, stagnant water. During the deposition of the sediment the largest grains settle out at the bottom and the finer grains higher up.

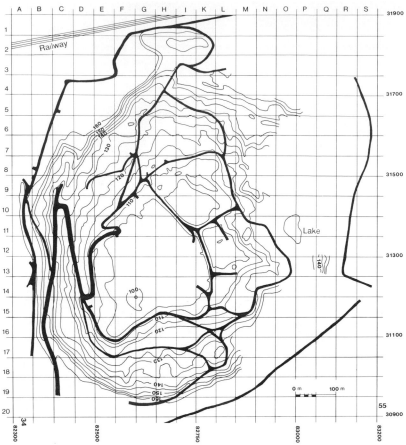

Fig. 8: Map of the Messel graben superimposed on a grid. This map is an important aid in geological interpretation. The grid facilitates both horizontal and vertical orientation in the excavation area for the geologist, as well as facilitating site descriptions for individual finds.

point. Subtropical to tropical climatic conditions made a strong increase of algae possible, which, during their decay in lower water layers, could de-oxygenate the environment and lead to incomplete decomposition of organic matter. This is how the kerogen was formed and how fossils remained, qualitatively, singularly intact. All theories must, however, attempt to resolve the remaining questions.

Model 1. A basin within an extensive river system

Almost 50 million years ago the northern valley of the upper Rhine did not, as yet, exist

in its present form. From the results of the Senckenberg excavations in Messel from 1975 to 1981 (Franzen *et al.* 1982), it can be concluded that the area under investigation in the Messel region was a flat, undulating landscape, through which flowed a large river, whose extent and direction are still unknown. At those sites where the Earth's crust collapsed locally when the upper Rhine graben began to sink, the course of the river became interrupted by individual lakes. These lakes, one of which was the Eocene Lake Messel, were, in effect, basins in which drifted animal corpses and plant debris accumulated. The tectonically sinking lake basin delayed the silting up of Lake Messel. In the still water of the lake, which was only a few square kilometres in area,

this cargo of corpses and plant debris was deposited together with the suspended clay. With the aid of directional measurements on fish, a flow model for the shale sedimentation was constructed, two tributaries pinpointed, and a lake outlet found (Figs 9 and 10). Evidence for such a model comes from the number of cases of freshwater caddis-fly larvae, as well as from the great frequency of the fish *Thaumaturus* and of small perch in the region of the putative tributaries.

Model 2. A small section of a large lake

The description of the formation of Messel has already indicated that the conditions for sedimentation in the Messel area changed repeatedly. This is not so surprising when one considers that, according to the findings of Irion (1977), some 100 000 years were required for its formation. In their investigations the geologists Möller, Schaal, and Schmitz-Münker, therefore, give first priority to sedimentology and, as a second step, examine the fossil contents. Detailed recordings of profiles show homogeneously formed layers across large areas of the site. The distribution of the sediments shows that the oil shales of Messel represent a small section of the central part of a large lake. The graben, which was formed step by step, acted as a sediment trap for clay particles which spread as clouds of mud over large areas in the lake, particularly at times of rains and during floods (Figs 11 and 12). Dead plants and animals of every kind drifted in this water and settled out together with the clayey particles. Sediments from the lake-side region are entirely missing, but landslides in the border regions of the graben and fissures in the layers interrupt the homogeneous aspect of the oil shale and are evidence for the existence of a distinct trough at that time (Reineck and Weber 1983). In this context it is assumed that the extent and volume of the lake have changed frequently. In times of drought Messel could have even served as a refuge for aquatic life-forms, as the graben never dried out. The chemical and biochemical reactions, particularly at the lake

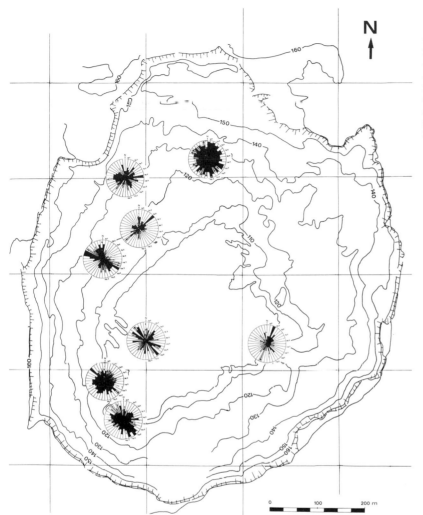

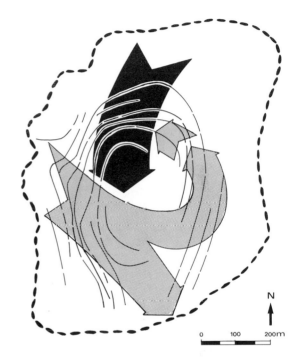

Fig. 9: Orientation data for fishes from different excavation sites. The data for fishes from each site are not derived from a single layer, but are the result of several years' excavation through sedimentary layers which record a time span measured in millenia.

Fig. 10: Reconstruction of flow directions of the Messel oil shales using directional data from fish (see Fig. 9) and further site measurements (see Model 1). Black continuous lines are reconstructed contours of equal lake depths (isobaths).

bottom, were drastically influenced by these oscillations and are mirrored by the corresponding precipitation of minerals.

The most frequent fossil finds are recruited, of course, from the inhabitants of the lake. In contrast the composition of fossil finds from the terrestrial flora and fauna is more subject to chance and the frequency of finds is relatively smaller. Not only the early horses and crocodiles, but also small perch, mudfishes, and the fish *Thaumaturus* were found just as often at the lake borders as at the centre of the site. Sometimes only a few decimetres of

sediment lie between two finds but these decimetres can reflect several thousand years of sedimentation. Assumptions that finds of certain groups of vertebrates are limited to particular sites have not yet been confirmed.

No statements can be made as yet as to the cause of the graben formation and it must be pointed out that the development of this sedimentation model is not yet conclusive.

Model 3. A deep crater lake in a subtropical forest

Rietschel (1987) explains the Messel graben as a maar-like crater of volcanotectonic origin. He considers the deficits in flora and fauna and reaches the following conclusions. After major subsidence the crater filled up with coarse, decaying debris and water '. . . until a surface outlet could be formed . . .' so that a deep lake with steep banks and predominantly hostile environmental conditions could become established. Fishes and reptiles were able to

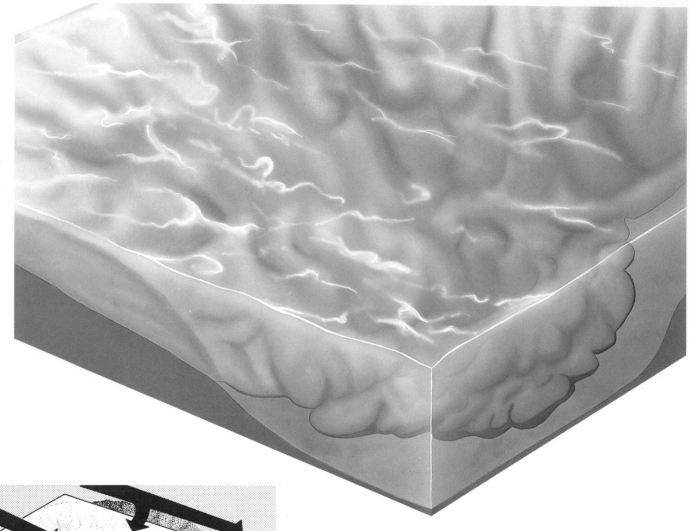

Fig. 11: A highly schematic sedimentation model for the deposition of clay particles in a shallow depression. A 'slurry' of greater density spreads slowly below the less dense water particles and inundates the depression, which traps the clay particles. It is this sunken area with its decreased flow rate which allows sedimentation to occur.

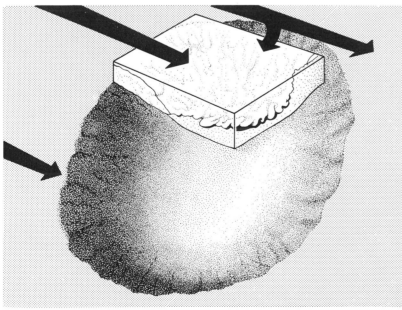

Fig. 12: Sketch of a medium-sized depression. The position of Fig. 11 is superimposed. The arrows indicate the flow directions of suspended sediment with a broad, leading front.

reach the crater via the lake outlets, and springs of groundwater flowing from the steep banks of the lake explain the fossil evidence of a freshwater fauna. Small streams carried a light clay suspension, which was deposited in the lake together with flowers and leaves blown into the water and with mammals that fell into the lake along its borders.

This model can be used to explain certain absences in the fauna including, for example, crabs, water insects, and water birds. Between the primary producers (green algae) and carnivorous fish, herbivorous fish are missing. In addition, there are floristic absences such as

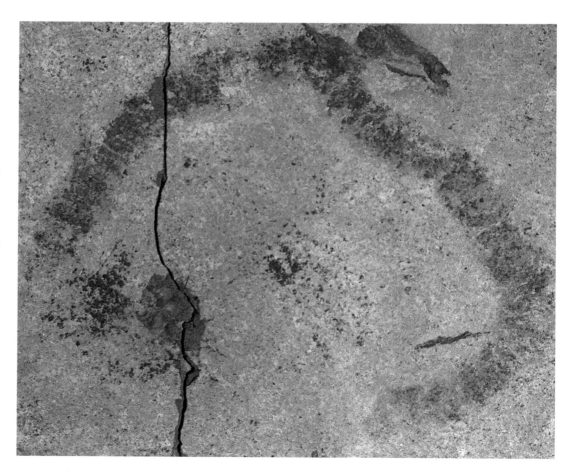

Fig. 14: A second find which complements the specimen shown in Fig. 13. The course of the line is less densely convoluted and appears to show a trail typical of a grazing snail.

Fig. 13: (below, right). This approximately 30 cm-long, apparently segmented form on the surface of a layer captures the imagination of the observer. Only with knowledge of the find depicted in Fig. 14 does it become obvious that the areas represent the pattern of a very tightly zigzagging path.

Fossil or pseudofossil?

The oil shales of Messel are quite sensational for the scientist, and sometimes finds are made which still cannot be classified with certainty. The structure in Fig. 13 was found when a geological profile was taken in 1986. It quickly captures the imagination of the observer and the theory that this might be a track, or even a worm, is attractive.

The length of the shape measures barely 30 cm and the width 1.5 cm, with insignificant variations in the measurement; both ends have indistinct limits. More than 70 distinct areas, reminiscent of segments, making up the shape, are arranged at approximately the same distance from one another, and are placed either like treads of a ladder vertically, or at a shallow angle to the longitudinal axis of the structure. The light-coloured material is

themselves provide an answer to all questions. The environments in which sedimentation occurs are very complex and very changeable. Attempts to explain the wealth and distribution of organisms in an ecosystem may fail, and unreliable simplifications may result from this. Thus an interpretation can appear reasonable and intellectually satisfying, but will quickly lose its value if the main points in a reconstructed sedimentation model are wrong. Future research will show which model most closely approaches reality.

roots, stems, branches, and twigs. Furthermore, besides water-lilies, the higher water plants and representatives of vegetation from the border of the lake are missing.

Rietschel assumes that the good fossil preservation could also be due to the presence of gases with conserving properties, which might even have contributed to the death of the animals.

A critical evaluation is necessary

Although it is the aim of research to find a good hypothesis with which to explain phenomena, a series of observations cannot by siderite ($FeCO_3$) and its thread-like shape is often compared with that of snail faeces (Rietschel, oral communication). The structure itself is thus an accumulation of threads of siderite which are ordered at regular intervals and are aligned at the specimen border parallel to the longitudinal axis. In 1987, a second, similar structure attracted the attention of the excavation team from the University of Tübingen (Fig. 14) but it consists, regrettably, only of fragments. Both finds are derived from the same or from closely juxtaposed horizons.

Are these fossil tracks, compaction marks, or grazing trails?

The environment—so hostile to life—that must have existed at the lake bottom during the time of deposition of the Messel formation almost excludes the formation of grazing trails. During a short interval of improved environmental conditions, snails could, however, have been the first to settle on the bottom of the lake. The snails found at Messel are, according to a communication from A. Zilch (in Franzen *et al.* 1982), pond snails (Viviparids); see Fig. 368). According to the advice of R. Janssen there might even be a correlation between the width of a 'segment' or, respectively, the line in Fig. 13, and the estimated width of the foot of a Messel snail. Some snails, for example, feed on algae and, during their zigzag or undulating grazing, leave a trail in the form of a densely convoluted meander, a two-dimensional pattern on the lake floor. Snails would not produce permanent traces on the very soft and fine-grained surface of the decaying mud, however, but might initiate the alignment of threads of siderite. Such considerations are true also for compaction marks. An object that falls to the ground, leaves a mark in the loose sand, and is later transported away from that spot would also produce no permanent traces in the decaying mud. Only finds of snails at the ends of such structures could be evidence that the trace is indeed a track.

Could this structure be a fossil worm?

Its segmentation and length shows some correlation with segmented worms. The absence of any soft parts and bristles makes a positive statement concerning this view impossible. The lake floor in the area of the Messel site was, at the time of its formation, neither inhabited, nor burrowed in, by filter feeders. Snail shells are known, but were probably washed into the lake deposits. The same explanation must be adopted for the presence of other, rare inhabitants.

Are these structures actually pseudofossils?

Sediments and sediment particles of all sizes can pass during their deposition, redeposition, and diagenesis through all kinds of deformation. In this process, structures develop which can be recognized on the bed surface, or in section. They all can be traced to physical or chemical processes. It is implausible to relate the alignment of siderite threads to water currents. Over the entire area of the oil shale outcrop no unified flow direction is recognizable. The possibility that this structure developed during diagenesis can be excluded from consideration. Local settlement or slides would be noticeable at the surface of a layer and would certainly not be uncommon in the Messel formation.

We have to wait for the results of future excavations if we wish to find an answer to this and similar riddles.

4

The vegetation: fossil plants as witnesses of a warm climate

There are only a few fossil sites in which such a rich fauna and flora exist side by side as at Messel. Plant fragments were found in the last century soon after the start of mining for oil shale, and Engelhardt described a rich leaf flora as early as 1922. From a comparison with recent plants he deduced that there had been a 'very warm climate' with a 'tropical–subtropical character' and envisaged plant geographical connections with South-east Asia. He also pointed out, however, that the flora of Messel cannot be compared with a 'contemporary time'. Although the material on which this work was based is still kept in the Hessen Landesmuseum in Darmstadt and in the Senckenberg Natural History Museum in Frankfurt, a revision of Engelhardt's work is difficult because the specimens are in a poor state of preservation. Nevertheless, in 1971 Sturm began a revision of this museum material, although he was able to complete work only on the Lauraceae. In the meantime, the excavations which have taken place over the last 10 years have led to the accumulation of a host of new plant material which provides rich information. The excavations have taken place under the auspices of the Senckenberg Museum and the other research institutes; in particular, those of the Hessen Regional Museum in Darmstadt, the Regional Natural History Collection in Karlsruhe, and the Royal Belgian Museum in Brussels.

The vegetation: fossil plants as witnesses of a warm climate

FRIEDEMANN SCHAARSCHMIDT

The foundation of the new plant collections comes from the excavations of the Senckenberg and the other research institutes, particularly those of the Hessisches Landesmuseum in Darmstadt, the Regional Natural History Collection in Karlsruhe, and from the Royal Belgian Museum in Brussels, and the greatest part is deposited in the Senckenberg Natural History Museum. In addition, excavations have been carried out with specific goals, in recent years by the palaeobotanic section of the Senckenberg Research Institute. For the study of pollen there are at least available samples from research borings into the oil shale made in 1980.

By these means material has been brought together which is not, as previously, limited to leaves but which also contains sizeable collections of fruits, seeds, and flowers. In addition, there are plant microfossils such as spores, pollen, and algae which have been obtained from samples from outcrops and from drilling cores. The large-volume of plant fossil material forms an ideal basis for a scientific treatment. The peculiarities of preservation of the plant remains in Messel have indeed required some years of experimentation and searches for specific methods of preparation, conservation, and graphic display of the results. As a result of such endeavours our knowledge concerning the vegetation in Lake Messel and its environment has increased significantly. Thus it was possible in the last few years to carry out some research projects partly supported by the German Research Council (spores and pollen by Thiele-Pfeiffer, Munich, 1988; leaves by V. Wilde, Frankfurt/M., 1989; fruits and seeds by M. Collinson, London, 1983, and 1986; and lake algae by K. Goth, Frankfurt/M., 1990). The description of the flora which follows and which makes reference to the most important groups is based mainly on these studies.

There exists in Messel an almost unique opportunity to compare the rich fossil finds of the individual groups of organs (leaves, fruits and seeds, flowers, spores and pollen). In doing this the correlations have in part proved very good, but there have also been significant differences. These differences can originate in the variable suitability for preservation of the different plant organs, but can also stem from the growth habit of the plant and the ability of plant organs to survive transport.

Ferns and fern allies (Pteridophyta)

The pteridophytes are even today a highly varied class of plants, with such different forms as the typical ferns, horse-tail (*Equisetum*), and club-moss (*Lycopodium*). Appropriately, they also have very different habitats. Many indeed are adapted to moist surroundings, but others can endure drought. In the warm regions of the world several genera of ferns even reach tree size. In previous geological epochs this variability was even greater and even horse-tails and club-mosses existed as trees!

In the Eocene the class of pteridophytes had already attained approximately the composition that it has today. This is also shown by the leaf debris (Wilde 1989), which was, however, found only rarely and then mostly in the form of small fragments. Sporangia, in which spores are formed and which enable a classification with the recent genus *Thelypteris*, were found on one fragment only. This fragment belongs therefore to the polypdiaceous ferns (Polypodiaceae *sensu lato*), the most modern family of ferns, and is related to the marsh fern of Germany. In this relationship *Rumohra recentior* (Fig. 15) also belongs, as well as a fragment of the 'mangrove fern' *Acrostichum*, the only fern which occurs in mangrove woods and which can tolerate salt water without requiring it. More frequent, by comparison, is a fern judged to belong to the

Fig. 15: Pinna of *Rumohra recentior*, a polypod fern, without sporangia. Each leaflet is 6 cm long.

Schizaeaceae and represented at Messel by 30 fragments. The main evolutionary period of this family was in the Mesozoic. The fronds of the species *Ruffordia subcretacea* occur, apart from Messel (Fig. 16), also in the Geiseltal near Halle and at other recovery sites from the Lower Tertiary. Their relatives are found today mainly in the warm regions of the world (Fig. 17). Some fragments of yet another family were found, namely the royal ferns (Osmundaceae) which also have their evolutionary climax in the Mesozoic. The species at Messel is *Osmunda lignitum*, which had a wide distribution in the Tertiary.

The finds of spores that can be associated with ferns are very much richer (Thiele-Pfeiffer 1988). They are easily recognized by either a three-rayed 'trilete' germination pore or a simple 'monolete' one. At Messel 12 genera with over 20 species belong in this relationship. Thus five species were found which belong to the 'lesser club-moss' (*Selaginella*) from the

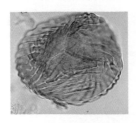

Fig. 16: Reinforced spore of the fern *Ruffordia* (Schizaeaceae). The spore is named *Cicatricosisporites paradorogensis*. Magnification, *c*.300×.

Lycopodiaceae. Among the fern spores a large part probably belong to the Schizaeaceae (Fig. 16), several to the Polypodiaceae, and one very rarely occurring form to the Osmundaceae. In spite of the wealth of the spore flora, its distribution over individual families appears to correlate approximately with the leaf fragments. Among the spores there is even a species (*Cicatricosisporites paradorogensis*) (Fig. 16) that was isolated by Barthel (1976) from sporangia of the fern *Ruffordia subcretacea*. A few belong to the Cyatheaceae with which tree ferns are classified nowadays. Their presence in Messel must therefore also be expected.

Conifers (Coniferae)

As gymnosperms the conifers are really a floristic element of the Mesophytic and appear with the first early representatives at the end of the Carboniferous. During the course of the Mesozoic they evolved and are today still represented by five families. The most species-rich and the youngest of these are without doubt the pines (Pinaceae), which even today

Fig. 17: Present-day distribution of the Taxodiaceae.

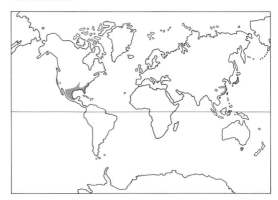

make up the northern softwood belt but which are restricted to mountain areas in the tropics. Apart from the pine, these include (*Pinus*), the fir (*Abies*), the larch (*Larix*), and the cedar (*Cedrus*). Of the other families, only two are represented in Europe, the cypresses (Cupressaceae) by the juniper (*Juniperus*) and cypress (*Cupressus*) and the Taxaceae by the yew (*Taxus*). The rest of the 50 genera with 550 species are—sometimes with only small geographical ranges—strewn across the globe. Among these, two more families should be familiar. The swamp cypresses (Taxodiaceae) include the redwoods (*Sequoia*, *Sequoiadendron*, and the swamp cypress (bald cypress) (*Taxodium*), and to the araucarias (Araucariaceae) belongs the Norfolk Island pine (*Araucaria heterophylla*).

Corresponding with the fragmented nature of the group, their climatic demands are also very different. While some advance almost to the arctic tree-line, others require a mild climate and can scarcely endure frost. The Podocarpaceae are thus largely limited to the warm areas of the world. There, however, they thrive mainly in the mountain forests: in the rain forests of the lowlands conifers are not important.

It is therefore not surprising that conifers are very rare at Messel. According to Thiele-Pfeiffer (1988), pollen, which she attributes to the pine (*Pinus*), occurs regularly and the pollen of *Cathaysia*, a conifer similar to the larch and still extant in China, occurs rarely. The swamp cypresses (Taxodiaceae) are even represented by three different species of pollen grains. One belongs to the Japanese umbrella pine (*Sciadopitys*) which is today represented by one species in Japan. A second species is attributed to the Chinese water pine (*Glyptostrobus*) (Fig. 18). A further form of

Fig. 18: Pollen grain with a slit-like tear from a member of the Taxodiaceae, probably of *Glyptostrobus*, named *Inaperturopollenites concedipites*. Magnification, *c*.500×.

pollen is regarded as belonging to the Cupressaceae but cannot be attributed to a definite genus.

Of these conifer pollen forms, only those of *Pinus*, Cupressaceae, and *Glyptostrobus* occur 'regularly' if not 'abundantly' in cores. All others are limited to single samples. When one considers in what quantities pollen is produced by the wind-pollinated pines, one must conclude that conifers cannot have played an important role on the shore of Lake Messel. This is made even more evident by the rarity of macrofossils. According to Wilde (1989) there are a few fragments of both the conifer *Doliostrobus* Marion 1884 (Fig. 20) and the Chinese plum yew *Cephalotaxus* Endlicher 1842. The genus *Doliostrobus* belonged presumably in the family Taxodiaceae and had a wide distribution during the Lower Tertiary, but its attribution to the Araucariaceae is still under discussion. Engelhardt described this conifer from Messel as early as 1922. More recent collections include some fragments of twigs, cone scales (Fig. 20), and a cone fragment. The cone scales are easily recognized by the light streaks of resin. The Chinese plum yews (Cephalotaxaceae) are still extant, with eight genera, from the Himalayas to southern

Fig. 19: Twig of the now Asiatic plum yew *Cephalotaxus*. The stomata are in the two pale stripes on each needle. 25 mm long.

Fig. 20: Cone scale of the conifer *Doliostrobus* cf. *certus*, probably a member of the Taxodiaceae which was widely distributed in the early Tertiary. 15 mm long.

have preferred swamp habitats, whereas other genera, such as *Pinus*, point to the existence of dry locations. According to their present climatic demands, the conifers which occur in Messel are evidence of a subtropical to warm-temperate climate. Mai and Walther (1985) take *Doliostrobus* as an indicator of a monsoon climate.

Plants resembling grasses

In today's flora, grass-like plants play a major role. In open woods and also in areas that, due to their low rainfall, are sparsely wooded or treeless, we find grass-like plants as ground cover. We call such arid areas steppes or savannahs. But also beyond the timber-line in the high mountains and in areas too cold for trees in the arctic or antarctic tundras, grass-like plants are a substantial constituent of the flora. Thus we can hardly visualize present-day ground-cover vegetation without grass-like plants, since it is, conceptually, very easy to

Fig. 22: Flowering grass-like shoots of a member of the Restionaceae from Australia.

Japan. Until now they were known to occur during the Tertiary only from the late Eocene to the Pliocene. The Messel samples dating from the Middle Eocene are thus the oldest finds. They consist of a few needles and leafy bits of twigs (Fig. 19) and are easily recognized by the stomata which are arranged in two bands on the lower side of the leaf. The stomata lie parallel to the longitudinal axis of the leaf and have two lateral subsidiary cells. Buds at the tip of the twigs indicate a temporary arrest of growth which could have been caused by a seasonal climate.

The rarity of macrofossils and the small proportion of conifer pollen found in Messel allow us to conclude that conifers were rare in the immediate vicinity of the lake or, more probably, did not exist there. Reconstructions with *Glyptostrobus* or *Taxodium* on the shore of the lake have used as models the brown-coal swamps with which the flora of Messel has nothing in common. The conifers that have been documented must have grown mainly in the hinterland where *Glyptostrobus* would

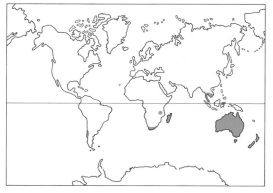

Fig. 21: Present-day distribution of the Restionaceae. Now limited to the Southern Hemisphere, they were of world-wide distribution during the Tertiary.

cover, in our imagination, areas without trees by a dense growth of 'grass-like' leaves from no specific family. Even today when we observe a grass-like vegetation it does not always consist of true grasses. Nevertheless, on steppes, meadows, and even in the reed-belts bordering

ponds, the true grasses (Gramineae) dominate. In these the flowers are reduced to single anthers or, respectively, to ovaries and protective glumes. These aggregate into small spikelets which form, in turn, larger units such as, for example, spikes, panicles, or racemes. Today grasses are one of the most morphologically varied families; they have conquered many very different ecospheres and their highly varied forms range from annuals to 40-metre-high woody bamboos.

Throughout the world there grow among these true grasses plants which, although they look similar, have a different floral structure and belong to entirely different families. The most important of these in our ecosphere are the rushes (Juncaceae), the sedges (Cyperaceae), the reed-maces (Typhaceae), and the bur-reeds (Sparganiaceae). The leaves of all these families, which today are mostly dominant in moist, boggy habitats or at the borders of water-bodies, or which form distinct ecotypes, can hardly be distinguished from one another or from true grasses, yet the pollen is generally easily recognized. This is true most of all for the grasses, whose pollen grains each have a single, distinct pore. Such pollen occurs very seldom in the Lower Tertiary and has not, for example, yet appeared in Messel. In the Geiseltal we meet grass pollen only very occasionally in the younger layers of the Upper Eocene, and from fossil data we know that only in the Upper Tertiary does it become frequent enough to allow us to postulate the existence of sizeable open grasslands (steppes).

Even if grasses are still missing in Messel we have, from the pollen, evidence of all the other families mentioned. Fossil fruits of the sedges (Cyperaceae) have even been found. These were, as single fruits, already known from the London Clay (from the Lower Tertiary) as *Caricoidea* (*Carex*: sedge). They occur at Messel still joined together in short infructescences (Fig. 23).

Apart from these families, still represented in our flora today, it was possible to document at Messel a further family at present only found in the Southern Hemisphere, the Restionaceae. The pollen of this family is very characteristic and samples dating from the Cretaceous and Lower Tertiary have already been reported from other sites (also European) in the Northern Hemisphere. It is described as the 'artificial genus' (form genus) *Milfordia* (Fig.

Fig. 24: Finely stippled pollen grain of *Milfordia*, a member of the Restionaceae, of which two species occur at Messel. Magnification, c.500×.

24). The genera of this family occur nowadays mainly in moist habitats in South Africa, New Zealand, and also in Malaysia, Chile, and Patagonia (Figs 21, 22). It appears that the Restionaceae occupied habitats world-wide during the Cretaceous and the Lower Tertiary, from which they were later partly displaced by grasses. In Messel two species of *Milfordia* were documented (Fig. 24) mainly at the base of three deep drill cores (Borings 1, 2, and 4). It appears that a special lake-side vegetation or an initial bog vegetation was present before the start of the oil-shale formation. Because of the absence of *Milfordia* from several drill cores, Thiele-Pfeiffer (1988), assumed either that individual shallow pools existed before the actual sinking of the lake bottom, or that the initial sinking was uneven. Because the pollen profile from that time does not differ from the later one, the surrounding forest vegetation must already have existed at that time and have hardly changed thereafter.

As the leaves and stalks of all the grass-like families mentioned are very similar, it is, unfortunately, not possible to decide to which one the 'culms' belong, which are not infrequently found in Messel. They are sedge-like culms, articulated by nodes, from which leaves arise at an upward angle. They may come from reed grasses but equally they may come from the Restionaceae. From their frequency in the oil shale one can conclude that they probably formed a 'reed-like belt' around the edge of the lake. They are definitely not true grasses.

Palms (Palmae, Arecaceae)

At present, palms, with a large number of genera (about 200) and species (2800), are at home in all warm areas of the world, and just reach Europe in southern Spain and southern Italy with the dwarf palm (*Chamaerops humilis*) (Fig. 25). Because the northern and southern extent of the area is climate-dependent, palm fossils are valuable climatic indicators.

Fragments of palm frond were already found in Messel in the last century and Engelhardt (1922) reproduced the first picture of a pinnate leaf. In subsequent years only a few fragments were recognized and incorporated in the collections. Only in the contemporary layers of the Prinze von Hessen mine, near Messel, did there occur (in a silicified horizon) petrified stem bases with a large number of roots which were correctly classified as pinnate palms by Müller-Stoll (1935).

Only very recently two sizeable pinnae (Fig. 26) as well as smaller leaf fragments have been found. Based on the wavy epidermal cells of the leaf, Wilde (in Schaarschmidt and Wilde 1986, p. 192) classified these with the species-rich scaly-fruit palms (Lepidocaryoideae), which are at present distributed mainly in the tropics from India to the Indo-Malaysian Islands. Their easily recognized fruits were found, not infrequently, in the European Tertiary. Today they include such important genera as the rattan palms (*Calamus*) and the sago palms (*Metroxylon*).

Surprisingly, some years ago insignificant bud-like objects were recognized as palm blossoms (Fig. 27), of which in the meantime

Fig. 25: Present-day distribution of palms in the tropics and subtropics, with extensions into warm–temperate areas.

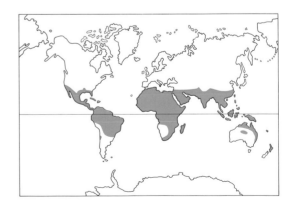

Fig. 23: Compound infructescence of *Caricoidea*, a reed grass (Cyperaceae). Length, 12 mm.

Fig. 26: Fragment of the pinnate compound leaf of a palm frond (*Phoenicites*) which, by the structure of its epidermis, belongs in the Lepidocarioideae. Reduced, *c.*0.5×.

Fig. 27: Partially opened male flower of a palm, with three of the six anthers visible, and three petals and basal sepals. Length, 8 mm.

flowers have a similar structure: some have a condensed structure and always contain anthers; others are more slender, contain no anthers, and show thickening in their lower third. Obviously these are monoecious male and female flowers.

In the male flowers the anthers can be visible in different ways. Frequently, the contours are visible through the petals, or they extend partly between the flower petals. Sometimes one sees them exposed on the inside of opened petals. Under the fluorescence microscope, one frequently still sees in the long anthers, the yellow glistening, monocolpate (one groove) pollen, which also occurs singly on the surface of the flowers and is evenly distributed in the sediment. The pollen has already long been known from other sediments, under the Latin name *Monocolpopollenites tranquillus* Pflug 1953 (Fig. 28).

The finds of palm blossoms finally allow the attribution to the palms of a type of fruit frequently found in Messel—longish-oval, often wrinkly structures that measure approximately 2 cm in length. Attached to their base is the same three-part calyx as in the palm blossoms (Fig. 29).

several thousand were procured (Schaarschmidt and Wilde 1986, p. 179). They are approximately 8 mm long and their structure corresponds with the basic plan of monocotyledons: three sepals, three petals, and two whorls of three anthers each. The ovular region is, unfortunately, as occurs for most buds, pressed flat against a black coal layer so that the structure of the female organs cannot be studied. It is apparent, however, that not all

Fig. 28: Palm pollen grain from an anther, with a deep germination groove. Scanning electron microscope: magnification, 1000×.

It is difficult to associate the flowers with a Recent palm genus, due to the wealth of forms in this family. It is subdivided into up to 15 different subfamilies. The relatively simple structure of the unisexual flowers supports its relationship with the betel palms (arecoid palms). This group of palms, the largest with 88 genera, is at present distributed throughout the tropics with its centre in East Asia. Well-known representatives are the royal palm *Roystonea regia* and the betel palm *Areca catechu*.

Evidence that both organs belong to one species comes from the finds of series of immature fruits and from the identity of the calyx with the corolla. The leaves are quite another matter. They were attributed to another subfamily; that means that they cannot

Fig. 29: Palm fruit with a distinct calyx at the base which coincides with the calyx. Length, 25 mm.

Fig. 30: Fruit of a Recent palm, sectioned longitudinally on the right hand side. The mesocarp is fibrous, as in the coconut. Natural size.

originate from the same plant. It would, though, also be possible to assume the existence of a tribe in which the characteristics of both evolutionary lines were united, that is, a tribe intermediate between the two and which is today extinct. There is no evidence for this. Further, rarer finds of palm fragments not described here—differently shaped blossoms, inflorescences, and fruits—show that there could not have been just one species of palm in the area surrounding Lake Messel. The probability of the survival of fossil evidence depends on various prerequisites. The blossoms must originate from a species which flowers abundantly and whose flowers are easily dehiscent from the axis. They must not, therefore, as in many palms, be inserted deeply into the axis. The leaves, on the other hand, can originate from any of the different species that grew close enough to the bank. The palm leaves, in the natural environment, usually stay on the trunk even after their death so that they are torn off only occasionally, for example, by the wind. This explains too the rarity of leaf fragments in Messel in comparison with those of flowers and fruit.

It can be observed during preparation that the palm blossoms frequently appear incomplete and threadbare. This appearance can be explained with the help of small pellets found frequently on the surface of such flower fragments. Under the fluorescence microscope these are shown to contain palm pollen. It is easy to explain the 0.2 mm diameter structures as fossil faecal pellets (coprolites). The producers were probably Microlepidoptera. There are still today several genera which live parasitically on palm blossoms. The 'date moth' *Batrachedra* often causes great damage when its larvae eat the blossoms and young fruits of the date palm. The finds from Messel are evidence that similar host–parasite relationships already existed in the Middle Eocene.

Even though the palms are today a very varied family one can still draw climatic conclusions from their occurrence because they are even today primarily restricted to the tropics and subtropics. There is no reason to

believe that it was ever different. Palms are the best evidence that Messel had a warm climate.

Aroids (Araceae)

The Araceae are a varied family with many, mainly herbaceous, forms. Many of their taxa have creeping axes or tubers, from which the erect shoots with often large leaves and inflorescences arise (for example, *Calla*). Other representatives are lianas, which either remain anchored to the ground by means of aerial roots (for example, *Philodendron*) or later become epiphytes when the lower part dies. Although a few genera (*Arum* and *Calla*) extend into the temperate zone, the bulk of species are distributed throughout the tropics. The flowers occur in great numbers on spadixes above a spathe. The latter can constrict in the middle in some genera, as in *Arum*, catch insects in the lower, bubble-like part, and force them to pollinate. The variably shaped leaves can become very large (in *Amorphophallus* 2–3 m) and, in contrast to other normally parallel-veined monocotyledons, frequently exhibit pinnate venation. The fine venation between the large veins is also characteristic: in contrast to the multicostate leaves of monocotyledons (e.g. banana) which show, at most, simple connections (commissures), the tertiary veins in the Araceae are reticulate. By this means

Fig. 31: Present-day distribution of the Araceae; most species are found in the warm areas of the world.

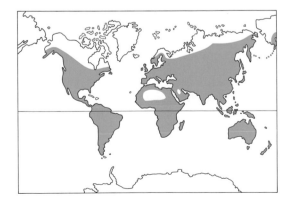

Fig. 32: Leaf of a member of the Araceae. Most extant representatives are found in the tropics and subtropics as herbaceous plants, lianas, or epiphytes. In Messel one finds several leaf forms; the one illustrated has a broad central vein and strong primary lateral veins. The intermediate veins are reticulate (not visible in photograph). Length, 23 cm.

fossil leaves of the family can be clearly recognized (Fig. 32).

While at other sites the Araceae are represented primarily by their seeds (*Epipremnum*), at Messel we find only leaf fragments. Some of these are in the form of large and beautiful leaves, but others are only small shreds that can, however, always be recognized by the above-mentioned venation (Fig. 32). While up to now Araceae have been described very rarely from the Tertiary, Wilde (1989) was able to distinguish three groups. In one the main vein and the strong lateral, primary veins dominate and are connected by the curving fine venation. In the two others only the central vein is apparent, while the thin connecting veins run closely parallel.

The laurel family (Lauraceae)

The laurels are mainly woody evergreens. With 32 genera and 2500 species they are spread throughout the world, with main centres of distribution in Asia and America. Here they form an important fraction of the forests and dominate most of all in the mountain forests at altitudes between 1300 and 2700 m. Their outliers are the laurel-forest relicts on the Canary Islands. Apart from the laurel (*Laurus nobilis*), the genus *Cinnamomum* with the cinnamon and camphor trees and the avocado pear (*Persea americana*) is well known. At Messel the leaves of this family are most frequently preserved—rarely the fruits as well.

Leaves of the Lauraceae occur in large numbers in Tertiary sediments. The best known of these are the three-veined leaves which were formerly regarded as leaves of the cinnamon tree (*Cinnamomum*) (Fig. 37). Since then, we know that not all three-veined leaves belong to the Lauraceae (in Messel similar forms occur also in *Ficus*), and a large part of the family has simple, not very characteristic, leaves with reticulate venation that cannot always be clearly distinguished, morphologically, from other form-groups (Fig. 34). Anatomical investigation of the leaves is therefore essential for their identification (Wilde 1989). They often contain oil-particles in their tissues (Fig. 36) and show characteristic stomatal structures: two subsidiary cells always lie parallel to the guard cells (Fig. 35). In addition, simple trichomes with a unicellular base are often present.

The nomenclature of the fossil laurel-like leaves has altered, historically, several times. Although the Recent genera cannot be definitively separated by their leaves, it has been useful to subordinate the fossil ones to two form genera: the reticulate-veined leaves are named *Laurophyllum* Goeppert 1854 and the three-veined leaves *Daphnogene* Unger

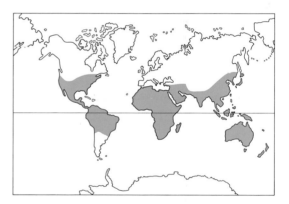

Fig. 33: Present-day distribution of the Lauraceae.

1850. Both can be separated only by examination of their cuticles.

Two questionable species of fruits of Lauraceae are mentioned by Collinson (1986). They are up to 2 cm large berry-like fruits, partly with a stalk (Fig. 38).

The pollen of the Lauraceae has such a thin outer membrane that no fossils survive. It was,

Fig. 34: *Laurophyllum lanigeroides*. Pinnate-veined leaf of a member of the Lauraceae, with circular holes in the tissue of the leaf which were presumably caused by a fungus, and are characteristic of this species (see the last section of this chapter, 'Interactions between animals and plants'). Length, 10 cm.

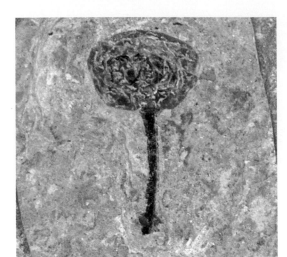

Fig. 38: Berry-like fruit with peduncle from a member of the Lauraceae. Length with peduncle, 23 mm.

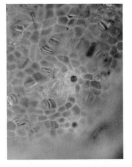

Fig. 35: Fluorescence micrograph of the leaf cuticle of a member of the Lauraceae. Epidermal cells, stomata with two parallel subsidiary cells, and two stumps which once bore trichomes are visible. Magnification, 150×.

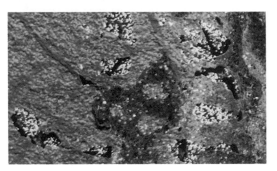

Fig. 36: (above). Leaf of a member of the Lauraceae; the oil particles in the mesophyll between the upper and lower epidermis are brightly fluorescent. Magnification, 25×.

Fig. 37: (left). *Daphnogene crebrigranosa*. Twig with several leaf fragments belonging to a member of the Lauraceae. The leaves have three main veins at the base. Length, 22 mm. Magnification 1.2×.

therefore, not possible to collect evidence of its presence in Messel.

Due to the work by Sturm (1971), which was restricted to the leaves of Lauraceae, an impression may have arisen at some time in the past that forests of laurel existed in Messel. Although this does not do justice to the variability of the Messel flora it is correct in so far as the Lauraceae were an important and species-rich element in Messel. As today, they would mainly have made up a large part of the lower woody scrub up to and including small trees. They are only one of the many angiosperm families in Messel, but serve as important evidence for subtropical conditions.

Water-lilies (Nymphaeaceae)

Although the water-lilies (Nymphaeaceae) today contain only nine genera (with 90 species), they assume so many forms that some authors split them into up to five families. We will unite them here in three subfamilies.

The most independent are the Nelumbonoideae, with the lotus flower (genus *Nelumbo*); this is represented by one species each in North America, Asia, and northern Australia. It is distinguished by the fact that the nut-like fruits are not fused, but are arranged spirally and submerged in the disc-like, flattened receptacle at the top of the floral axis.

The blossoms of the hairnix flower (Cabomboideae), with their few constituent parts, also have free carpels. Of the two genera, *Brasenia* is distributed throughout the Northern Hemisphere, with the exception of Europe, and *Cabomba* occurs in America.

The largest subfamily is that of the true water-lilies (Nymphaeoideae), in which the usually circularly arranged carpels are free from one another, but are frequently submerged in the receptacular tissue. The subfamily contains the well-known water-lilies of the genus *Nymphaea*, which, with 40 species, has an almost world-wide distribution, and the yellow water-lily *Nuphar*, which occurs with a few species in the Northern Hemisphere. The other species have a restricted area of distribution: the spiny genus *Euryale* is, with one species, restricted to Asia, and the large-leaved *Victoria*, with two species, to tropical South America. The submerged *Barclaya* has a tube fused from petals and stamens, and occurs with three species in Indo-Malaysia, while *Ondinea*, which has no petals, has one species and is restricted to western Australia.

The strong morphological and geographical disjunction of the family indicates that it must be quite old in evolutionary terms, and that the individual groups may even be of different origin. For such a family, early fossil finds such as those from Messel are of particular importance.

Water-lily leaves are found frequently at Messel, and their existence there has been known for a long time (Engelhardt 1922, p. 85, Plate 28, Fig. 3). They have a strong central

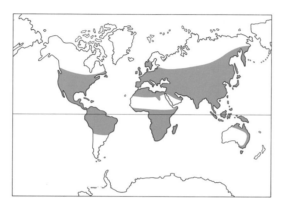

Fig. 39: Present-day distribution of the Nymphaeaceae.

Fig. 40: Recent water-lily (*Nymphaea*) in flower. The floating leaves are not fused at their bases.

vein and reticulate secondary veins (Fig. 41). At the base the leaves are deeply notched but not auriculate. According to Wilde (1989) they are morphologically very similar to the leaves of the recent water-lilies of the genera *Nymphaea* and *Nuphar*. The leaf fossils are often difficult to prepare, because only a little organic substance from the thin laminae remains. They are easily recognized, however, because they are always covered by a light film which produces as a strong yellow glow under the fluorescence microscope. The film probably originates from a dense growth of epiphytic algae, which leads to the conclusion that the leaves had a submerged or floating habit.

Apart from the leaves, Nymphaeaceae 'blossoms' have also been found (Fig. 43). Their most striking feature is a bold pattern of reticulation on *c.* 4 cm long, oval shapes. Occasionally there are, at the base, still some broad elliptical to lanceolate leaves, which can easily be classified as sepals. At the opposite end of the oval shape there is a tuft of stamens in which, under the fluorescence microscope, pollen grains can still be detected. These were also studied using light and scanning electron microscopy (Figs 44 and 45). Thiele-Pfeiffer (1988) also found this pollen distributed in the oil-shale cores and described it as *Monocolpopollenites crassiexinus*. The pollen grain has an elongated opening sealed by a lid-shaped structure.

The external view already suggests that these are floral organs. But how can the various shapes be explained; which organs do they represent? Initially, the central reticulate structure was particularly puzzling. If this was the ovary then its position below the stamens is unusual, because it is usually located above. Epigynous ovaries may exist in water-lilies (*Nymphaea*), but the carpels are always arranged in whorls and are coincident with the petals. Only in the lotus (*Nelumbo*) do irregularly arranged carpels occur, but they are inserted on flat receptacles. The solution to the puzzle was found during the preparation of yet more 'blossoms'. In these a central pattern of reticulation was never found, but rather an irregular, wrinkled surface. The reticulation

Fig. 41: Leaf with distinct venation, belonging to a member of the Nymphaeaceae. Laminar tissue is almost entirely absent, lighter spots presumably caused by algal film (leaves are often completely covered with this and appear white, without distinct venation); leaf base not fused. Length, 16 cm.

must therefore be an internal structure which is covered by the wrinkled layer. Between both layers, cases of insect larvae were found regularly (Fig. 43), which were composed of little stones just as in cases from extant caddis-fly larvae (Trichoptera). Furthermore, seeds, which definitely are those of Nymphaeaceae, were discovered in some samples between the stamens or more towards the centre (Fig. 46). These 'blossoms' belong, therefore, without doubt to that family. Although their structure departs in major aspects from that of Recent Nymphaeaceae, common features can still be recognized, particularly with the Indo-Malaysian *Barclaya*. These bear five sepals at the base (which are also identified as bracts). At the base the petals have fused with the sepals forming a tube, at the bottom of which there is a circle of 10 carpels.

The structure and biology of these flowers of Messel can be reconstructed as follows (Fig. 42). On a long stalk above some large sepals, a form of container arises, which is formed either by the stamens alone or with petals as well. At the upper rim this 'cup' bears a whorl of stamens. The ovary enclosed within was probably much smaller during the flowering period, and the free carpels were arranged spirally on a conical receptacle. Following pollination, the ovary matured into a compound fruit, entirely filling the space in the cup. At maturity the carpels might have become mucilaginous and dissolved, just as in several Recent Nymphaeaceae. This mucilage could then be food for the larvae of Trichoptera which probably also aided the discharge of seed. The cases have up to now been found in every completely preserved fruit and it can, therefore, be assumed that the process was not haphazard, but the rule. A precondition for this is, however, that the fruits, when ripe, were in the water so that, therefore, as in *Barclaya*, the flowers developed underwater or, more probably, the pollinated flowers or mature fruits reached the water, for example, by the bending of the peduncle. This is a further example of plant–animal interaction.

The fruits of Nymphaeaceae are also important in another respect because they represent a group (perhaps a new subfamily) of extinct Nymphaeaceae which unites primitive characteristics (spirally arranged receptacles) with advanced ones (fused stamen tube). They thus show that the diversity of Nymphaeaceae

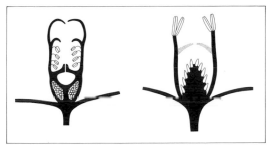

Fig. 42: Reconstruction of the Messel Nymphaeaceae flower (right), compared with the flower of the Indo-Malaysian *Barclaya* (see text for explanation).

Fig. 44: Fluorescence micrograph of pollen grains in an anther of a fossil flower belonging to the Nymphaeaceae; longitudinal germination groove partially closed. Magnification, 150×.

Fig. 45: Scanning electron micrograph of a pollen grain from a flower belonging to the Nymphaeaceae. The longitudinal groove is closed by an obvious lid. Magnification, 1000×.

was greater in the Eocene than today and that their origin lies much earlier, possibly in the Cretaceous.

Knowing so much about Nymphaeacean fruits, we must ask ourselves where they grew. It is certain that their habitat was not in the area of oil-shale formation. Such a site was too hostile for them, and the water certainly too deep. Also, rootstocks, which are not infrequent in other areas, have never been

Fig. 43: Messel water-lily flower fruiting. Below and right are two sepals; above, the longitudinally split area of the receptacle after the disintegration of the carpels. Above left, undamaged external area laid open with a line of anthers at the top. Between them, as well as at the base of the receptacle, there are caddis-fly larval cases (Trichoptera). The larvae fed, presumably, on the mucilaginous mass of the disintegrating carpels. Natural size.

found. Thus, the only conclusion that remains is that, outside the area of oil-shale formation, there must still have existed shallower, flooded areas with water-lilies and a different bank vegetation. The water-lilies cannot, of course, give us any information concerning the extent of such areas.

Climatic information also cannot be expected from them, because the family is today cosmopolitan and occurs in all climatic zones apart from the coldest areas.

Fig. 46: Aggregation of seeds of Nymphaeaceae, possibly freed by the decay of the carpels. Length, 6 cm.

The moon-seed family (Menispermaceae)

This family contains mostly lianas and only a few bushes and trees. The leaves are often large. They frequently have palmate venation and are sometimes lobed. They are widely distributed throughout tropical rain forests, and only a few representatives live in subtropical or temperate climates (Fig. 47). By means of unilateral growth the stone fruits often become kidney- or horseshoe-shaped (moon-seeds).

In Messel the above-mentioned fruits are the principal finds. Collinson (1986) lists seven genera, of which the *Tinospora* are distributed nowadays primarily in Asia. The others cannot be correlated with recent genera. The preservation in Messel is so excellent that we find not only hard stone kernels but also occasionally the soft parts of fruits (Fig. 48).

Three-lobed leaves are also classified with this family (Fig. 49). They have stomata on both sides of the leaf and not, as is more usual, only on the underside.

The pea family (Leguminosae)

The Leguminosae are, with 17 000 species, one of the largest and most variable families on earth. Common to all species are the fruits, 'pods', which form from a folded carpel and bear a row of seeds along their dorsal suture. Their habitat can vary significantly and we find

Fig. 47: Present-day distribution of the Menispermaceae.

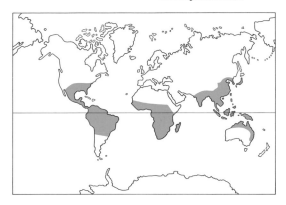

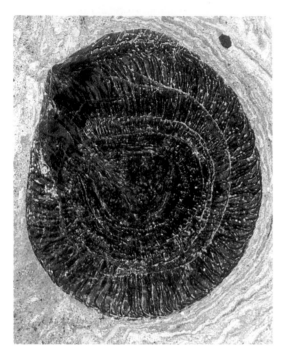

Fig. 48: Stone fruit of a member of the Menispermaceae. To the left are hilum and raphe; externally the soft parts of the carpel are completely preserved. Length, 25 mm.

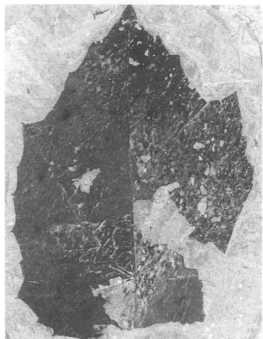

Fig. 49: Leaf of a member of the Menispermaceae, which mainly occur as lianas in the tropics. Lamina fused at base. Natural size.

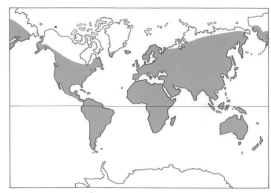

Fig. 50: Present-day distribution of the Leguminosae.

Fig. 51: Pinnate leaf of a member of the Leguminosae. Probably part of a double-pinnate leaf of a member of the Mimosoideae, in which such leaves occur frequently; central section of the pinna not preserved. Natural size.

among them trees, bushes, shrubs, herbs, and lianas. The leaves are mostly simply or doubly pinnate.

The majority of the well-known 'butterfly' flowers are contained in the subfamily Papilionoideae. The flowers of the Caesalpinioideae can have a radiate structure, while those of the Mimosoideae are reduced and cluster in small heads. The two latter subfamilies thrive mainly in warm areas, while the Papilionoideae are also distributed in the temperate zone and even extend into the frigid zone. Leguminosae can be found in extreme, dry habitats, but also as tall trees in the tropical rain forest.

At Messel we find pods and also pinnae or single leaves. Wilde (1989) divides the latter into five species, for only one of which, however, is there extensive fossil material (Fig. 52). Because hitherto only single leaflets were available, this species was classified till now with the Apocynaceae (dogbane family). In Messel a larger pinnate leaf was found to which the single pinnae still adhere by swollen stems as so frequently observed in Leguminosae leaves. This negates a classification with the Apocynaceae, which always have single leaves, and increases the likelihood that these are Leguminosae leaves. The occurrence of Leguminosae in Messel is supported by pods which were portrayed by Engelhardt (1922) and have since been found repeatedly.

The tea family (Theaceae)

From this family the tea bush (*Camellia sinensis*) in particular is of great economic importance and is grown over wide areas of

Fig. 52: Fluorescence micrograph of a pinnate leaf belonging to the Leguminosae. The petioles of the pinnules, as in Recent Leguminosae, are thickened, which is probably evidence of nyctinasty. Leaf length, 10 cm.

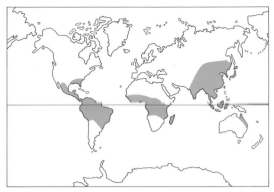

Fig. 53: Present-day distribution of the Theaceae.

Fig. 54: Fruits of a member of the Theaceae with many stalked berries and leaves—a rare case of direct connection between two organs. Length of rock slab, 17 cm.

Asia (Fig. 53). Like most of the trees and bushes of species in this family it is characterized by leathery, evergreen leaves. The insect-pollinated flowers are so striking, that the related *Camellia japonica* is grown as an ornamental plant. The family is distributed throughout the warm zones of the world. Here it mostly avoids moist habitats, but extends, by contrast, into the broad-leaved forests. The family is known to have existed since the Cretaceous and was already species-rich in the Lower Tertiary.

Until now only leaf fragments have been found at Messel, which Wilde (1989) classified as belonging to two genera. Of these *Polyspora* is extant today in South-east Asia and has beautiful striate epidermal cells. One of its two species has, until now, been found only in the stomach contents of *Ailuravus macrurus*. The other genus is *Ternstroemites* and is unknown in the Recent flora. During the printing of this book a sizeable cluster of fruits—perhaps the most beautiful plant fragment from Messel— was found (Fig. 54). The leaves, which still adhere between the fruits, enable an exact identification because the characteristics of their epidermis have been superbly preserved.

The icacina family (Icacinaceae)

The reasons that this family is only little known are its low economic importance and its distribution which, with only a few hundred species extant, is essentially limited to the warm areas world-wide (Fig. 55). Trees, bushes, and lianas occur here mainly in rain forests and swamps. The flowers are mostly small and are grouped in inflorescences; the fruits are stone fruits.

Fossil fruits are widely distributed in the Tertiary and down to the Cretaceous, and two genera from Messel were listed by Collinson (1986) (Fig. 56). Wilde (1989) could, by contrast, only classify, with reservations, a single, albeit large, leaf with the Malaysian genus *Stemonurus*. Since, however, pollen from two genera is not infrequent, we must assume that the tropical family accounted for a significant portion of the Messel flora.

The grape family (Vitaceae)

The grape vines are distributed world-wide in the tropics and subtropics. They have their centre of distribution in the Indo-Malaysian zone and extend with, however, only a few species into the temperate zone (Fig. 57). Most of the 700 species (12 genera) are lianas; only a few are erect bushes. Most important of all for man is the grape vine (*Vitis vinifera*) and some similarly exploited North American species of *Vitis*. Several other genera are valued as

Fig. 55: Present-day distribution of the Icacinaceae.

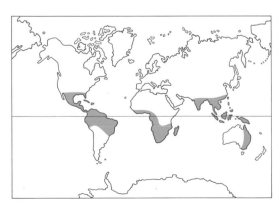

Fig. 56: Kernel of a stone fruit of a member of the Icacinaceae, a tropical family which includes numerous vines. Length, 12 mm.

ornamental plants as, for example, the Virginia creeper (*Parthenocissus*) and the Russian vine (*Cissus*).

In the Tertiary the seeds, in particular, were widely distributed and occur, for example, frequently in the London Clay flora (Collinson 1983). Collinson (1986) mentions three species from Messel, of which two belong to *Vitis* and *Tetrastigma*. The drop-shaped seeds (Fig. 58) are easily recognized: dorsally they have a round or elongated depression (chalaza) and carry ventrally two grooves, one on either side of the raphe. *Vitis* seeds occur occasionally in the oil shale, but are present in large numbers in the sands at the north-eastern perimeter of the pit. They occur here mainly in association with the seeds of the citrus family (Rutaceae). The sands are a formation close to the shore and the seeds of both families were, presumably, washed up there.

The description of fossil leaves of the Vitaceae has been infrequent until now. Wilde

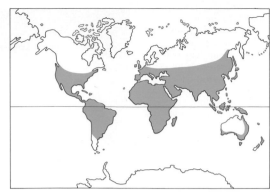

Fig. 57: Present-day distribution of the Vitaceae.

(1989) attributes a few, shallow-lobed leaves to *Ampelopsis* (Fig. 60). Although similar leaves also occur in other families, the fine striation that originates from the stomata of the lower cuticle appears to be characteristic for *Ampelopsis*.

Two pollen forms of the Vitaceae also occur at Messel (Thiele-Pfeiffer 1988). This is the

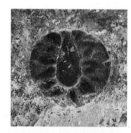

Fig. 58: Seed from an extinct genus related to our grape family (Vitaceae); hilum at the centre of the seed. Diameter, 6 mm.

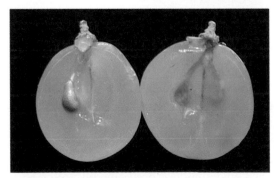

Fig. 59: Longitudinal sections through present-day grapes.

first evidence of *Vitis* pollen from the Eocene (Fig. 61). The second form probably belongs to *Cissus*.

Based on various organs, the following genera of Vitaceae have been identified in Messel: *Vitis*, *Tetrastigma*, *Ampelopsis*, and *Cissus*. Of these, *Cissus* occurs today world-wide in the tropics, *Ampelopsis* and *Tetrastigma* have their distributional centres in the tropics and subtropics of South-east Asia, and *Vitis* is distributed throughout the world in the subtropical and warm-temperate belts.

From the frequency of their fossils one can conclude that the Vitaceae played an important role among the probably generally high proportion of lianas in the forest of Messel.

Evidence that the grape-vine was already a dietary favourite in the Eocene are seeds of *Vitis* found in the stomach of an early horse (*Propalaeotherium parvulum*) in the Hessen Regional Museum in Darmstadt (von Koenigswald and Schaarschmidt 1983). It is

Fig. 61: Fusiform pollen grain of a vine (Vitaceae) named *Tricolporopollenites marcodurensis*, with three grooves and three pores. Magnification, *c.*500×.

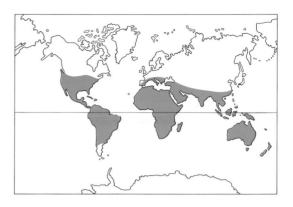

Fig. 62: Present-day distribution of the Rutaceae.

unlikely that the berries were eaten directly from the plant. It can be observed in vines today how berries fall off when ripe and accumulate in large numbers on the ground. Here the small horses would have gained easy access and may have welcomed the berries as a sweet change in their diet.

The citrus fruit family (Rutaceae)

The Rutaceae are primarily smallish trees or bushes; some climb with the help of spines, as does the genus *Toddalia* whose fossil distribution is widespread. The main area of distribution of this species-rich family includes all warm areas world-wide (Fig. 62); only a few genera extend into the temperate zone, as, for example, the diptam (*Dictamnus*), which grows wild in German dry woods. It is, as are most Rutaceae, rich in aromatic oils. The economically most important genus is *Citrus* which contains the familiar lemon and orange.

Among the plants of the Messel oil shale, the Rutaceae play only a moderate role. Only one leaf shape attributable to this family has been discovered until now. These leaves with an entire margin belong to the genus *Toddalia* (Fig. 64). This climbing bush is today represented by only one species in South-east Asia. The fossil of a similar leaf is already known from the Middle-German Upper Eocene and seeds are found quite frequently in the Tertiary.

On the basis of the pollen, four species of the form-genus *Tricolporopollenites* probably belong to the Rutaceae. The only moderate

Fig. 60: Leaf of a member of the vine family (Vitaceae), probably related to the East Asian genus *Ampelopsis*; identification of the multiform leaves is based on the structure of the epidermis. Actual size.

Fig. 63: Inflorescence and compound leaves of *Euodia hupehensis*, an arborescent member of the Rutaceae from China.

frequency of fossil pollen from this insect-pollinated family does not permit a conclusion as to what proportion of the general vegetation was made up of this family. But some samples of a five-petalled flower were found (Fig. 65) which probably belong to this family. Strikingly, 10 stamens of each flower show a voluminous pollen content in fluorescent light.

The firm seeds are discovered occasionally

of *Vitis* (Fig. 58). From this evidence we can conclude that the Rutaceae made up, at least locally, a significant proportion of the vegetation.

fruits. The nut-tree thrives nowadays in the warmer regions of Germany but, following the ice age, it was probably reintroduced by man to the regions north of the Alps. The family today contains only 50 species and seven genera which have a peculiarly disjunct distribution: only the walnut, *Juglans*, is distributed across the largest part of the warm-temperate Northern Hemisphere, down to the tropical Jamaica. The other genera have only limited, natural distributions: *Pterocarya* in East Asia, *Engelhardia* in India and southern China (Fig. 67), *Carya*, the hickory nut, in North America, and the genera *Oreomunnea* and *Alfaroa* in

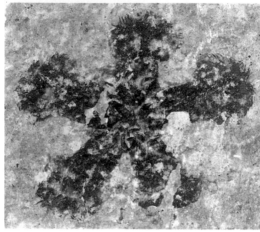

Fig. 64: (left). Leaf of *Toddalia*, a climbing bush belonging to the Rutaceae with a Recent distribution in Asia. Only one species is extant today. Length, 5 cm.

Fig. 65: (above). Flower, presumably from a member of the Rutaceae. The outer border of the petals is fringed; 10 anthers are identifiable, but no carpels. Diameter, 10 mm.

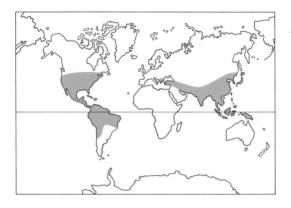

Fig. 67: Present-day distribution of the Juglandaceae.

Fig. 66: Seed of a member of the Rutaceae, with a finely patterned surface and a hilum at the top left. Length, 3 mm.

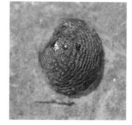

Central America. This list alone shows the different climatic conditions which extend from the tropical/subtropical to the temperate climate.

All genera have large, compound leaves. The male flowers are arranged in a catkin. The fruits are stone fruits, as in the walnut, or nuts, which in some genera, as in *Engelhardia*, are surrounded by wing-like membranes.

The Juglandaceae are represented at Messel primarily by a great variety of leaves and forms of pollen. According to the variability of leaves Wilde (1989) distinguishes, for the moment, four groupings according to shape, which show relationships with the true walnut (*Juglans*),

in the oil shale (Fig. 66). At the northern perimeter of the pit, seeds of the fossil genus *Rutaspermum* (Collinson 1986, p. 219) occur, however, in large numbers together with seeds

The walnut family (Juglandaceae)

The Juglandaceae are known generally, because some important species bear tasty but also with the Central American *Oreomunnea*, and with both Asiatic genera, *Engelhardia* and *Platycarya*. Usually only single leaflets are uncovered, but, occasionally, an intact compound leaf has been found in Messel (Fig. 68). Apart from the mostly entire border of the lamina, the stomatal structure as well as the multicellular peltate trichomes of the leaves are characteristic (Fig. 70).

Some pollen forms, for which relationships with the Juglandaceae are assumed, are particularly richly represented in Messel. To be sure, distinguishing between the simple triporate structure of these short-axis pollen grains and similar pollen-grain structures from the Myricacea (bayberry family) and other families is not always easy (Fig. 69). However, some of the eight different pollen genera have characteristic, bulge-like folds, as in the primitive 'Normapolles-Element'. Such pollen shapes were found by Friis (1983) in flowers of the Cretaceous, which have been classified with the Juglandaceae. They are also correlated with such Asian genera as *Engelhardia* and *Platycarya* and, furthermore, with the American *Oreomunnea* and *Carya*. We must, however, remember that the Juglandaceae are a very old family, which could have contained even more genera in the Eocene than today. In addition, the only catkin found in Messel to date appears to belong to the Juglandaceae (Fig. 71).

The fruits of this family are, unfortunately, very rare in Messel—especially the large stone kernels of the genera *Juglans* and *Carya*. Only the winged fruits of *Engelhardia* were apparently carried in on the wind; these have already been illustrated (Fig. 72) by Engelhardt (1922). The absence of large fruits is evidence that the trees in question did not grow close to the open water of the lake.

The forest of Messel

The Messel formation contains a large number of plant fossils; we have here introduced those which play an outstanding role. The appended list (Fig. 86) may give an impression of the real number of families for which evidence has been found to date at Messel. No monograph,

Fig. 68: Compound leaf of a member of the Juglandaceae. On the basis of the structure of the epidermis it is closely related to *Engelhardia*. In the Eocene the leaf had only one leaflet at the top, while in Recent *Engelhardia* there are two; partially double-dentate. Length, 15 cm.

Fig. 69: Pollen grain with three pores belonging to a member of the Juglandaceae, named *Caryapollenites triangulus*. Magnification, *c.*500×.

however, can give an impression of the collected vegetation as it originally grew in the environment around Lake Messel. The reason for this is due to the form of the preservation. Lake Messel was a catchment area into which plant fragments drifted in various ways.

Fig. 70: (below, left). Fluorescence micro-graph of a leaflet of a member of the Juglandaceae related to *Engelhardia* (small yellow dots: peltate trichomes). Length, 6 cm.

Fig. 71: Male catkin from a member of the walnut family (Juglandaceae). Magnification, 5×.

Fig. 72: (below). Winged fruit of *Engelhardia*. Size, 17 mm.

1. By detachment of plants from vegetation on the bank. This means of preservation can only have been important for water and shore plants.

2. By drifting in with tributaries or on a periodically rising lake surface. As larger tributaries were certainly absent, the drifting can have taken place only via small brooks or during inundations of the whole site. From the investigation of recent conditions we know, however, that bank vegetation had such a strong filter action that fragments could hardly pass.

3. Wind-blown fragments. This means of preservation applies only to small plant fossils. A natural sorting, according to size

and ability to be carried by wind, occurred. Thus the larger fossil fruits could only have originated from the lake-side vegetation and some fruits such as large fruits of the Juglandaceae are entirely absent. Leaves could have been wind-blown across shorter distances and pollen over larger ones.

The larger plant fragments must, therefore, have, for the most part, fallen directly into the water and only small plant materials, notably pollen and spores, could have been blown in over wide distances. The pollen of wind-pollinated (anemophilous) plants predominates since it is produced in larger amounts than that of insect-pollinated (entomophilous) plants. At Messel we have therefore to consider a fossil assemblage (taphocoenosis) which originates from different areas, and on the basis of which conclusions as to the exact distribution of plants can be drawn only with difficulty. Only the variable transportability of different plant organs can tell us whether certain plant groups grew closer to, or more distant from, the lake. It is obvious that species occurring in the closest vicinity should be represented more frequently, in such a fossil deposit, than those which had to be transported in from a larger distance. Because most fossil deposits of plants were accumulated in lakes or bogs, we mainly discover the floras of such damp habitats. Evidence for the vegetation in the vicinity of Lake Messel consists, therefore, not only of the presence of certain families, but also of the absence of some groups or organs. To explain this further, we still have to consider some of these critical families.

The Mastixiaceae occur comparatively rarely at Messel. The leaves (so difficult to identify) are completely absent and only a few of the distinct fruits were found (Collinson 1986). But, although pollen occurs in two forms (Thiele-Pfeiffer 1988), it is not numerous. The very rare occurrence of this family at Messel is curious, because the Mastixiaceae fruits are usually among the most frequently found fruits of the warmth-requiring floras of the European Tertiary. Kirchheimer (1943, for example) had even introduced the term 'Mastixioideae-floras'. The family has today a very restricted distribution in Eastern Asia with its centre in the Malaysian Archipelago. This is, however, surely the only remaining area of a once widespread distribution. There their habitat consists of tropical, evergreen rain forests without a distinct dry season. The rare occurrence at Messel could be related to their almost

Fig. 73: Pollen grain of a member of the Mastixiaceae named *Tricolporopollenites satzveyensis*; fruits widely distributed during the Tertiary. Magnification, *c.*500×.

complete absence from the lake-side. The heavy fruits cannot be transported by the wind. The infrequent finds of pollen for these entomophilous plants are not surprising.

The position of the beeches (Fagaceae) is even more extreme. Neither their leaves nor their fruits were found in Messel, but the pollen is richly represented. All three groups of organs are easily recognized. No fewer than eight different forms of pollen can be distinguished (Thiele-Pfeiffer 1988), which can

Fig. 74: Present-day distribution of the Mastixiaceae.

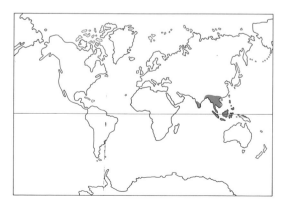

be classified with different, Recent genera as, for instance, the chestnut (*Castanea*) and oak (*Quercus*). The heavy fruits are here, again, absent as are also the leaves of the form genus *Dryophyllum* (Wilde 1989) which are so frequent at other fossil sites. Since the large organs are evidently entirely absent and the pollen of the entomophilous plant could have been carried in from a great distance, the only possible conclusion is that the Fagaceae did not grow in the immediate vicinity of the lake, but in more distant localities. These might have been drier areas. Here is one of the few hints, pointing towards a differentiation of the flora that could have been due to a varying elevation of the landscape.

The occurrence of fossils of screw pines (Pandanaceae) is rare but noteworthy. Only some of their leaf fragments and a number of

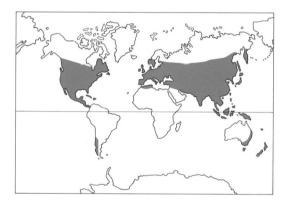

Fig. 75: Present-day distribution of the Fagaceae.

shreds of cuticle have been found in the rock. But these, nevertheless, can always be clearly recognized by the structure of the epidermis (Wilde 1989). The fact that no larger leaf fragments have been found is not surprising because, as in other monocotyledons, the narrow, sword-like leaves are not shed but disintegrate on the parent plant. The few genera of the family are made up mainly of trees with striking aerial roots, and, rarely, also

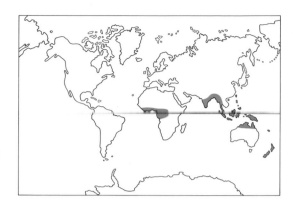

Fig. 76: Present-day distribution of the Pandanaceae.

of climbing plants. They are distributed today from Asia to Equatorial Africa. They are either forest-dwellers or grow on river banks and sea coasts. Not many fossils have been described from the Pandanaceae, but they are nevertheless noteworthy as indicators of a warm climate.

What conclusions can now be drawn from the fossil collection of Lake Messel as regards the vegetation in the lake and its surroundings? According to Rietschel (see Chapter 3, 'The genesis of the Messel oil shale'), the basin of the lake subsided and reached its final depth in a short time. This means that it must have been surrounded from the beginning by steep banks. That tallies with a large number of macrofossils which could not have been wind-blown from afar but must have fallen into the water from the lake-side forest. It is striking that the lowest samples of some deep drill cores contain a larger proportion of Restionaceae (Thiele-Pfeiffer 1988). This allows us to assume that there was a swamp phase at the start of the sedimentation, which must have sustained underground movement at the bottom of the depression.

But, during the period of sedimentation, the lake cannot have been surrounded by steep banks alone; series of water plants that require shallow water are evidence for this. In some localities, at least, shallow water must have existed where water-lilies rooted in up to 2 m depth (Wilde 1989). Behind this, in still-shallower water and in swampy areas, a belt of monocotyledons probably became established. Because there could not yet have been grasses and hence no true reeds, Cyperaceae (sedges) and Restionaceae assumed the role of succession plants. The occurrence of such grass-like plants has been already mentioned above. In this reed-like belt other plants could also have been represented, for example, the Araceae which are strikingly frequent in Messel, and ferns, especially the salt fern *Acrostichum* and the royal fern *Osmunda*.

A border to this swamp belt might have been formed by dense shrub in which screw-pines (Pandanaceae) were represented and which merged with forest slopes covered in Lauraceae, Juglandaceae, Theaceae, and Leguminosae. This forest may have been, as is usual in rain forests, stratified according to altitude and, at the higher elevations, may have had a somewhat more open character with an admixture of Fagaceae. Towards the water it must have had a dense canopy.

In the interior of the forest, and perhaps even more so at its perimeter, lianas obviously played a major role, as a series of families is known from Messel that either consist exclusively of lianas or at least contain numerous climbing species, for example, Menispermaceae, Leguminosae, Rutaceae, Icacinaceae, Vitaceae, Araceae, and Pandanaceae.

In places where the bank was steep, the forest probably bordered the water directly so that it formed, together with the bushes, a dense bordering thicket at the lake-side. Leaves, fruits, and small twigs could here have dropped into the lake from branches that reached down to the water, or could have been blown down from higher branches. The absence of larger trunks preserved as fossils is due to the fact that fallen trees decayed in the forest or on the shore and were not transported because of the lack of currents. For this reason, collections of wood are generally restricted to forest swamps and river sediments and, as a rule, are absent from still-water sediments. Certainly, the large stands of swamp cypress, so beloved of illustrators in their

reconstructions of brown-coal swamps and also of Lake Messel, did not exist anywhere at the immediate lake-side.

We are accustomed to study present-day vegetation by areas, and we can recognize without difficulty its 'banding' according to climatic zones. This appearance can, however, be disturbed by several factors. We can, for instance, find a flora on the southern slopes of a mountain that is characteristic of a far more southern climate. The opposite is true for the northern slope, and certain ecological habitats, such as swamps, can have a stronger influence on the composition of a flora than the geographical latitude.

At fossil sites this study by areas is not possible. In reality, there are only occasional windows which allow a glimpse of certain local situations. Every site, therefore, has its peculiarities which can arise from different causes and which make the comparison of the sites with one another more difficult. The flora of Messel was characterized by Mai (1981) as 'paratropical rain forest', a term that was introduced by Wolfe (1969) to replace the older term 'subtropical forest'. It is supposed to indicate moist, warm rain forests that contain evergreen, large-leaved representatives of the families Lauraceae, Icacinaceae, Mastixiaceae, Menispermaceae, Sterculiaceae, Theaceae, and palms. That this is no true tropical vegetation is indicated in the first place by the lack of purely tropical families such as the Dipterocarpaceae, Melastomataceae, and Musaceae and also by the presence of deciduous families, such as the Betulaceae, Hamamelidaceae, Juglandaceae, and Platanaceae. This must, therefore, have been a species-rich rain forest, which, however, thrived in northern latitudes where temperate floras grow today.

If this rich flora is to be compared with another, then that of the Geiseltal near Halle is most suitable. When the Middle Eocene portion of that flora is compared with its contemporary one at Messel, the correlation observed is very limited (Wilde 1989). But, by contrast, the flora of the Geiseltal shows many similarities with the Helmstedt flora with which it was also contemporary. What are the reasons for these differences? The answer is simple:

both the Geiseltal and Helmstedt have brown-coal floras which evolved in swamps; Messel, however, is a lake deposit. But coal floras are strongly locally influenced floras which cannot be compared directly with Messel. Their fossils have generally not been transported over long distances and mirror, juxtaposed, several local floras from, for example, individual brooks. The Messel flora, by comparison, draws on a larger region for sites of origin and is not subject to the special conditions of a swamp. Thus, the Messel flora gives a more exact picture of the Eocene flora of Western Europe than that of the Geiseltal and is, therefore, a 'zonal flora'.

The only other similarly rich flora known is that from the London Clay (Collinson 1983). That flora, however, consists almost entirely of fruits and seeds. In addition, owing to its coastal position, it has a unique character and, for example, includes mangroves.

Interactions between animals and plants

At Messel many leaves are preserved incompletely, and it is frequently uncertain whether an injury was sustained. In some cases force may have contributed when the leaf was 'torn' off. This should be true most of all for groups which do not experience natural leaf abscission, as do most deciduous trees (even when the leaves remain on the tree for several years). Particularly in monocotyledons (as, for example, reed-like plants, palms, and screw pines) and also in the ferns, the leaves dry out on the parent plant and finally fall off through external agents such as wind or decay.

Damage to leaves can come about in purely mechanical ways. Animal agents cause certain eating patterns or changes in the leaves such as callus formation or galls. After being damaged, the leaf in Fig. 77 continued growing and was therefore certainly damaged during its growth period. Presumably an animal ate part of the leaf. Particular patterns of eating, such as the mines of insects, have, up to now, been observed only rarely at Messel. Occasionally, characteristic patterns of dashes are found which stem, presumably, from accumulations of insect eggs.

Irregularly destroyed leaves that result from the partial decay of dead leaves or from fungal attacks on living ones are more frequent. However, galls caused by animal or plant

(Fig. 34). This regular occurrence in almost all of the leaves of that species is reminiscent of the black spots that can be observed every year in the maple, and are caused by the fungus *Rhytisma acerina:* Like these, the *Laurophyllum* spots also certainly originate

so well preserved is due to the fact that certain plant cell-wall substances are indigestible and are also not destroyed by the process of fossilization. They are the cutin of which the externally protective cuticle is composed and the sporopollenin in the membrane of the

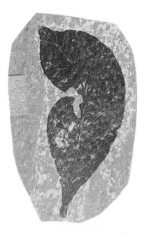

Fig. 77: Leaf, clearly damaged by a herbivore when immature. The curved shape is evidence that it continued growing after being damaged. Natural size.

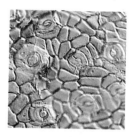

Fig. 79: Cuticle of a leaf from the stomach contents of an early horse. Magnification, 180×.

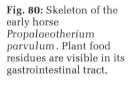

Fig. 80: Skeleton of the early horse *Propalaeotherium parvulum*. Plant food residues are visible in its gastrointestinal tract, mainly in the form of seeds of the vine *Vitis* which were presumably collected from the ground. Length of the skeleton, c. 50 cm.

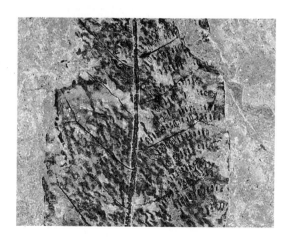

Fig. 78: Eggs laid by an insect on a leaf surface. The long oval eggs were deposited in crescent-shaped bands. Magnification, c. 2×.

parasites are rare. A striking example, though, is furnished by the almost round holes in the leaves of *Laurophyllum lanigeroides* which are surrounded by a conspicuous swelling

from a species-specific parasite.

There have been various reports on plant remains found in the stomach contents of vertebrates. Thus, as early as 1922, using chemical techniques, Kräusel obtained pine needles from the stomach contents of the saurian, *Trachodon*. The remains of leaves in the gastrointestinal tract of the early horse *Propalaeotherium* have been documented by Franzen and Richter (in Franzen 1977). Just as in the remainder of the oil shale, the protective layers of leaves (the cuticles) can be easily isolated and studied under the microscope. The ability of these plant fragments to remain

Fig. 81: Seed of the vine *Vitis* in the gastrointestinal tract of the early horse *Propalaeotherium*. This is evidence of the variety of foods eaten by early horses. Seed c. 6 cm long.

spores and pollen grains. They are both organic compounds and consist, like those in man-made plastics, of long-chain molecules which are chemically almost indestructible. Lignin too has great resistance. Apart from wood it occurs in other woody tissues, for example, in the cell walls of fruits and seeds which are eaten mainly by rodents and which can later be recovered from their stomachs.

Since the shape of the epidermal cells is visible on the cuticle, the family and genus from which such plant fragments in stomach contents derive can be identified. Apart from remains of laurels (Lauraceae), Sturm (1978) identified five more plant families, including walnut (Juglandaceae), in the stomach of an early horse. This serves to demonstrate that these early horses, previous to the evolution of grasses, browsed on leaves from trees and bushes. Evidence that their food was still more varied came from the find of a *Propalaeotherium*, in whose gastrointestinal tract grape seeds (*Vitis*) were found (von Koenigswald and Schaarschmidt 1983). The ripe berries had dropped, presumably, from the high lianas and were easily taken up by this small animal.

Entire or chewed seeds occur to varying degrees in the stomachs of birds and rodents (Fig. 82). They allow conclusions to be drawn concerning the way of life of these animals. It is, however, quite unusual to also find vegetable matter in the gastrointestinal

Fig. 82: Seeds in the gastrointestinal tract of a member of the woodpecker family. Seeds 4 mm long. (Whole animal illustrated in Fig. 217.)

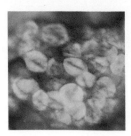

Fig. 84: This aggregation of pollen grains is the remains of food from the gastrointestinal tract of a jewel beetle (Fig. 83) and appears to derive from a variety of different flowers. Fluorescence micrograph: magnification, 180×.

Fig. 83: Jewel beetle with wing case spread. The yellow material within the body is pollen (Fig. 84). Length, 1 cm.

contents of insects. Thus, in one beetle, yellow material, which turned out to be pollen when studied under the fluorescence microscope, can be recognized under a carapace (Figs 83 and 84). The beetle was obviously a pollinating vector for which pollen was a food source. In the abdomen of a leaf-eating beetle, fragments of different cuticles were discovered.

Evidence that various relationships between insects and plants already existed in the Eocene come from the example of microlepidopteran coprolites on palm blossoms and from the finds of caddis-fly cases on water-lily blossoms. Advanced states of parasitism and symbiosis were here already in operation.

Fig. 85: Schematic reconstruction of the vegetation in the vicinity of Lake Messel: a paratropical rain forest, rich in vines. In some places the rainforest reaches the water, in others there is intermediate vegetation represented by Araceae and Pandanaceae in shallow water. There are palms at the border of the forest. Mixed in, at a greater distance from the border of the lake, are Taxodiaceae and Fagaceae. (Bushes and herbs are excluded from the diagram.)

Pteridophyta (ferns and fern allies)		P/S	F/S	LVS
LYCOPSIDA (club mosses)				
	Selaginellaceae (*Selaginella* fam.)	4	—	—
FILICES (ferns)				
	Osmundaceae (royal fern fam.)	1	—	1
	Schizaeaceae (curly grass fern fam.)	4	—	1
	Polypodiaceae (polypod fern fam.)	11	—	4
Gymnospermae (seed plants without pericarp)				
CONIFEROPSIDA (softwoods)				
	Cephalotaxaceae (plum yew fam.)	—	—	1
	Cupressaceae (cypress fam.)	1	—	—
	Taxodiaceae (swamp cypress fam.)	3	1	1
	Pinaceae (pine fam.)	2	—	—
Angiospermae (seed plants with pericarp)				
Monocotyledonae (with one cotyledon)				
ALISMATIDAE				
Hydrocharitales	Hydrocharitaceae (frogbit fam.)	1	—	1
COMMELINIDAE				
Restionales	Restionaceae (restio fam.)	2	—	—
Cyperales	Cyperaceae (sedge fam.)	2	1	—
Typhales	Typhaceae (reed mace fam.) or			
	Sparganiaceae (bur reed fam.)	1	—	—
	Zingiberaceae (ginger fam.)	—		1
ARECIDAE				
Arecales	Palmae (palm fam.)	1	1	1
Pandanales	Pandanaceae (screw-pine fam.)	—	—	1
Arales	Araceae (aroid fam.)	—	—	3
LILIIDAE				
Liliales	Smilacaceae (green-brier fam.)	—	—	3
Dicotyledonae (with two cotyledons)				
MAGNOLIIDAE				
Magnoliales	Magnoliaceae (magnolia fam.)	1	2	—
Illiciales	Illiciaceae (anise-tree fam.)	—	—	1
Laurales	Chloranthaceae (*Chloranthus* fam.)	1	—	—
	Lauraceae (laurel fam.)	—	2	22
Nymphaeales	Nymphaeaceae (waterlily fam.)	1	1	2
Menispermales	Menispermaceae (moonseed fam.)	—	8	1
HAMAMELIDAE				
Hamamelidales	Cercidiphyllaceae (katsura-tree fam.)	—	—	1
	Platanaceae (plane-tree fam.)	—	—	1
	Hamamelidaceae (witch-hazel fam.)	3	1	—
Eucommiales	Eucommiaceae (*Eucommia* fam.)	1	—	—
Myricales	Myricaceae (sweet gale fam.)	3	—	2
Fagales	Betulaceae (birch fam.)	3	—	—
	Fagaceae (beech fam.)	8	—	—

Fig. 86: Survey of the plant families found in Messel. The numbers refer to the numbers of species identified on the basis of pollen and spores (P/S) and leaves (LVS), and to the number of genera identified by fruits and seeds (F/S). (After Collinson 1986; Thiele-Pfeiffer 1988; and Wilde 1989.)

DILLENIIDAE

Theales	Symplocaceae (sweetleaf fam.)	1	—	1
	Theaceae (tea family)	—	—	5
Malvales	Tiliaceae (lime fam.)	2	—	—
	Sterculiaceae (sterculia fam.)	1	—	1
	Bombacaceae (silk cotton fam.)	3	—	—
Urticales	Ulmaceae (elm fam.)	4	—	2
	Moraceae (mulberry family)	—	—	1
Salicales	Salicaceae (willow family)	1	—	—
Ericales	Cyrillaceae (*Cyrilla* fam.)	2	—	—
	Ericaceae (heath family)	2	—	—
Ebenales	Sapotaceae (*Sapodilla* fam.)	2	—	—

ROSIIDAE

Rosales	Rosaceae (rose family)	1	1	1
Fabales	Leguminosae (pea family)	—	2	5
Myrtales	Lythraceae (loose-strife fam.)	3	1	—
	Thymelaeaceae (mezereum fam.)	1	—	—
	Myrtaceae (myrtle family)	—	—	1
Cornales	Nyssaceae (tupelo fam.)	2	—	1
	Cornaceae (dogwood fam.)	—	1	—
	Mastixiaceae (mastic-tree fam.)	2	1	—
Santalales	Olacaceae (*Olax* fam.)	2	—	—
	Loranthaceae (misteltoe fam.)	—	—	2
Celastrales	Icacinaceae (*Icacina* fam.)	2	2	1
	Aquifoliaceae (holly fam.)	3	—	—
Euphorbiales	Euphorbiaceae (spurge fam.)	—	1	—
Rhamnales	Vitaceae (grape family)	2	3	1
Sapindales	Staphyleaceae (bladdernut fam.)	—	1	—
	Aceraceae (maple fam.)	—	—	1
	Anacardiaceae (cashew fam.)	1	1	—
	Simaroubaceae (assia fam.)	—	1	—
	Rutaceae (citrus fruit family)	4	4	1
Juglandales	Juglandaceae (walnut fam.)	16	2	4
Polygalales	Malpighiaceae (*Malpighia* fam.)	—	1	—
Umbelliferae	Araliaceae (ivy fam.)	2	—	2

ASTERIDAE

Gentianales	Apocynaceae (dogbane fam.)	—	—	1
	Oleaceae (olive fam.)	1	—	—
Dipsacales	Caprifoliaceae (honeysuckle fam.)	—	1	—

5

Giant ants and other rarities: the insect fauna

Until the appearance of a new spate of research reports in the 1970s, hardly anything was known about the insects of the Messel oil shale. The only source of information was the work of Meunier (1921), who described 19 species of beetles, bugs, and cockroaches. A few papers made the first fragments of moths and one ant known between 1970 and 1980. Thus, for many decades our knowledge of this group of animals was limited to a few finds only. For some years now this situation has distinctly improved and even 'on the spot' mostly unimpressive finds receive the consideration which is their due. Meanwhile, the collection of the Senckenberg Natural History Museum alone includes several thousand items and each dig brings new material to light. Besides 'old friends', this frequently includes representatives of groups whose fossils have never previously been found.

Giant ants and other rarities: the insect fauna

HERBERT LUTZ

It is already evident that the insect fauna of Messel has been preserved with significantly more species than contemporary insect-bearing layers from the Geiseltal and many other sites. On a world-wide basis the Messel pit can be assessed as one of the most important sources of information about insects of the middle Eocene. The scientific importance of the Messel finds is based not least on the fact that relatives are frequently uncovered of groups that live today in subtropical and tropical forests and whose phylogeny could not, until now, be documented through any fossils.

The most frequent finds at Messel are of beetles. Among the material of the Senckenberg Museum from 1979 to 1984, they comprise 63 per cent of all finds (all further data concerning the percentage frequencies of the individual orders also relate to this collection). Unfortunately, the state of preservation often does not allow a reliable identification of the samples. In spite of this, a significant number has in the meantime been correlated to various families (the percentages relate to the given proportion within the sum total of the beetle finds).

Click beetles (Elateridae)	15.8	%
Weevils (Curculionidae)	12.8	%
Jewel beetles (Buprestidae)	8.4	%
Dung beetles (Scarabaeidae)	3.9	%
Stag beetles (Lucanidae)	1.7	%
Ground beetles (Carabidae)	1.4	%
Water beetles		
(Dascillidae (Eubrianacinae in the		
US))	1.4	%
Longhorn beetles (Cerambycidae) ..	0.5	%
Rove beetles (Staphylinidae)	0.26	%

Fig. 87: Leaf beetle (Chrysomelidae) with very well-preserved structural colours. This family is very species-rich today and includes at northern latitudes numerous metallic coloured species. Magnification, 14.3×.

Among the remaining finds are a multitude of mostly metallic, colourful leaf beetles (Chrysomelidae; Fig. 87), species that probably belong to the nocturnal ground beetles (Tenebrionidae), and many small representatives of other families that have not yet been investigated.

In this the Messel beetle fauna shows a certain similarity with that of the Geiseltal and that from the Upper Palaeocene lake deposits of Menat (Puy de Dôme, France) where jewel and click beetles are also species-rich and numerous. Jewel beetles, particularly in Messel, have lived up to their reputation. All species still glow in their original, almost fantastic array of colours. Some are shiny metallic blue and green (Fig. 88), others gold-green and coppery with dark-brown longitudinal streaks and spots (Figs. 89–91). Until now the only comparable finds described were those from the leaf-coal of the Geiseltal.

It is noteworthy that a large proportion of the Scarabaeidae found are most probably species feeding on dung ('dung beetles'). They certainly had a close relationship with the herbivorous mammals. Leaf-eating representatives, for example, relatives of the cockchafer are, by comparison, found far less frequently.

Fig. 88: Jewel beetle (Buprestidae). This family, mostly rare as fossils, comprises exclusively species needing a warm climate. Their commonness in the sediment of Lake Messel suggests that the climate was distinctly warmer during the middle Eocene. Magnification 4×.

Fossil finds of the stag beetle are among the greatest rarities. It is all the more pleasing that this family, closely related to the dung beetles, is represented in Messel with two confirmed and one questionable species (Fig. 92). The larger of the two certain species (Fig. 94) is distinguished by a sparkling blue and green metallic coloration. Its mandibles are, however, rather small and insignificant and do not yet form such imposing 'antlers' as those of the European stag beetle. The probably nearest, Recent relatives of the Messel stag beetle live in the jungles of South-east Asia.

In the case of two fossil ground beetles, the species belong to the subfamily Scaritinae whose Recent representatives are flightless, nocturnal predators which dig deep tubular burrows for themselves. As they mostly prefer loose, sandy soils close to the water it appears possible that both species at Messel also lived in the lake-side area and that they were perhaps washed into the lake by flood-water.

Also of particular scientific importance are the few finds of the family of longhorn beetles (Cerambycidae; Figs 93 and 95). This group is only rarely documented by fossils and then mostly by small species. Complete finds of very large representatives, as illustrated by the 50 mm long specimen in Fig. 93, were unknown until now.

The larvae of the water-beetle genus *Eubrianax* Kiesenwetter 1974 are undoubtedly one of the great surprises among the insects found to date. Only a single fossil specimen from the Middle Eocene sediments of southern France was known. This is because these animals can only live in highly oxygenated water, generally in the area of a waterfall. It was all the more surprising then, that all at once 12 complete specimens were found in the sediments of Lake Messel (Fig. 96). The conclusions to be drawn from this will be discussed in detail later (Lutz 1985).

Among the similarly frequent Hymenoptera (17 per cent) the ants (Formicidae; Figs 97–99 and 101) dominate with 87.5 per cent. It is not surprising, therefore, that the first documentation for this order in Messel comes from this family (Gahl and Maschwitz 1977). They are represented exclusively by winged, reproductive animals which can be classified into five different subfamilies: the Ponerinae, Myrmicinae, Dolichoderinae, Formicinae, and Formiciinae. In almost all instances they are unspectacular animals, close relatives of still-extant groups.

The representatives of two new species of ants can, however, be counted as the most exciting insect finds of Messel. The winged females ('queens') of the larger of the two species attain a wing-span of up to 16 cm. With this they surpass not only all other ants, but also, by far, all other known representatives of the order. They become even larger, and with an estimated weight of up to 10 g, heavier than some hummingbirds which weigh little more than 2 g. Even the common wren grows only marginally larger.

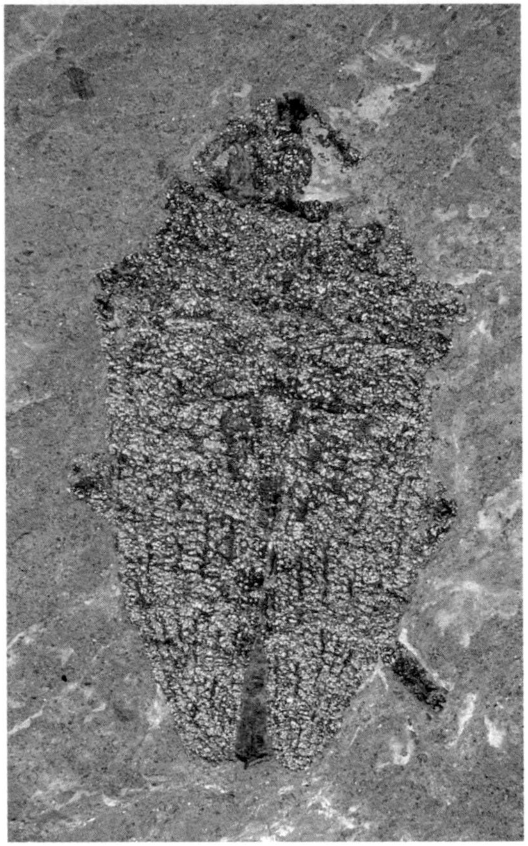

Fig. 89: Jewel beetle (Buprestidae). All specimens of this species show the same brown stripes on a metallic gold-green background. This is evidence that we are looking at the original colour pattern. Magnification, 4.7×.

Fig. 90: *Scintillatrix rutilans* (lime jewel beetle), a Recent central European species which has almost the same colouring as the specimens from Messel. They are not closely related, however. Magnification, 11.5×.

With a collection of, by now, approximately 200 finds, it was possible to collect such a wealth of information as seemed possible hitherto only from fossils preserved in amber. Even for some Recent species hardly more information is available. Thus, it was possible not only to differentiate between two species of a genus (*Formicium* Westwood 1854), but we also succeeded in matching in each case males (Fig. 98) and females (Fig. 97) with one another.

Thanks to a method of preparation developed specifically for this purpose, it is possible under the light microscope to observe and study the internal organs of the body. In order to do this the finds must, in a lengthy process, be freed completely from the surrounding matrix and transferred to resin. Only then can we, for instance, recognize the very fine sclerotic stiffening on the inner side of the trachea wall, formed like a spiral spring to reinforce the wall, or the dust filters of the spiracles.

It was also possible to reconstruct the stinging apparatus. This was very tiny in relation to the size of the entire abdomen and contained only a short, weak sting. This confirms without any doubt that these animals could no longer sting, but had to defend themselves by the spraying poisons—evidence for a high measure of specialization.

On the other hand, no evidence has been found yet of vestiges of a crop-closing mechanism, the so-called proventriculus. This very probably indicates that these species had a type of proventriculus that was not suitable for fossilization. The chitinized, decay-resistant parts of this organ were probably already reduced.

Comparable features are only known from a few Recent groups of ants and permit preliminary conclusions as to the probable feeding habits of the giant ants of Messel.

In all probability they did not feed on liquids that had been 'interim-stored' more or less long in the crop as do their closest living relatives (e.g. the red wood-ant, *Formica rufa* Linnaeus 1758). It is much more likely that they were highly evolved food specialists which could do without the storage of food reserves in their

Fig. 91: The elytra (wing cases) of this jewel beetle were torn off when the oil shale was split so that the blue tergites of the abdomen are visible. In the living animal these are only visible during flight. This is probably a species of the same genus as the one in Fig. 89. Magnification, 7.8×.

Fig. 92: An as yet unidentified beetle, probably a stag beetle (Lucanidae). In this specimen too the strong mandibles, which are not covered by a frontal shield (clypeus), are striking. This is evidence against an affinity with the dung beetles (Scarabaeidae). Magnification, 2.8×.

Fig. 93: Longhorn beetle (Cerambycidae, Prioninae). This subfamily contains many large, nocturnally active inhabitants of subtropical and tropical forests. Magnification, 1.2×.

Fig. 94: Stag beetles (Lucanidae) are very rarely found as fossils. This metallic species is found fairly commonly in Messel. Comparable Recent forms are known only from the tropics of South-east Asia. The only comparable specimen was discovered in the last century in Baltic amber. Magnification, 2.9×.

Fig. 95: Longhorn beetle (Cerambycidae, Lamiinae). This species still shows clearly pigmented lines on the elytra. The larvae of most species live in wood. Magnification, 3.9×.

Fig. 96: *Eubrianax* larva (Coleoptera, Dascillidae). Nowadays these larvae live only in clear oxygenated water; this suggests that the Messel finds were washed into the lake. Magnification, 9×.

crop. Such strategies are known from some Recent groups which feed on solid food particles, for example, seeds or fungal cultures (harvest and leaf-cutter ants of the subfamily Myrmicinae) or which regularly undertake huge hunting trips (driver ants of the subfamilies Dorylinae and Ecitoninae).

All these results permit a sure determination of the systematic position of this genus (Lutz 1986). It belongs in a new subfamily, the Formiciinae, the extinct sister group of the Formicinae that includes the well-known 'red wood-ant'.

Furthermore, the finds at Messel allow for the first time the classification of three other Eocene species which had been described long ago on the basis of isolated fore-wings. These are members of the same genus. This discovery makes it possible to separate these species from the Myrmiciidae (= Pseudosiricidae) and makes it easier to reassess this Cretaceous family which is found in almost all palaeontology textbooks, but whose systematic position has been controversial until now.

Besides the ants, one regularly finds species of other groups. Usually, however, these are single finds and their identification is therefore very problematical. Among this material, representatives of the following families certainly occur: parasitic wasps (Ichneumonidae), chalcid wasps (Chalcididae), tiphiid wasps (Tiphiidae), potter wasps (Eumenidae), scoliids (Scoliidae), and flower bees (Anthophoridae). In other cases the finds are probably representatives of the spider hunting wasps (Pompilidae) or the digger wasps (Sphecidae) (Fig. 102). These finds in particular lead us to expect that many important discoveries will be made regarding the phylogeny of some groups of the Hymenoptera.

Notably, no representatives of plant wasps ('Symphyta') have been found so far. In view of the occurrence of the majority of their Recent representatives in the temperate latitudes of the Northern Hemisphere, this absence from Messel can probably be taken as indicating that a similar climatic preference already existed in the middle Eocene.

The bug fauna of Messel (Heteroptera; 12.5 per cent) appears to be singularly lacking in variety; more than 80 per cent of finds are of burrowing bugs (Cydnidae). The first finds in Messel were also of this family (Kinzelbach 1970b; Meunier 1921). Among the remaining material are representatives of the capsid bugs (Miridae), assassin bugs (Reduviidae), and, probably, shield bugs (Pentatomidae).

With our present knowledge we must start from the assumption that the majority of the

Fig. 97: Female *Formicium giganteum*, the Messel giant ants. Resin-transfer preparation of an almost complete specimen. Magnification, 0.7×.

Cydnidae lived in the more immediate vicinity of the lake. Their distribution in the Messel pit, however, still allows us to also consider the possibility that these animals were washed into the lake from tributaries.

Rietschel (oral communication) was able to observe in the Peruvian rain forest that Cydnidae deliberately fly to corpses of larger mammals and suck on them, even though, according to references in the literature, they are plant suckers. Since animal corpses were normally available at Messel, either washed on to the shore or drifting, it seems likely,

Fig. 98: Male *Formicium giganteum*. All Messel ant finds are winged specimens which perished during their nuptial flight. No representatives of the flightless worker class have been found as yet. Magnification, 2.6×.

therefore, that these species of Cydnidae came to the lake intentionally and fell regularly on the water surface.

On the other hand, many of the Recent species prefer dry and loose ground where they dig. Such conditions occur not only in dry warm climates, but also in moist climatic conditions, for example, on open sun-exposed and sandy shores of rivers and lakes. Certainly, it is not known whether the species of Messel preferred such a habitat or if they were ground-dwelling inhabitants of the forest; nevertheless,

the possibility that the frequency of burrowing bugs could indicate the existence of open, dry surfaces should not be disregarded.

Finds of cicadas (Auchenorrhyncha, 'Homoptera'; 1.5 per cent) are in Messel significantly rarer than those of all the orders discussed up to now. Besides smaller species, including the representatives of the frog-hoppers (Cercopidae), isolated wings of the Ricaniidae and/or Flatidae, and, perhaps, the remains of a spiny frog-hopper (Membracidae), fossils of several species of the

cicada (Cicadidae) have also been found. These consist not only of isolated wings but also of complete specimens. Some are distinguished by wings with dark bands and spotted wings (Fig. 100). One species attains a wing-span of at least 160 mm, a size which is known today only among tropical representatives of the family. From the group of the lantern-flies (Fulgoridae) and closely related families we have, as yet, only a single find (Fig. 103).

As regards the cockroaches (Blattodea; 1.5

per cent) several large species were found which can reach lengths of up to 50 mm. Some beautiful and completely preserved specimens were uncovered, particularly in the youngest layers at the centre of the Messel pit (Fig. 105). However, this order is altogether far rarer at Messel than would have been expected

according to the old assertions of Meunier. As far as can be assessed at present, the specimens belong probably in all cases to the family Blattidae.

From the group of the crickets (Saltatoria–Ensifera; 0.5 per cent) only a few species are recognized at present. Apart from one

specimen which could belong to the Gryllacrididae (Fig. 106), these are exclusively bush crickets of the family Tettigoniidae (Fig. 104). Relationships to tropical groups are apparent for most of these finds.

Two winged specimens of the very rare (as a fossil) ghost cricket (Phasmatodea) have

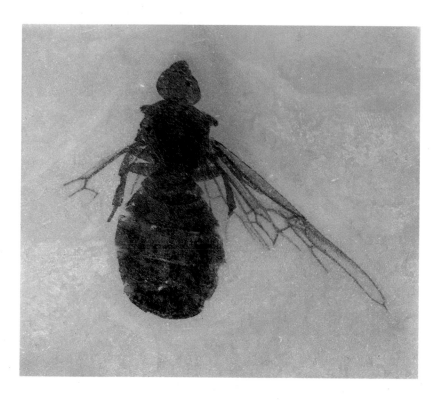

Fig. 99: (left). Female of the smaller *Formicium simillimum*. Magnification, 1.4×.

Fig. 102: (right). As yet unidentified wasp of the form of a spider hunting wasp (Pompilidae) type. Magnification, 1.2×.

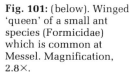

Fig. 101: (below). Winged 'queen' of a small ant species (Formicidae) which is common at Messel. Magnification, 2.8×.

Fig. 100: Cicada (Auchenorrhyncha, Cicadidae). As well as this rather small species there are also larger forms in Messel with wing-spans of almost 160 mm. Magnification, 0.9×.

already been found at Messel. The species of these primitive stick-insects still remain unidentified: they are very slim with body lengths of, respectively, *c.* 80 and 120 mm (Fig. 107). They are deposited in the Landessammlung für Naturkunde in Karlsruhe. Apart from some very small specimens found in amber, no other finds are known throughout the world which could approach the Messel specimens with regard to size and completeness. Besides, these are the oldest finds which belong unequivocally to the stick-insects.

Flies and mosquitoes (Diptera; 0.4 per cent) are extraordinarily rare in Messel. When compared with most other sites for Tertiary insects, this is a very surprising result. In spite of intensive searching, only five specimens were uncovered between the years 1979 and 1984. These are one species each related to the crane-flies (Tipulidae) or the gnats (Limoniidae), and, probably, the non-biting midges (Chironomidae); one horse-fly (Tabanidae); one species that probably belongs

Fig. 103: Lantern fly (Auchenorrhyncha, Fulgoroidea). Magnification, 3.4×.

Fig. 105: Cockroach (Blattodea). Usually only isolated wing fragments are found at Messel. Magnification, 1.5×.

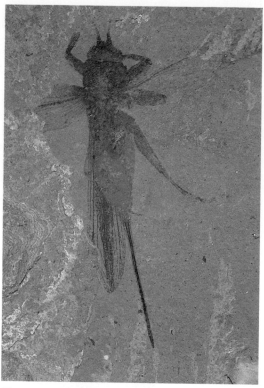

Fig. 106: Bush cricket (Saltatoria). Female with an ovipositor that is almost as long as her body. Magnification, 1×.

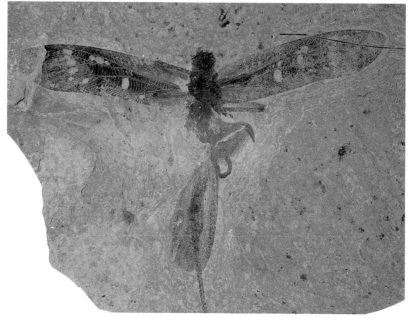

Fig. 104: Huge bush-cricket female (Saltatoria, Tettigoniidae). The pattern on the fore-wings and the sickle-shaped ovipositor are striking. Magnification, 0.7×.

with the athericids (Athericidae); and a further representative of the suborder Brachycera, which has not been classified as yet. Only recently a further find, a hover-fly (Syrphidae), was added. The rarity of this order at Messel can only be explained by the consideration that this group—favoured by the fact that their bodies would float for an extended period—were carried to the lake shore by currents near to the water surface, before they could sink to the bottom of the lake.

Similar conditions presumably held for the Lepidoptera (0.25 per cent) that came into the lake. Only three badly preserved remains of these have been recovered to date. These belong, with a high degree of certainty, to the moths. However, before the first finds of wings and an almost complete specimen were available, moths could already be documented by means of isolated wing scales from the stomach contents of bats (Richter and Storch 1980; Richter 1987). The mine in a leaf described by Kinzelbach (1970a) was presumably caused by a representative of the Microlepidoptera.

From the termites (Isoptera, 0.08 per cent), so characteristic of today's tropics, only one specimen has been found to date (Fig. 109). This is a winged adult that, as was also true for the ants of Messel, came to grief during its short nuptial flight and ended up in the lake. Termites generally form colonies with extraordinarily large populations and produce a correspondingly large number of winged, mating adults. With the evidence of such rich vegetation in the environment of the lake and of the very warm climate, we would have

expected more numerous finds of these wood-consuming insects.

Just as surprising is the extensive absence of many aquatic insect species which should have been preserved in the sediments of the lake (see below). Dragonflies (Odonata; 0.16 per cent) have been recorded up to now only as a wing fragment and one or two nymphs of the hawker dragonflies (Anisoptera); stoneflies (Plecoptera; 0.8 per cent) are represented by one partially disintegrated larva. The caddis-flies alone (Trichoptera) are found in great numbers but only in the form of their empty cases, never as imagos. Besides cases formed from grains of sand, there are occasionally some that consist only of spun silk, to which the smallest particles of the sediment have become attached. Now and then, many cases

regularly in water-lily patches and perhaps even used them as food.

Stylops (Strepsiptera), an order with relatively few species, belong to the most remarkable insects of all. All of them develop as parasitoids in the body cavities of other insect groups—usually of Hymenoptera and cicadas—and leave their host only as fully formed insects. This, however, is only true of the males and females of two families (Mengeidae and Mengenillidae). In all other families only the tiny, winged males leave the host, while the females remain for life and only poke the fused head–thorax part out from between the abdominal segments of the host's body. All species give birth to live young. The tiny, fast larvae can jump, they actively penetrate the new host and so complete the life cycle. Since all the

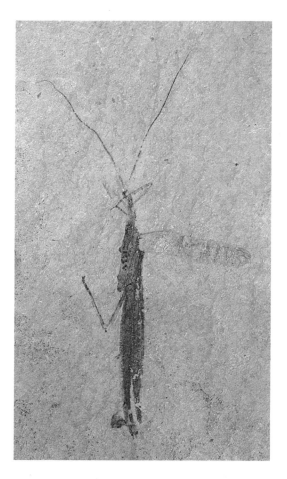

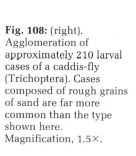

Fig. 107: (left). Stick-insect (Phasmida). Fossils of this order are extremely rare. The Messel species were probably inhabitants of the jungle. Magnification, c.0.5×.

Fig. 108: (right). Agglomeration of approximately 210 larval cases of a caddis-fly (Trichoptera). Cases composed of rough grains of sand are far more common than the type shown here. Magnification, 1.5×.

were embedded side by side (Fig. 108), because they were held together by the spun threads. Occasionally cases of larvae are wedged between the stamens of water-lilies so that we can presume that this species lived

males measure only a few millimetres and live only a few hours, it is not surprising that this order had been known previously only from a few specimens in fossil resins.

The first find of an only 0.12 mm long larva

(primary larva) that had penetrated a beetle just before its death was made in the Geiseltal (Kinzelbach and Lutz 1985). It was all the more surprising that a small ant—probably related to the wood ants—was found at Messel, between whose abdominal segments protruded two inconspicuous oval, chitinous structures. These could only be the penultimate developmental stages (puparia) of Stylops. This constitutes the first fossil documentation of a visibly infected (stylopized) host.

In the Recent fauna only one family is known that develops within ants. This is the phylogenetically youngest subgroup, the Myrmecolacidae. It is a peculiarity of this family that the tiny males develop exclusively in ants and the very large females, by contrast, develop in praying mantises (Mantodea) or in grasshoppers (Saltatoria). The fossils at hand must, accordingly, be male puparia of a hitherto unidentifiable species of the Myrmecolacidae. The Geiseltal specimen, moreover, also belongs to this family.

These finds are evidence not only that in the Middle Eocene the same parasite–host relationship already existed as exists today, but simultaneously, that we must assign a far greater age than has been usual to the entire order Strepsiptera, because the Myrmecolacidae are evolutionarily the youngest family. In addition, both of these finds are good indicators of a tropical–subtropical climate during the epoch of the European Middle Eocene.

But now for some general aspects of the insect fauna of Messel which is, more precisely, designated as a thanatocoenosis. This specific term signifies that we must differentiate between the association of dead organisms that have been embedded together in a specific locality ('association in death' = thanatocoenosis) and the original living association (biocoenosis) of a given locality. A spectrum of fossil species does not necessarily give a faithful picture of the once-existing biocoenosis. We must rather expect that, for whatever reasons, the fossil preservation of certain groups of animals is less likely than that of others. The striking peculiarities of the spectrum of insect species from Messel are

Fig. 109: Termite (Isoptera). The only find from Messel. This winged, sexually mature specimen, like the ants, drowned in the lake by mischance during its nuptial flight. Magnification, 4.1×.

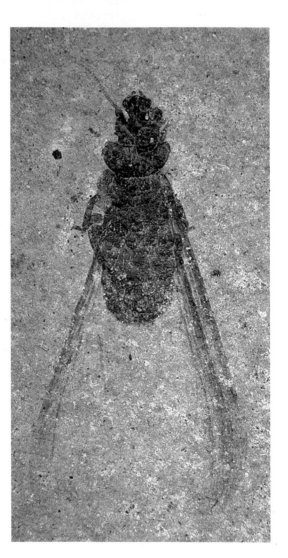

evidence that here too the original fauna has not been preserved 'true to life'.

Presently, representatives of 14 orders are known. As already indicated, the largest part of the finds by far is recruited from the Coleoptera, Hymenoptera, and Heteroptera. Also frequent are the cases of Trichoptera, which, as a rule, consist of grains of sand. All other orders are very rare indeed. This is particularly noticeable in the case of Diptera. Today this order not only makes up a large part of the insect fauna, but is usually significantly more frequent at other fossil insect recovery sites than at Messel. Often it is even the dominant element of the thanatocoenoses there.

When this fact is more closely investigated the common parameter for many representatives of the rare or absent groups turns out to be that they are small and of ephemeral build and have, in relation to body size and weight, large wings and sometimes long legs. Beetles, bugs, and the majority of the recovered Hymenoptera, on the other hand, are heavy, compact animals whose ratio of body volume (≈ weight) to surface area is rather large.

Experiments with Recent insects show that these differences influence the length of the floating phase of an insect corpse. This suggests attributing the different frequencies of the fossil occurrence of the individual orders at least in part to a selective separation of 'freights' in the surface transport and not to a gap in the fauna caused, for example, by predators. This is confirmed by the observation that there were apparently no insectivorous animals in Lake Messel.

Of the animal species that have been documented in the lake, the following must be considered first: knife-fish (*Thaumaturus*), perch-like fish (Percidae), immature bowfins, and gars, as well as frogs (Palaeobatrachidae). The latter have apparently lived only sporadically in the lake (Wuttke 1986) and, due to their rarity, should not have had a large influence. The Thaumaturi and a portion of the perch were, according to the investigations of Franzen *et al.* (1982), largely resident in the tributaries and can therefore hardly count as regular residents of the lake. Accordingly, they also play a lesser role as insectivores. Of the two other species of fish—bowfin *Cyclurus kehreri* Andreae 1893 and the gar *Atractosteus strausi* (Kinkelin 1884)—which rate as 'true' inhabitants of the lake, young animals below 100 mm in length are, noticeably, almost completely absent. (Only the excavations in the youngest layers at the centre of the pit have brought many small specimens to light—the insect fauna of this area is, however, still too little known to permit the recognition of possible changes in the composition of the thanatocoenosis.) With an average size of 250–300 mm among the recovered specimens, fish cannot be considered as insect predators and particularly not of small species as, for example, flies and mosquitoes.

The rarity of small insects cannot, therefore, be explained by their having been eaten by predators. The large number of insect fragments in a cubic metre—according to Franzen *et al.* (1982) one can count on *c.* 700–1500 finds and locally maybe even more—can rather be taken as confirmation of the results available at present which suggest that the lake's suitability as a habitat was limited. The absence of water insects, which are, so to speak, the true (autochthonous) lake residents, can also be explained in this manner.

The striking and unexpected absence among the fossils of species that are tied to a life in the lake as larvae and/or imagos can only be explained by assuming the almost complete absence of these species from the lake, at least during the period of deposition that has been investigated hitherto, i.e. that of the oil shale in the area of the key horizons α, β, and γ. We would expect, among others, the following groups: (1) consumers of detritus or vegetable matter as, for example, mayflies (Ephemeroptera), many caddis-fly larvae with cases composed of plant fragments (Trichoptera), various diving beetles (Hydrophilidae), non-biting midges (Chironomidae), mosquitoes, gnats (Culicidae), soldier flies (Stratiomyidae); (2) predators such as, for example, dragonflies (Odonata), alder flies (Sialidae), various water bugs (Nepidae, Belostomatidae, Notonectidae, and others).

The oxygenated shore zone and the water layers close to the surface should have, in spite of the generally inimical conditions in the deep water, offered sufficient living space to all these groups.

The few known finds of water insects or, respectively, of their larvae—12 larvae from the beetle genus *Eubrianax* (Fig. 96), one stone-fly larva, one or two nymphs of dragonflies—come from the north-western area of the Messel pit and, along with the countless cases left by Trichoptera larvae, show a significantly increased frequency in this area. As regards their environmental requirements, the *Eubrianax* larvae are the best suited for informed conjecture.

All known Recent species of this genus live in extremely oxygenated and disturbed water, for example, at waterfalls, in rapids, and in the surf zone of large lakes (Lakes Tanganyika and Nyassa in East Africa). Judging from the entirely similar form, this must also have been true for their middle Eocene relatives from Messel. But there cannot have been a comparable environment in Lake Messel. In explaining the occurrence of this genus in Messel, we can only surmise that *Eubrianax* was washed in via a tributary. Thus these fossil remains also demonstrate that a tributary entered at the north-west of the pit, as Franzen *et al.* (1982) have already postulated in their model of the site. Cases of Trichoptera larvae, built of large grains of sand, and stonefly larvae are also found more frequently in running than in quiet water. It can therefore be stated that this tributary must have been a clear, well-oxygenated body of water. Presumably there was, not too far from the lake, even more turbulent water as, for example, in rapids.

According to all the information accumulated to date, one must start from the assumption that as yet unknown environmental factors prevented a colonization of the lake by water insects and many other groups of animals. The following possibilities deserve particular consideration:

1. Extreme pH-values are worthy of discussion: here both very high values caused by phytoplankton blooms and very low values which usually occur in bog waters and tropical black waters must be considered. Algal blooms at Messel have, in the meantime, been proven and were probably caused primarily by representatives of green algal cocci, which, by way of periodic mass reproduction, caused the fine lamination of the sediment (Goth 1986). As fossils giving evidence of chemical conditions, dinoflagellates have a particular significance; the cells of this group of algae have not to date been documented at Messel. Thus, a periodic rise of the pH value of the lake water to values of up to pH 10–12 appears conceivable.

On the other hand the chemical analysis of coprolites has furnished evidence of their often strikingly high heavy-metal content (Schmitz-Münker 1986); this suggests the presence of acidic water which increases the mobility of metals. The reconciliation of these two results must await the further geochemical analysis of the sediments and the coprolites.

2. In his work on fish, Micklich (1985) considered brackish conditions. These would account for the absence of many water insects, but, at present, there is no further evidence for this hypothesis.

3. High contents of dissolved, poisonous substances: this applies to toxic metabolic products of the phytoplankton (particularly during the algal bloom, known in the marine environment as 'red tides') but also as soluble substances from washed in or blown in plant material. The poisonous action of tannins on water insects is known. It was observed locally in the Amazon Basin and on the Rhine (leaf litter of poplars in the old tributaries of the river).

4. Extremely high water temperatures can probably be disregarded in this case.

A decision as to whether one or several of the above points apply is not yet possible. Ultimately, it is not inconceivable that the absence of water insects might be due to causes different from the above. For this purpose, finds of species that certainly lived in the lake would be very important. They could to a particular degree—by comparison with their mostly well-known Recent relatives—serve as indicators of the water composition in Lake Messel.

The rarity or the complete absence of some groups that should occur, given the wealth of confirmed plant species and the presumed forest-like character of the vegetation, cannot be explained at present. This is true, among others, of the longhorn beetles (Cerambycidae) and the termites (Isoptera). Possibly presently unknown selection factors played a similarly important role here as they do for flies and gnats. The absence of the wood wasps ('Symphyta') can perhaps be explained by the fact that Recent representatives of this group are distributed primarily throughout the temperate zones of the Northern Hemisphere so that their absence in Messel constitutes an indirect indicator of a tropical/subtropical climate.

In any case, all insect species that can to date be reliably interpreted point toward this conclusion. In the first case, there are the species-rich jewel beetles that are represented by many individuals and which belong with their metallic green, blue, and red-gold gleaming colours to the most striking insects of Messel; further there are representatives of the leaf-beetle genus *Eosagra* Haupt 1950, large species of cockroaches, the representatives of cicadas with a wing-span of up to 160 mm, and, last but not least, the stylops of the family Myrmecolacidae.

Besides all the information concerning the differing insect species *per se*, the structure of their bodies, their habitat, and their minimum geological age, we can also gain in direct insight into almost 50-million-year-old relationships between insects and the environment they lived in.

Significant clues to the relationships between insects and plants arise, for example, from the occasional preservation of gut contents which allow us, in the case of pollen-feeding jewel beetles, to make statements concerning the flowering plants that were visited. In other instances, relations between an, as yet unidentified, caddis-fly and water-lilies can be traced. It also was possible to document, through tiny faeces on flower petals, that palm blossoms already suffered from a pest, presumably a microlepidopteran. Occasionally one also finds leaves with galls caused either by gall midges or gall wasps; in other leaves tiny larvae of flies or micromoths have eaten mines ('leaf mines'), whose shapes are species-specific, through the mesophyll of the leaf (Schaarschmidt 1986).

The investigation of the fossil stomach contents of mammals showed that insects were in part eaten coincidentally with other food but may have also comprised the main food item for individual species. The first vestiges of what were probably moths rather than butterflies were found, in the form of wing scales, in the stomach contents of bats. This is of importance in so far as it is evidence that the middle Eocene bat species already fed primarily on flying insects. One can even surmise, but not yet document, that in tracking and recognizing prey they used echolocation, as do their modern relatives (Richter and Storch 1980).

The first evidence of interrelationships among the insects, the already mentioned stylops puparia in the abdomen of an ant and the above-mentioned connections between insects and plants and, respectively, insects and mammals, lead us to expect still further direct evidence for such relationships in the future.

These few hints clarify the 'supradisciplinary' significance of the further research into the insect thanatocoenosis of Messel. One of the most difficult problems of the Messel research, as always, still proves to be the construction of a valid model of Lake Messel itself. Here, in particular, we may hope that through future finds of autochthonous residents of the lake—if they ever existed—to obtain important information concerning the composition of the lake water. In addition, we may justifiably hope to uncover further washed-in representatives of fauna from flowing water, i.e. torrents and streams, because Messel offered particularly good preconditions for their fossilization. Every additional find leads us to expect important insights into a biocoenosis whose members are normally not preserved but are destroyed in the course of drifting in the water long before their embedding becomes possible.

Even in future, of course, additional finds of partly known species will be of importance, because in many instances only then can a correct answer be given to questions concerning their systematics as, for example, in the case of the finds classified as stag beetles (Lucanidae) which constitute, along with those found in the Baltic amber, the oldest fossil evidence of this family.

The spiders: infrequent predators at Messel

Fossils of this class are very great rarities at all recovery sites of fossil insects; this is also true for Messel. Until now only vestiges from representatives of two orders have been found, namely from the orbweb spiders (Araneidae = Araneae auct.; Figs 110 and 111) and from the harvestmen (Phalangiida = *Opiliones* auct.). The Senckenberg collection contains only five specimens of the order Araneida, which are all only a few millimetres in size. Three of the individuals investigated by Wunderlich (1986) can be classified as belonging to the family of orbweb spiders, but the recognizable characteristics are not sufficient to make more detailed statements about their taxonomic relationships. Wunderlich, however, thinks that they might possibly be related to the orbweb spiders of the genus *Singa* C. L. Koch 1836 which live today in moist habitats.

Fossil harvestmen of the superfamily Phalangioidea have become known from a few species only and almost exclusively from fossil resins (amber); by contrast, from freshwater deposits only one species from the Lower Oligocene at Florissant, Colorado, was described. The only proven specimen from this group at Messel as yet is the oldest find to date.

In the biocoenosis of Messel, spiders, as hunters of insects and of other arthropods, must have played a role comparable in importance with their role today. The explanations for their rarity are based, on the one hand, on the fact that their thin exoskeletons do not preserve very well. On the

Fig. 110: Spider (Araneae) preserved in pyrite. Spiders and harvestmen are very rare at Messel. Apart from a few, mostly badly preserved, orbweb spiders, only one daddy-long-legs has been found (Phalangiida). Magnification, 6.3×.

other hand, one can observe in Recent spiders that most species, when they accidentally fall into water, can run quite well on the water surface. It must therefore be presumed that the majority of specimens do not drown but can reach safety at the nearest shore.

Fig. 111: Unidentifiable male orbweb spider (Araneidae). Species of similar appearance live in moist habitats today. Magnification 8×.

6

Ancient knights-in-armour and modern cannibals

The majority of the vertebrate fossils of Messel—almost a by-product of the search for higher vertebrates—are, as always, fishes. The bowfin (*Cyclurus ('Amia') kehreri*) alone accounts for almost 90 per cent of the finds at some excavation sites. Fish fossils are known from all hitherto excavated areas of the fossil site. Nevertheless, one can observe significant differences in the composition of the species spectrum of the individual sites. In this, the conditions of sedimentation as well as the stratigraphical classification (the geological age) of the clayey shale probably play a role. This appears not to matter to the same extent for all groups of fishes found in Messel, but rather to apply particularly to the 'true' bony fishes (Teleostei).

Ancient knights-in-armour and modern cannibals

NORBERT MICKLICH

All representatives of the fish fauna of Messel are advanced 'bony fishes', which are characterized, among other shared parameters, by the simplified structure of their unpaired fins. They are, therefore, as Neopterygii (fishes characterized by the structure of their unpaired fins), distinguished from other more primitive bony fishes (Class Osteichthyes). However, even within the Neopterygii, the Messel species must be classified according to their rather variable organizational or, respectively, evolutionary stages. The systematic spectrum is correspondingly wide. To date, seven genera and species from six different families and five orders are known. As shown in Fig. 151, this list will soon be enlarged. Thus, the so-called 'juvenile' perch is certainly a new genus with at least one further species and, again, in *Palaeoperca* we may have to differentiate between more than just one species.

Atractosteus, a primitive gar

At Messel the gars are represented by *Atractosteus strausi* (Kinkelin 1884) and are among the most frequently recovered fish species. The percentage of bowfin (*Cyclurus*) finds is, however, still greater. The gars are distributed more or less evenly among all excavation sites and all oil-shale horizons. The length of the majority of specimens lies between 20 and 30 cm. Significantly smaller animals (less than 10 cm; cf. Fig. 119) as well as giants, almost 1 m in length, have also been discovered. Due to the scarcity of such 'lucky finds', it is, of course, impossible to make statistically reliable assertions regarding their distribution in the entire excavation area. It is, however, striking that the majority of smaller specimens comes from the older layers in the north-east of the Messel pit, while the most spectacular specimens can, apparently, be expected in the younger layers to the north-west.

Atractosteus finds are easily recognized at first sight (Figs 113 and 145). The body is clad in massive, rhomboid scales which are also shiny due to their coating of ganoin—a substance similar to tooth enamel. These scales are connected one under the other in a peg-and-socket manner and thus form, like chain mail, an armour plating, which, although it has almost no gaps, is still relatively flexible. In

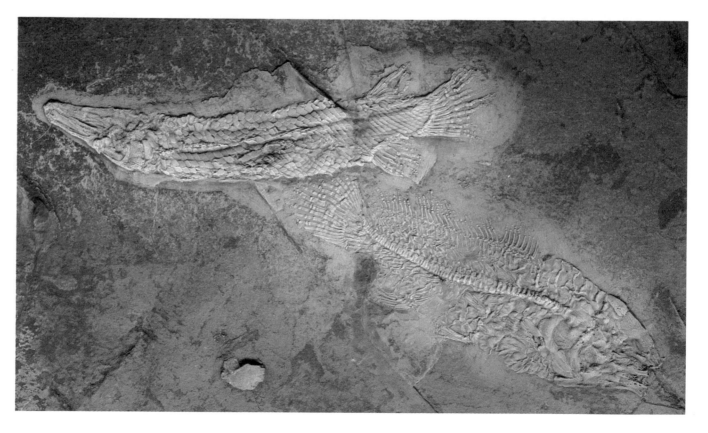

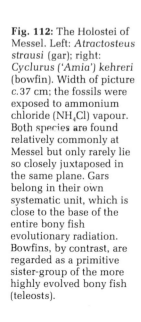

Fig. 112: The Holostei of Messel. Left: *Atractosteus strausi* (gar); right: *Cyclurus ('Amia') kehreri* (bowfin). Width of picture *c.* 37 cm; the fossils were exposed to ammonium chloride (NH_4Cl) vapour. Both species are found relatively commonly at Messel but only rarely lie so closely juxtaposed in the same plane. Gars belong in their own systematic unit, which is close to the base of the entire bony fish evolutionary radiation. Bowfins, by contrast, are regarded as a primitive sister-group of the more highly evolved bony fish (teleosts).

Fig. 113: *Atractosteus strausi*. Length, *c.*80 cm. Messel gars were fish-eating predators and could attain body lengths up to 1 m. Their shiny ganoid-scale armour, the beak-like elongated snout, and the internally asymmetrical caudal fin make them easily distinguishable from other fish species in Messel. In particular, their skull structure furnishes some important clues for their systematic classification: besides the so-called 'primitive' characters, several other special structures can be recognized that occur only in gars.

addition, the snout of the animal is prolonged like a beak and numerous bones of the skull exhibit a surface strongly ornamented by a coating of enameloid (a substance similar to

the ganoin of the scales, from which it can only be distinguished by its fine structure). This gives gars a primitive aspect reminiscent of crocodiles (Fig. 115), and they are, therefore, also dubbed 'cayman fish'.

Indeed, the gar-like fishes (order Lepisosteiformes) are classified on the basis of the above-mentioned features as well as on some other 'primitive' characters as being among the earliest members of the Neopterygii. The numerous small bones behind the eyes (the so-called suborbitals; cf. Fig. 115) are considered to be just as primitive as the construction of the lower jaw, composed of numerous bones, the internally asymmetrical, semiheterocercal caudal fin, recognizable in specimens with incomplete ganoid-scale armour, or the enlarged ridge scales (fulcral scales) in front of the median fins. At the same time the skeletal structure of gars shows peculiarities that exclude them from a closer relationship with the bowfin-like fishes (order Amiiformes) as well as with all the other Teleostei. This has prompted the taxonomists to place them in a separate category, the Ginglymodi (Wiley 1976). One example of these peculiarities (autapomorphies) is the

Fig. 114: *Atractosteus strausi*. Detail from the jaw region. Length *c.*3.5 cm. The margin of the upper jaw is fringed by numerous tooth-bearing infraorbitalia which only occur in gars. Just as singular is the ridged enamel crown of the teeth.

arrangement of bones along the periphery of the upper jaw. These 'infraorbital' bones are numerous and positioned one behind the other; they bear small teeth topped by pleated enamel. Other specializations include 'opisthocoelous' vertebrae, which are convex anteriorly and concave posteriorly (Figs 114 and 116).

The order of gar-like fish (Lepisosteiformes) includes only one family, the Lepisosteidae. Here two genera can be distinguished— *Lepisosteus*, Lacépède 1802 with four extinct and four extant species, and *Atractosteus* Rafinesque 1820 with five fossil and three Recent species (Wiley 1976). These are in some ways so similar that scientists have sought to reflect these similarities in the systematic subdivision of both genera (cf. Wiley and Schultze 1984). The identification of the genus rests primarily on fine details of the gill skeleton or of the structure of the teeth and their distribution among the individual jaw bones. Such details can be only rarely discerned in the fossil, making the identification very much more difficult. Up to now it appears that all Messel gars that have been examined in detail are 'true' Atractostei. However, the presence of the genus *Lepisosteus* in Europe is also substantiated by Eocene finds and its presence in the Geiseltal has recently been verified (Gaudant 1988). *Lepisosteus* and *Atractosteus* species even occur together in other Eocene fossil sites (e.g. in the Green River Shales from Wyoming; Grande 1980). It is, therefore, not impossible that such finds may yet be made at Messel or that they are already present, unrecognized, among the fossils already collected. In differentiating between the *Atractosteus* species and also in assessing their family relationships, Wiley (1976) made considerable use of the pattern and rugosity of the skull-bone sculpture.

According to these features *Atractosteus strausi* is the only species that still possesses completely developed enameloid ridges on the operculum and is therefore regarded as the most primitive representative of the entire genus. Interestingly, one can recognize in some individual Atractostei from Messel differences in the pattern of the enameloid ridges on some skull bones (cf. Figs 117 and 118) which, however, might be due merely to infraspecific variation.

Fossil gars are found throughout Europe, North America, Africa, and India. The oldest remains in Europe and North America date from the Cretaceous. While the gars apparently died out in Europe during the Oligocene/early Miocene (Weiler 1963), they still survive in North America. As they differ only slightly from the gars of the Cretaceous, the extant gars

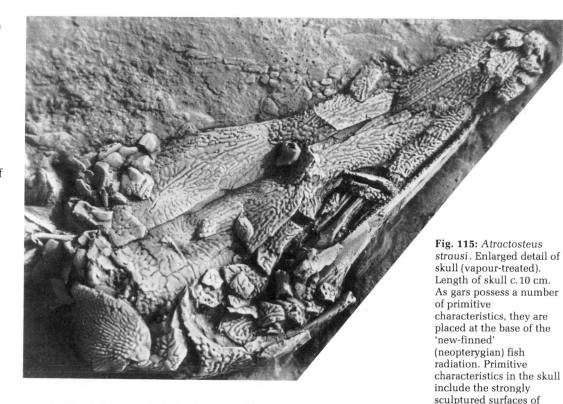

Fig. 115: *Atractosteus strausi*. Enlarged detail of skull (vapour-treated). Length of skull *c*. 10 cm. As gars possess a number of primitive characteristics, they are placed at the base of the 'new-finned' (neopterygian) fish radiation. Primitive characteristics in the skull include the strongly sculptured surfaces of some bones (due to coating with a ganoid-like enamel substance), the numerous suborbital elements behind the eye, and the still relatively complex structure of the lower jaw.

Fig. 116: *Atractosteus strausi*. Detail of trunk vertebrae (vapour-treated). Width of picture *c*. 3.9 cm. Only gars develop opisthocoelous (anteriorly convex, posteriorly concave) vertebrae. These are flexibly joined to each other, completely ossified, and firmly fused with their processes.

Fig. 117: (left), Fig. 118: (right). *Atractosteus strausi*. Detail of opercular bone (vapour-treated). Width of each picture *c*. 1.5 cm. Gar specimens from Messel sometimes show distinct differences in the sculpturing of some bones in the head and in the gill covers (opercula). This probably results from natural variability rather than species differences.

can be described as 'living fossils'. The individual Recent species, however, do not go back very far: *A. spatula* (Lacépède) is, at most, of Miocene origin. The family relationships to fossil species, however, lead us to suspect a far older age in some individual cases. Thus, *A. tropicus* Gill may possibly extend back into the Cretaceous (Wiley and Schultze 1984). From this viewpoint the 'life-span' of some systematic subgroups

Fig. 119: *Atractosteus strausi*; young animal, length *c.*8 cm. In young gars the scales are frequently absent from the anterior end of the body. Recent Lepisosteids also develop ganoid scales in the tail region first.

corresponds also to the expectation of 'living fossils'. We can conclude from the distribution of Recent gars that they (more precisely, that their ancestors) occurred originally throughout the entire 'supercontinent' Pangaea (which was formed from the then still-attached land-masses of the Northern and Southern Hemispheres). The separation into today's genera and species should therefore have taken place at least 180 million years ago (Wiley 1976). The fact that the distribution of the Lepisosteidae has remained limited to North America and Europe since the start of the Caenozoic, suggests a close (land) connection between both continents (Gaudant 1988). But at least one of the oldest gar fossils originates putatively from marine sediments. Even some present-day gars migrate occasionally to brackish water or coastal zones. Therefore one cannot altogether exclude the possibility that fossil Lepisosteidae were also able to spread across smaller ocean straits (cf. Jerzmańska 1977*b*).

Recent Lepisosteidae prefer shallow waters with vegetation. As mentioned above, they migrate occasionally into brackish or coastal waters. The largest species, *Atractosteus tristoechus*, can attain a length of 3 m (it is not, however, dangerous to man). The particular construction of the air-bladder enables gars to also breathe air (as, for example, in periods of water scarcity that can occur particularly during the hot season). In the winter time many gars remain in the deeper parts of the river-bed and the food uptake is then minimal. In spring they visit the river bank for spawning. The eggs are deposited on water plants or on stones. Gars are typical predators. They lie in a hiding place waiting for their prey which they then grab in a lightning ambush. Until they reach a length of *c.* 5 cm, young gars feed on invertebrates. After that, they prefer fish. To supplement their diet they also eat crustaceans and insects.

Due to the very slight morphological differences between Recent and fossil gars, we can assume that the Atractostei of Messel fed in a similar predatory manner. Some specimens of *Atractosteus simplex* (Leidy), a closely related species from the Green River Shales of Wyoming, are fossilized together with their prey still held in their long jaws (Grande 1980).

Cyclurus (*'Amia'*) resembles the Recent bowfin

The genus is represented in Messel by *Cyclurus kehreri* (Andreae 1893). As mentioned above, this is the most frequently encountered species of fish in the Messel fauna. The length of most specimens is very similar to that of *Atractosteus*, about 20–30 cm. In this case, exceptional finds of animals with an overall length of *c.* 5 cm or of up to 70 cm are also known. Interestingly, the sites in the Messel pit where large bowfins are found mostly coincide with the sites where correspondingly large gars are found.

The bowfins, too, can already be easily separated, by simple characters, from the other species of fishes in Messel (Fig. 120): the elongated dorsal fin is composed of numerous fin-rays and extends almost to the caudal fin. The latter, as in gars, externally appears to be symmetrical. Because the bowfins lack the massive ganoid scales of gars, the asymmetrical internal structure can be easily recognized (Fig. 122). In this case the type of symmetry can be described as hemicercal: the most distal part of the vertebral column is (as in *Atractosteus*) bent upwards but still reaches the uppermost rays of the caudal fin. That is why the bowfins, in contrast to the gars, really do have upper (epaxial) caudal fin rays, which are separated from the lower ones by elongation of the vertebral column. In addition, under the bent portion of the vertebral column, numerous flattened, and at times elongated, supporting bones (hypurals) for the caudal fin rays can be seen. The large, more or less oval, and, on their upper surface, finely grooved scales (Figs 123 and 146) are also characteristic, as is the skull enclosed by numerous massive and sculptured dermal bones (also characteristic of *Atractosteus*).

Besides the hemicercal tail and the strongly ossified skull, the bowfin still displays several other primitive characteristics: the vertebral centra are still not completely ossified and are connected only by cartilage to the neural and haemal arches. The lower jaw is, furthermore, composed of a relatively high number of individual bones. For a long time such primitive characteristics were used to place bowfins and gars in a separate group, the so-called 'ganoid bony fishes' (Holostei). It is now realized that the bowfins share more important characteristics with the Teleostei and therefore belong in the same group, the Halecostomi (cf.

Patterson 1973; Schultze and Wiley 1984). Among other characters, the ganoin cover of the scales and skull bones is lacking and so are the suborbitals of the gars and the 'chondrosteans'. While these could still be parallel characteristics, evolved independently in bowfins and teleosts, there are other characters that obviously represent true specializations common to both. The maxilla is no longer solidly integrated into the massive complex of cheek-bones, and this has led to the partial mobility of the upper jaw. In the jaw and the gill-cover apparatus (operculum) there are new bones which are not present in the 'primitive' ray-finned (actinopterygian) fishes (Fig. 123). Finally, there are still some skeletal peculiarities that are found only in the bowfin. In this context we must mention in particular the so-called diplospondylous nature of the distal part of the vertebral column. Here each body segment contains two vertebrae, but of these only the front one bears neural arches (Fig. 122). A similar importance is attached to certain fusions of bones in the lower segment of the true caudal fin. Such peculiarities exclude a close relationship with the teleosts. It appears reasonable, therefore, to consider the bowfin-like fishes (order Amiiformes) as the primitive 'sister group' of all other teleosts.

Approximately seven families with a number of genera are assigned to the bowfin-like fish. The Amiidae (with *Amia calva* Linnaeus 1758 as the single extant representative) embrace about eight different genera (cf. Schultze and Wiley 1984). Only the genera *Amia* Linnaeus and *Cyclurus* Agassiz possess a very long extended dorsal fin (dorsalis), making them easy to distinguish from other bowfins. Yet *'Amia' kehreri* of Messel can no longer be classified with the genus *Amia* (Gaudant 1980; Schultze and Wiley 1984) on the basis of the presence of unusual, 'stylus-like' teeth (Fig. 121) on some jaw and palatal bones (which represent a specialization). It has instead been classified with *Cyclurus* Agassiz (this genus designation has taken precedence over *Kindleia* Jordan) (Gaudant 1987). For the identification of species in both genera, one must invoke, among other characteristics, the body proportions, the extension and formation of various bones of the skull, differences in dentition, as well as counts of vertebrae and fin-rays (Boreske 1974; Jerzmańska 1977*a*). Overall, we can recognize about seven species of *Cyclurus* and about six species of *Amia*, where individual morphological differences can be very small. The Messel species shows deviations (among others in the number of vertebrae and in the shape of the toothed area on the underside of the palate) from those amiids found in the Geiseltal and described by Jerzmańska (1977*a*) as *'Amia' kehreri*.

Fossil Amiiformes are known since the Triassic (Andrews *et al.* 1967, p. 652). Particularly during the Jurassic and Cretaceous, they were clearly distributed throughout Europe and also North America, Asia, and Spitsbergen. The earliest reliable records of the genera *Amia* and *Cyclurus* are from the Cretaceous of North America, but the first occurrence in Europe dates from the Palaeocene. Since the distribution of this family is most probably restricted to fresh water (the fossil species were also almost certainly freshwater dwellers), two former land connections are discussed as possible migration routes (cf. Schultze and Wiley 1984): either a route via Greenland and Spitsbergen to North Scandinavia (the so-called De Geer Route), or a south-eastern one via Greenland to England ('Thule' Bridge). Eocene *Amia*

Fig. 120: *Cyclurus ('Amia') kehreri*; length 23.7 cm. This was probably also a voracious predator. The very long back fin is typical, and also distinguishes *Cyclurus* from other fossil Amiidae. The absence of massive ganoid scales and the rounded shape of the caudal fin are also typical. On the basis of some 'primitive' characteristics, bowfins are placed at the beginning of the higher bony fish radiation. For example, the vertebrae are only incompletely ossified, and not fused with their neural arches.

Fig. 121: *Cyclurus ('Amia') kehreri.* Detail of palate, or, more precisely, of the upper jaw. Width of picture *c.* 2.5 cm. The formation of pillar-shaped, 'styliforus' teeth, which are slightly bent and have a blunted tip, is important in the differentiation between species of *Amia* and *Cyclurus.* The evolution of such teeth in representatives of the genus *Cyclurus* is usually interpreted as a specialization for eating hard-shelled food (e.g. crustaceans, molluscs). However, bowfins at Messel must have also eaten fish.

Fig. 122: *Cyclurus ('Amia') kehreri.* Detail of caudal fin. Length *c.* 12 cm. In contrast to the gars, the vertebral column of the Amiidae extends into the bases of the caudal fin-rays. The diplospondylous structure of the distal vertebrae close to the caudal fin is characteristic of bowfins: in every segment of the body there are two vertebrae, only one of which bears a neural arch.

Fig. 123: *Cyclurus ('Amia') kehreri.* Detail from the gill-cover region (vapour-treated). Overall length *c.* 11.2 cm. In this case also the skull bones of larger specimens are, in part, strongly sculpted. However, the outer layer of enameloid is missing. The development of an interoperculum (the lowest element in the series of gill covers) is regarded as an important indicator of a relationship with the teleosts.

species are, however, also known from Asiatic deposits. Because Europe and Asia were separated by the Turgai Strait in the early Tertiary, a further, more easterly migration route (across the Bering Strait) is assumed for these forms (Jerzmańska 1977b). In the Tertiary, *Amia* and *Cyclurus* probably had a wider distribution in the Northern Hemisphere. *Amia* itself had probably disappeared from Europe by the Eocene but has survived in North America to the present. The genus *Cyclurus*, in contrast, is still documented from Europe during the Oligocene but apparently became extinct in North America during the Eocene (Schultze and Wiley 1984; Gaudant 1988).

The area of distribution of the only extant bowfin species, *Amia calva* Linnaeus 1758, is limited to the eastern and eastern central areas of the USA and Canada and reaches from about the area of the Great Lakes to the river system of the Mississippi. It can be dubbed a 'living fossil' both because of the above-mentioned slight morphological deviations from other fossil species and also because of the long 'life-span' of the genus (it is at least 70 million years old). The Recent bowfin, *Amia calva*, lives mostly in clear, running water with rich vegetation. It has a maximum age of up to 30 years and can attain lengths of over 60 cm (Lee *et al.* 1980). Just as in the gar, the air-bladder, characterized by an enlarged net-like surface, can be used for breathing air and thereby enable survival even during de-oxygenation and high water temperatures in the summer. Bowfins spend the day in deep-water zones where they partly hide in the soft mud at the base of patches of vegetation. During the night they move to shallow water. They are mainly active during the hours of dawn and dusk, but feeding can also take place at other times. They hibernate in shoals (in depressions or on the pebble floor of lakes). In spring, at spawning time, they sometimes seek out extremely shallow water. There, mostly in hidden positions (below overhanging trees, sunken tree trunks), bowl-shaped nests are constructed. The juveniles, when hatched from the egg, remain for a while in the nest or in the immediate vicinity. They swim in shoals and are guarded by the male. Their food consists at first of small crustaceans. Larger juveniles, up to an overall length of *c.* 10 cm, also still remain in guarded shoals but swim, during the last stage, through areas of more than 100 m (Reighard 1903). Fully grown specimens are greedy predators. Although fish are preferred as the main source of food (they are sometimes cut in two by the very strong bite), almost anything is eaten.

We can recognize in *Cyclurus kehreri* (as also in other fossilized Amiidae) at most only slight differences from the adaptive type of the present-day bowfin. Therefore, ways of life corresponding with those of the Recent *Amia calva* can be presumed. Grande (1980) and Boreske (1974) believe that the 'styliform' teeth of *Cyclurus* (*'Amia'* or, to be precise, *Kindleia*) *fragosa* (Jordan), a species (again from the Green River Shales of Wyoming) very close to the bowfin of Messel, represent an adaptation to 'hard-shelled' food, for example, crustaceans or molluscs. Although *Cyclurus kehreri* from Messel has the same type of teeth, this species certainly also consumed fishes: some specimens have a prey fish in their jaws and even the (putative) food pellets contain corresponding vestiges of fish.

Thaumaturus: 'true' knife-fish or advanced Mooneye?

Finds of the *Thaumaturus* species that occurs at Messel (*Thaumaturus intermedius* Weitzel 1933) are rather numerous in the more north-western area of the pit but are otherwise scarce. In the very oldest and in the youngest layers it appears to be largely absent. Most recovered animals have a length of 3–6 cm. The largest specimens can attain a length of about 9 cm and originate from the more centrally situated areas of the pit. Dwarf specimens of 1 cm are also known.

Apart from the diminutive size, the identification is made easier by a dorsal fin set in the hind part of the body; the lack of fin-spines, and also the lack of massive sculptured bones in the skull; the rim of the operculum, which is complete; and the straight to forked caudal fin (Fig. 127).

In contrast to the previously described gar and bowfin, the Thaumaturi of Messel are true teleosts: the amphicoelous vertebrae (concave on both sides) are completely ossified and firmly fused with their arches; the caudal fin is homocercal, that is, the distal section of the vertebral column is only weakly bent upwards and no longer rises up between the tail rays (Fig. 126); the scales are rounded and devoid of ganoin (so-called cycloid scales; Fig. 147); the maxilla and premaxilla are mobile.

From a systematic view, the Thaumaturi of Messel have been on an epic voyage! They were classified originally—particularly due to the erroneous assumption that they had an adipose fin—with the salmon (Salmonidae) Shortly afterwards, Voigt (1934) noticed, during the examination of a closely related species from the Geiseltal, that characteristics, such as the position of the dorsal fin at the back of the body and the absence of an adipose fin, distinguished them clearly from the Salmonidae. He suggested, therefore, the establishment of a separate family, the Thaumaturidae, which he placed in the

suborder Salmonoidei. Later authors (Gosline 1960; Weitzman 1960), however, suggested that a relationship with the pike-like fishes (suborder Esocoidei [Haplomi]) would be more correct. More recent investigations of the skeletal structure suggest, in contrast, a connection to a quite different group of fishes (Gaudant 1981): the absence of certain bones in the upper jaw as well as the absence of epipleural elements beneath the vertebral column, the size reduction in the region of the operculum, and details of the structure of the caudal fin and its supporting skeleton (Fig. 126) are typical of the bony tongue-like fishes (order Osteoglossiformes).

Within the Osteoglossiformes up to three suborders are distinguished; here particular emphasis is placed on the soft anatomy. But, there are also some skeletal features which seem to be useful for the classification of fossil forms (Nelson 1968, 1969; Taverne 1977, 1978, 1979). Thus, the formation of a characteristic toothed plate in the lower section of the hyoid arch (on the so-called basi- or glossohyal), characteristic in *Thaumaturus*

Fig. 124: (left); **Fig. 125:** (right). *Thaumaturus intermedius*. Detail of the skull (vapour-treated). Width of picture Fig. 124: 2.4 cm; Fig. 125: *c.*2.2 cm. The massive, sculptured covering bones are absent from the skull of *Thaumaturus*, and the upper jaw is more mobile than in the Holostei. The reduction of certain elements of the skull and the development of teeth on the basihyal (Fig. 124, lower left) suggest a relationship with the knife-fishes.

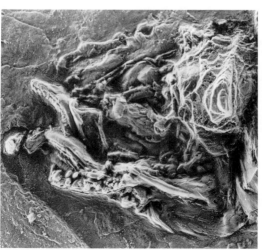

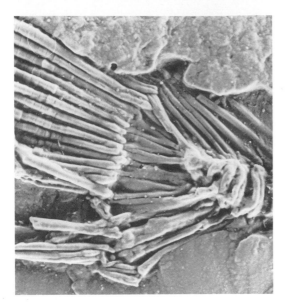

Fig. 126: *Thaumaturus intermedius.* Detail of caudal fin (vapour-treated). Width of picture *c.*1 cm. In contrast to bowfins and gars the Thaumaturi have a homocercal caudalis. The most distal part of the vertebral column is only weakly bent upwards and no longer reaches into the bases of the caudal-fin rays. At the same time some characteristics of the osteoglossomorphs (such as the complete neural spine on the antepenultimate vertebra) are also developed in this species.

Czechoslovakia do, nevertheless, exhibit unique peculiarities in the structure of the roof of the skull as well as in the arrangement of the marginal jaw elements (Voigt 1934; Obrhelová 1975). On the basis of these osteological peculiarities Gaudant assumes that the genus *Thaumaturus* Reuss is the only representative of a special Notopteroidei line, which can possibly be inserted between the 'mooneyes' (family Hiodontidae) and the more specialized knife-fishes. As a qualification to such debates, one must mention, however, that the otoliths of *Thaumaturus* species (which are known, for instance, from the Geiseltal and from Miocene deposits from the rift valley of the Rhine) resemble much more those of smelts (family Osmeridae). Thus, otolith specialists include the genus *Thaumaturus* in the family Osmeridae (Nolf 1985).

To date, no specimens of bony fish from Messel are known in which the otoliths have been preserved. However, the *Thaumaturus* species of Messel shows most of the skeletal characteristics of Osteoglossiformes just cited above (Figs 124–6) and should, therefore, in light of the foregoing discussion, be classified together with the other Thaumaturi in the suborder Notopteroidei.

The oldest finds of osteoglossomorphs (Lycopteridae) come from Asiatic deposits of

spannuthi Voigt from the Geiseltal and also in today's North American 'mooneyes' (family, Hiodontidae, e.g. *Hiodon alosoides*; the fossil genus *Eohiodon* Cavender from the Green River Shales must, however, also be included here), suggests a classification with the knife-fish relatives (suborder Notopteroidei;

Gaudant 1988). However, 'true' knife-fishes (family Notopteridae) display 'advanced' characteristics (for example, a stronger fusion of certain bones in the supporting skeleton of the caudal fin). The Thaumaturi of the Geiseltal, as well as the finds from the diatomaceous earth of Kučlin in

Fig. 127: *Thaumaturus intermedius.* Length *c.*8.6 cm. Messel Thaumaturi probably ate small animals. They are easily recognized by the very distally positioned dorsal fin or by the forked caudal fin. Furthermore, the fins are spineless and the adipose fin typical of salmon is absent. The vertebrae are amphicoelous (concave on both sides) and firmly fused with the neural arches. The absence of epipleurals is regarded as characteristic of osteoglossomorphs.

the Lower Cretaceous. The Osteoglossiformes apparently still had a wide distribution in the Tertiary. However, there are only a few records of the suborder Notopteroidei. Two fossil 'mooneyes' are described from the Eocene of North America (cf. Grande 1980); a 'true' knife-fish originates, putatively, from the Upper Tertiary deposits of Sumatra (Andrews *et al.* 1967). Gaudant (see above) hypothesizes the dispersal of the fossil Notopteroidei from their former Laurasiatic distribution whereby, on the basis of the distribution of Recent species, a Gondwana-based origin can also be envisaged. The genus *Thaumaturus* Reuss first appeared in the French Palaeocene (Jerzmańska 1977*b*; Gaudant 1979). Further Eocene finds are, as already mentioned, known from the Geiseltal and—if more recent dates are confirmed—from Bohemian sediments. Although *Thaumaturus* otoliths have been found in the Upper Tertiary of the rift valley of the Rhine and in the Mayence Basin even in slightly brackish zones (Weiler 1963), *Thaumaturus* appears to be a freshwater inhabitant. *Thaumaturus* was obviously restricted to Europe and became extinct here as early as the Miocene.

In the Recent Notopteroidei we differentiate between 'real' knife-fishes (family Notopteridae) and 'mooneyes' (family Hiodontidae). Knife-fishes—three genera with some six different species are known—occur in Africa and South-east Asia and prefer dense vegetation at the borders of slow-flowing rivers. Some can survive in the brackish zone. The partitioned and branched air-bladder can serve as an additional air-breathing organ. Generally, during the day the animals stay hidden and search for food at dusk. The smaller species eat insect larvae, worms, and other small animals. Larger species can reach a length of up to 1 m and feed on fish. Mooneyes, in contrast to the Notopteridae, occur only in fresh water. At present only one genus with two to three species is known; their distribution is entirely restricted to central and eastern North America (including the Mississippi Basin). While one species clearly prefers the clear, highly oxygenated water of larger lakes and (particularly during spawning) of rivers,

the other species prefers still water and even displays greater tolerance to turbidity (Lee *et al.* 1980). Mooneyes can attain a length of up to 50 cm. Their food consists predominantly of aquatic and terrestrial (also flying) insects and other invertebrates (such as crustaceans and molluscs) and, infrequently, of smaller fishes.

The Messel Thaumaturi have, in their appearance, stronger resemblances to recent mooneyes (Hiodontidae). The position of fins, as well as the jaw morphology and the dentition, suggest at first sight a hunter of small animals (insects, small crustaceans) of the river bank zone which either searched for food among the clumps of water plants or swam in the more open water close to the surface.

The Messel eel: unique find and 'lone-wolf'?

The 'Messel eel', *Anguilla ignota* Micklich 1985, is at present the only example of a separate species of the genus *Anguilla* Shaw. The specimen is *c.* 60 cm long and important eel characteristics can be recognized at first glance (Fig. 128). The body is extended and without pelvic fins; the vertebral column is composed of numerous, elongated, hourglass-shaped vertebrae; the skin appears naked (eel-scales are very small and can hardly be recognized with the naked eye).

The X-ray photograph of the skull (Fig. 129) shows that the upper jaw consists, in addition to the toothed maxillaries, solely of a unified plate (also with teeth), which has developed from the fusion of various bones of the upper jaw. This represents one of the most important common characteristics of the eel-like fishes (order Anguilliformes). Eel-like fishes as well as the tarpon-like fishes (order Elopiformes) pass, in the course of their development, through a curious larval stage, characterized by a lanceolate body form, the so-called leptocephalus stage. Due to this they are united in a separate systematic unit, the Elopocephala. Many aspects of the classification of the eel-like fishes are still under discussion. But the above-mentioned absence of pelvic fins, the formation of tiny cycloid scales in the skin (cf.

Fig. 148), and especially the characteristic pattern of dentition in the upper jaw of the Messel eel are good reasons to classify it with the genus *Anguilla* Shaw 1803 (suborder Anguilloidei). This is, therefore, probably the oldest find of a complete skeleton of a fresh-water eel: only eels of the genus *Anguilla* spend the greater part of their life in inland waters.

Members of the Anguilliformes (eel-like fishes) are known already from marine sediments of the Cretaceous and were found in such deposits in North America as well as in Europe (England) and Asia Minor (Lebanon). They are represented, particularly in the Eocene sediments of Monte Bolca in northern Italy, by many species (Blot 1978, 1980). The genus *Anguilla* evidently appears first in the Eocene but could not yet be documented in Monte Bolca (a very similar genus and species occurs there, however). At the moment, the genus contains, all together, 16 extant and several fossil species; the latter are partly based on recovered otoliths. From the Messel eel, as also from the other bony fishes of this site, no otoliths have been preserved. Comparisons with other species that have been described on the basis of their otoliths are therefore excluded. But even when the skeletal structure is more or less known completely, a differentiation of the Messel eel from other fossil eel species is not that simple. We must, on the basis of our knowledge of extant eels, take into consideration, for example, the changes during growth which can sometimes be considerable; further sexually based differences in size and proportion (secondary sexual dimorphism); or the tendency of eels to evolve differing types based on feeding patterns (for example, the broad-headed and the pointed-headed forms of today's river eel). The low total number of vertebrae (here also in conjunction with some characteristic details of the arrangement of bands of teeth on the upper jaw bones) as well as particular head and body proportions make it seem reasonable to classify the Messel eel in a species separate from those of the other representatives of the genus *Anguilla*. Of possibly closely related Recent eels, we must consider first of all Indo-Malaysian species (*A. borneensis, A.*

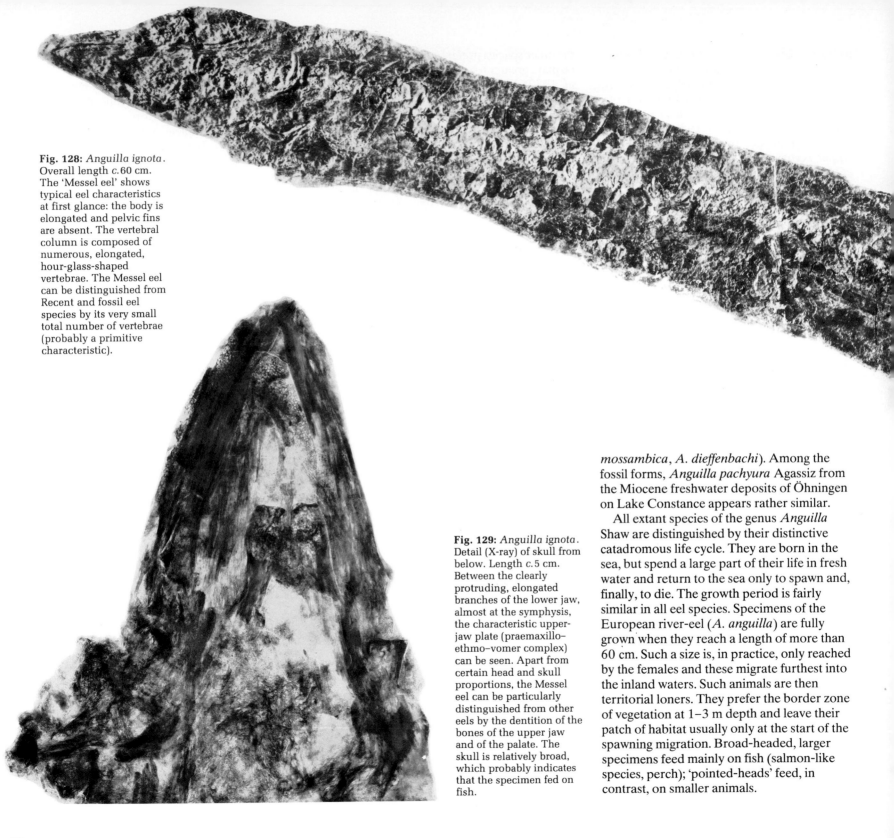

Fig. 128: *Anguilla ignota.* Overall length *c.* 60 cm. The 'Messel eel' shows typical eel characteristics at first glance: the body is elongated and pelvic fins are absent. The vertebral column is composed of numerous, elongated, hour-glass-shaped vertebrae. The Messel eel can be distinguished from Recent and fossil eel species by its very small total number of vertebrae (probably a primitive characteristic).

Fig. 129: *Anguilla ignota.* Detail (X-ray) of skull from below. Length *c.* 5 cm. Between the clearly protruding, elongated branches of the lower jaw, almost at the symphysis, the characteristic upper-jaw plate (praemaxillo–ethmo–vomer complex) can be seen. Apart from certain head and skull proportions, the Messel eel can be particularly distinguished from other eels by the dentition of the bones of the upper jaw and of the palate. The skull is relatively broad, which probably indicates that the specimen fed on fish.

mossambica, *A. dieffenbachi*). Among the fossil forms, *Anguilla pachyura* Agassiz from the Miocene freshwater deposits of Öhningen on Lake Constance appears rather similar.

All extant species of the genus *Anguilla* Shaw are distinguished by their distinctive catadromous life cycle. They are born in the sea, but spend a large part of their life in fresh water and return to the sea only to spawn and, finally, to die. The growth period is fairly similar in all eel species. Specimens of the European river-eel (*A. anguilla*) are fully grown when they reach a length of more than 60 cm. Such a size is, in practice, only reached by the females and these migrate furthest into the inland waters. Such animals are then territorial loners. They prefer the border zone of vegetation at 1–3 m depth and leave their patch of habitat usually only at the start of the spawning migration. Broad-headed, larger specimens feed mainly on fish (salmon-like species, perch); 'pointed-heads' feed, in contrast, on smaller animals.

In spite of its bad preservation the eel specimen of Messel exhibits the most important characteristics of Recent freshwater eels and must undoubtedly be associated with the same adaptive type. Thus, *Anguilla ignota* was also probably a catadromous migrating fish. But, yet again, this is possible only if Lake Messel was really—as postulated widely up to now—at least periodically joined to a river- or lake-system that had a connection to the sea.

The Messel eel can, therefore, serve as indirect 'evidence' of such a connection. The proportions of the skull or, respectively, the jaw, suggest furthermore the so-called 'broad-headed type', i.e. a predator of fishes. On the one hand, they agree very well with the measurements of corresponding specimens/types in *A. anguilla*. From a different point of view, a particularly broad skull (with a concomitant body diameter) seems disadvantageous for eels in terms of their other life habits (cf. Micklich 1983). A correlation with the feeding habit appears, therefore, the most plausible. Besides, all groups of animals preyed upon by Recent 'pointed-' and 'broad-heads' were already present in the Eocene. The

evolution of two 'variants' that are such opposites in terms of their nutritional biology would not be too surprising even in Eocene species of eel.

It is astounding, by contrast, that, to date, only one eel has been found in Messel. If the number of eels in today's lakes is taken as a yardstick, we should expect more of these animals to have lived in a water body of similar size (even if calculations are based only on the area of today's oil shale pit), even taking into consideration the Central European conditions that are rather unfavourable for eels. During an examination of various lakes in the Brandenburg Marches it was shown, for example, that at least half of those lakes that sustain, at least occasionally, hostile conditions on the lake bed (expressed as de-oxygenation or H_2S-enrichment which was, of course, also postulated for Lake Messel) hold a quite

satisfactory population of eels (Bauch 1961, pp. 164–83). Of nine other lakes, by contrast, that suffered no occasional de-oxygenation or H_2S-enrichment, only two were favourable for eels. An explanation for the rarity of eel finds at Messel could therefore lie in the 'type' of lake that Messel was. At the same time we must also consider the particular biology or, more

precisely, the probable (as in Recent eels) astonishing physiological capabilities or a higher susceptibility to processes of decomposition (fossil eels are certainly not such rarities at other sites).

Palaeoperca: the 'double-finned' Messel perch—a specialized temperate bass?

The species *Palaeoperca proxima* Micklich 1978 which occurs in Messel, is at the same time the only example of this genus of perch-like fish. Finds are recorded from various sites of the Messel pit. But these are mostly only single specimens. *Palaeoperca* is relatively 'frequent' only in older layers in the north-east of the pit. As far as can be judged, the length of most specimens varies between 17 and 20 cm. Animals shorter than 14 cm are actual rarities and finds under 10 cm are completely unknown. This is true also for specimens with lengths of over 25 cm. Interestingly, some particularly large samples (overall length 23–24 cm) come from layers that were excavated in the more southern area of the pit.

Palaeoperca proxima has a fusiform elongated body with spiny rays in the dorsal-, ventral-, and anal-fins (Figs 130 and 131). The dorsal fin is divided into a spiny-rayed proximal and a soft-rayed distal part (therefore 'double-fin'). The caudal fin is forked (Fig. 131). Perch of this kind are therefore distinguished easily from all other bony fishes at Messel. Even fragments can still be relatively reliably recognized by the characteristically broad area of scales in the distal half of the body which is composed of numerous tiny teeth (ctenii) (Fig. 149).

The margin of the upper jaw of the 'double-finned' Messel perch is, furthermore, formed exclusively of the toothed and mobile premaxillaries; the ventral fin contains one fin spine and five soft rays and is directly connected to the shoulder-girdle (cleithrum). It can, therefore, only be a representative of the perch-like fishes (order Perciformes). Based on the special composition of the musculature in the area of the gill arches, these are classified together with representatives of several other orders as the 'modern' Teleostei and are therefore, in turn, distinguished from the other teleosts as 'Neoteleostei'.

Perch-like fishes are, with *c.* 147 families and innumerable genera and species, the most numerous and most variform order of all vertebrates. The splitting into systematic subgroups is, as in the eels, consequently difficult. This is true both at the subordinal and familial levels. Thus, for a long time, those

Perciformes which did not show sufficiently distinctive characteristics to attribute them to a different suborder, were classified with the suborder Percoidei (the perch-like fishes *sensu stricto*), while the sea-bass family (Serranidae) filled a similar 'dust-bin' function. Once again, those Percoidei were placed in the Serranidae that could not be classified with any of the other families.

Further classification of *Palaeoperca proxima* is correspondingly difficult. Among other criteria, it is distinguished from the 'true' perch (family Percidae) by the higher number of spines in the anal-fin and the significantly lower number of vertebrae (cf. Collette and Banarescu 1977). At least from a systematic viewpoint, *Palaeoperca* cannot, therefore, be called, as one has read frequently in the past, a 'true' freshwater perch. The distinctive characteristics of the bones of the skull and cleithrum suggest that it is a distinct genus and species of the Percoidei, whose closest relatives should be found in the domain of the Percichthyidae in the widest sense. The important factors in this classification include, for example, the number and structure of the opercular spines (cf. Fig. 132), the particular distribution of the fin-spines and their support in the proximal part of the dorsal fin, details of the supporting skeleton of the caudal fin (Figs 134 and 135), and the number of vertebrae as

well as their distribution throughout the individual segments of the body. But the classification of this family also is still not entirely resolved. Some groups of genera have been newly defined and must now be distinguished from the original 'Percichthyidae' *sensu lato* (cf. Gosline 1966; Arratia 1982; Johnson 1984). But the 'double-finned' Messel perch corresponds in several characteristics quite well with the Recent North American genus *Morone* Mitchill 1914. Since the genera grouped about *Morone* have recently achieved family status, the 'double-finned' Messel perch may have to be regarded as a temperate bass species (family Moronidae). However, to date, not all distinguishing characteristics of this family could be verified in the Messel perch. At least some of the observed deviations from these characteristics could, nevertheless, be interpreted as possible specializations, but others seem, by contrast, rather 'primitive' (cf. Micklich 1985, pp. 74–6, 1989). When all facts are considered, *Palaeoperca* can possibly be regarded as a temperate bass (moronid) that was adapted to a diet of smaller animals and to an agile life in open water.

Certain characteristics are, at least in some Recent 'Percichthyids' in *sensu lato*, extraordinarily constant and remain so during growth. This is true particularly for the formation of spine-like processes at the distal

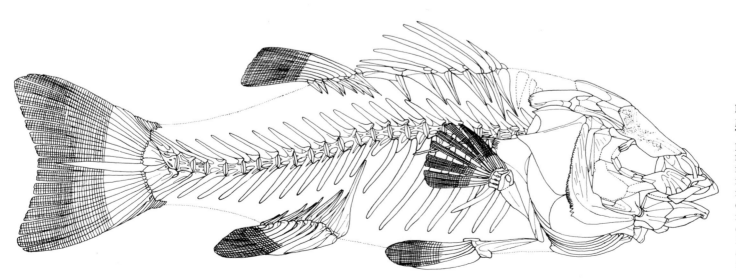

Fig. 130: *Palaeoperca proxima*. Reconstruction. The double-finned Messel perch is identified by its fusiform, elongate body, its two-part dorsal fin, and its slightly forked caudal fin. Fin spines are characteristic of perch-like fishes. In contrast to our extant river perch, *Palaeoperca* has three anal fin spines; it also has fewer vertebrae.

Fig. 131: *Palaeoperca proxima.* Overall length *c.*24 cm. Both the pectoral fins, which are positioned relatively high on the sides and the composition of the pelvic fin, as well as their means of attachment to the body, are important for systematic classification. The number and distribution of the vertebrae or the arrangement of the fin spines and their supporting structure in the proximal part of the dorsal fin are, by contrast, rather more important in the classification of this species. There is some indication that the double-finned Messel perch were partially specialized temperate bass.

rim of the principal operculum. Some Messel specimens of the genus *Palaeoperca* show, particularly in this area, variations that occur in Recent perch-like fishes only between representatives of different species (Fig. 133). Still, up to the present, no further reliably distinguishable differences can be proven among the specimens in question, but this may result from the rarity of such finds.

The geologically oldest finds of perch-like fish date from the late Cretaceous or, to be precise, from the Palaeocene. Perch-like fishes have, therefore, an approximate age of at least 65 million years. Because a large number of different genera and species have been recorded as early as the middle Eocene, their evolution must have been unusually fast. However, among the earliest Perciformes rather specialized forms (for example, Menidae; cf. Patterson 1964) can already be found. The sparsity of the pre-Eocene finds could therefore also be caused by gaps in the fossil record, so that the perch-like fishes might indeed have already gone through a far longer period of evolution.

Fossil bass (Moronidae) (mostly finds in the form of otoliths) were described from numerous localities in Europe that date mainly from the Miocene. Because the genus *Allomorone* Frizzel and Dante from North America (which is also based on otoliths) must clearly be classified with another perch family (Nolf 1985, p. 31) and because *Paramorone* David is only based on finds of scales, the corresponding evidence is, at the very least, questionable. In contrast, possible affinity to this family and a closer relationship with *Palaeoperca* must be discussed for other fossil 'Percichthyidae' whose area of distribution at times goes beyond Europe (Micklich 1988). Recent moronids occur only in Europe (*Dicentrarchus* Gill) and North America (*Morone* Mitchill or, to be precise, *Roccus* Mitchill), if one disregards the genera *Siniperca* Gill and *Lateolabrax* Bleeker for which the systematic affinity has not been entirely reliably established. Biogeographical interpretations are, at present, almost impossible if only because of the uncertainties in the classification of fossil forms. The Recent Moronidae should, more or less, derive almost directly from marine ancestors which were probably distributed throughout the Northern Atlantic (Woolcott 1957).

The area in which present-day European temperate bass (genus *Dicentrarchus* Gill) are found extends from the coasts of Norway to the Mediterranean. The animals attain a length of up to 1 m and prefer open, coastal water. During spawning, however, they migrate into the lower courses of rivers. Juveniles feed mainly on soft-bodied prey and crustaceans; fully grown specimens mainly eat smaller fishes. The North American species of the genera *Morone* Mitchill and *Roccus* Mitchill are distributed throughout wide areas of the Atlantic coast and also in inland waters from the area of the Great Lakes to the Mississippi Valley and grow only a little larger than their European relatives. For spawning, the more ocean-going species also enter the rivers. Fully grown animals feed, in similar fashion to the European temperate bass, on small crustaceans, molluscs, and smaller fishes. During spawning they cease to take nourishment.

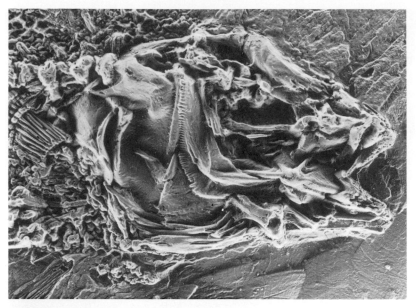

Fig. 132: *Palaeoperca proxima*. Detail of skull (vapour-treated). Length *c.*5.5 cm. The skull also exhibits important perch-like characteristics. Thus, the perimeter of the upper jaw is bordered only by the premaxilla, which can be moved forward. The small gape and details of the snout region suggest that this fish ate small animals. The distal margin of the proximal gill-cover is finely toothed and the ventral margin is completely smooth, in contrast to that of other Messel perch. Both the flattened, spine-like processes at the end of the main operculum are important for classification at family level.

Fig. 133: *Palaeoperca* sp. Detail of operculum (vapour-treated). Width of picture *c.*8 mm. The main operculum of some *Palaeoperca* specimens shows variation in shape; here, for example, the lower spine is split.

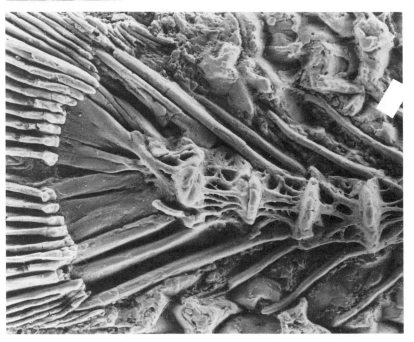

Fig. 134: (left); Fig. 135: (right). *Palaeoperca proxima*. Details of the skeleton supporting the caudal fin (vapour-treated). Width of picture Fig. 134: 3 cm; Fig. 135: 5 mm (reversed). The development of a small, dorsally positioned, additional bone plate (Fig. 134) and a tiny ventral hook (Fig. 135) are regarded as primitive features of perch-like fishes.

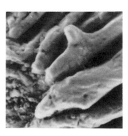

Perch of the genus *Amphiperca*: Australians at Messel?

The species *Amphiperca multiformis* Weitzel 1933, which was described from Messel, is found somewhat more frequently than the 'double-finned' perch, but was looked upon as rare for a long time. In the meantime, however, quite massive deposits have been discovered further south of the Messel pit, from the lowest level of certain layers that had formerly been uncovered by open-cast mining. Most *Amphiperca* finds have an overall length of 13–18 cm. Of specimens under 10 cm (the so-called juvenile forms) only a few finds could up to now be identified with certainty as genus *Amphiperca*. The particularly large specimens of over 22 cm in length are, yet again, of similar rarity.

Just as in the 'double-finned' Messel perch, the fin spines of the dorsal, ventral, and anal fins can, in *Amphiperca* specimens also, serve as the simplest character for differentiating them from other bony fishes at Messel. In contrast to *Palaeoperca*, the spiny- and soft-rayed sections of the dorsal fin are not separated and the caudal fin is rounded posteriorly (Figs 136 and 137). Further important differences with regard to *Palaeoperca* lie, among others, in the number and distribution of the vertebrae within the individual sections of the body, in the differential serration of the proximal gill-cover (preoperculum), as well as in the development of additional bones in the area of the upper jaw (cf. Figs 138 and 139) or in the skeleton supporting the caudal fin. Fragments can, again, be recognized quite easily from the very much smaller (in comparison with the *Palaeoperca*) field of ctenii in the distal sector of scales (the ctenii are, in these cases, occasionally entirely absent) (Fig. 150).

Due to the presence of all other typical characteristics of perch-like fishes which have already been listed in the discussion of the 'double-finned' Messel perch, *Amphiperca multiformis*, too, is undoubtedly a representative of the Percoidei (order Perciformes). Furthermore, a classification with the percichthyids in the widest sense (Micklich 1987) is here again supported by the number and distribution of the vertebrae, by details in the skull (shape of operculum, form of the hyoid arch, and in the skeleton supporting the caudal fin (Figs 138–140)), and also by the number and distribution of fin-spines and their supports (pterygophores) in the proximal section of the dorsal fin. In this case, however, the most closely comparable Recent perch must, possibly, be looked for

1978; Turner 1982). Even so, a possible common ancestor would have to have been a representative of a euryhaline species, tolerant of changes in the salt content of the water, which probably occurred and were distributed throughout the Tethys Sea (cf. Micklich 1987; Gaudant 1988). Such a species can hardly be dubbed 'Australian'. By far the largest number of shared characteristics is observed, however,

percoids that are found frequently in the brown coal of the Geiseltal near Halle. However, a new investigation of the Geiseltal perch (Micklich and Gaudant 1989) has clearly shown that these, as suggested by Voigt (1934) in his first description of these fossils, belong to a separate genus and species (*Anthracoperca siebergi* Voigt).

In the *Amphiperca* materials some striking

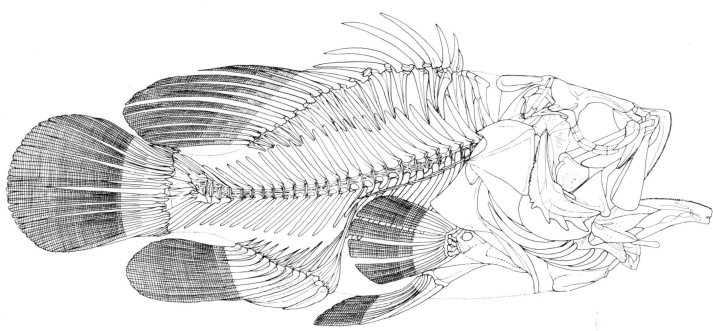

Fig. 136: *Amphiperca multiformis*. Reconstruction. This is again a typical representative of the perch-like fishes, with spines on the dorsal-, ventral-, and anal fins. The dorsal fin, in contrast to *Palaeoperca*, is continuous and the caudal fin is distally rounded. The shape of the body, the structure and dentition of the jaws and palatal bones, as well as stomach contents so far investigated, suggest that these were fish-eating predators which ambushed their prey.

among the 'Australians' (for example, species of the genera *Macquaria* [*Percalates*] Cuvier and Valenciennes 1830 or *Maccullochella* Whitley 1929). This is, however, more or less pure speculation at present. For one thing, a large part of the similar characteristics may result from parallel evolution (Micklich 1987, pp. 74–85); for another, the genus *Percalates*, to be sure, has already been described from (Early) Tertiary sediments (*Maccullochella* is, in contrast, only known since the Pleistocene) but all records, fossil or Recent, are restricted to Australia (Andrews *et al.* 1967; MacDonald

in the comparison with another fossil genus and species. This, too, is found in the probable Eocene, diatomaceous shales of Kučlin in Northern Bohemia that were mentioned above; it was described by Obrhelová (1971) as *Bilinia uraschista* Reuss 1844. The similarities in some important characteristics are so striking that the question arises as to whether *Bilinia uraschista* and *Amphiperca multiformis* are not just different species of one and the same genus. Such a similarity at generic level was postulated for a long while for *Amphiperca multiformis* Weitzel and for the

morphological differences also occur (Figs 138 and 139). These can only in part be explained as (possibly growth-associated) intraspecific variations (Micklich 1987). Almost always there are details which can only be observed in certain individual specimens. At present, it does not appear reasonable, therefore, to distinguish between several species of *Amphiperca*.

Present-day *Macquaria* [*Percalates*] and also *Maccullochella* species live in inland waters near the coast, the lower courses of rivers, and in south-eastern Australian estuaries. They

Fig. 137: *Amphiperca multiformis*. Length *c.* 19.5 cm. Specimens of the genus *Amphiperca* can also be distinguished from the double-finned Messel perches by their more numerous vertebrae. There are further significant differences in the skull. The lower margin of the preoperculum is strongly toothed and there is an additional bone (the supramaxillary) in the upper jaw (see also Figs 138 and 139). Some important characteristics (as, for example, the two-pronged main operculum or the number and distribution of the vertebrae) correlate well with the 'percichthyids'; others (e.g. the number and arrangement of the proximal spines of the dorsal fin and their supports) bear a closer resemblance to the Serranidae.

Fig. 138: (left); **Fig. 139:** (middle). *Amphiperca multiformis*. Details of skulls. Width of pictures, Fig. 138: 7.6 cm (vapour-treated); Fig. 139: 8.1 cm. (Plaster cast.) Not all of the morphological variation can be irrefutably traced to infraspecific variability or to changes during fossilization. In Recent Percoidei the slope of the supraoccipital crest (a pointed process of bone on the back edge of the roof of the skull) is sometimes important in the identification of species.

Fig. 140: (right). *Amphiperca multiformis*. Detail of caudal fin support. Width of picture *c.* 12 mm. In contrast to the caudal-fin support of the double-finned Messel perch (Fig. 134), the additional small dorsal bone plate is missing. The upper and lower hypurals (flat, small triangular bones, below the prolongation of the vertebral column) are, furthermore, very closely juxtaposed even at the mid-line.

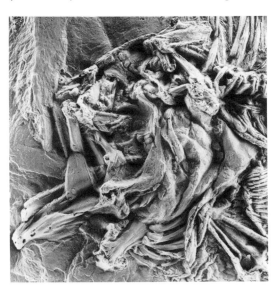

reach a length of 43–58 cm (*Macquaria*), respectively, up to 1.8 m (*Maccullochella*) and mostly prefer quietly flowing (also still) and partly muddy waters. For spawning some species return from the rivers into brackish water or the sea; others remain permanently in fresh water. The principal food sources are either various crustaceans or smaller fishes (*Maccullochella*) or insects and their larvae (*Macquaria*; cf. Scott *et al.* 1980).

With regard to body structure and jaw shape *Amphiperca* appears to be adapted to a predatory life in an environment full of obstacles. The jaws are elongated, the palate (palatinuum) set with teeth and, probably, double anchored to the brain-case (Micklich 1987), the body has a relatively high back, and the caudal fin is rounded. The animals lurked in a hiding place (for example, between water-plants or branches) and overpowered their prey (which probably consisted of small fish) in a lightning attack with a fast snap of the powerful jaws. To date, stomach contents could be documented only sporadically (the results are, however, based only on observations with the light microscope). The few findings, however, correspond well with the nutritional strategy that has been suggested. In one specimen lumps of fragments of numerous soft fin-rays can be recognized; in another, two tiny ganoid scales were found lying in the area of the abdominal cavity. In several specimens small, rounded, weakly conical structures are observed between or on the anal fin-rays. These are with a high degree of certainty the pharyngeal teeth of the so-called 'juvenile forms'. These can hardly have been washed selectively into the vicinity of only one fin and, therefore, probably derive from the gut.

'Juvenile forms'—small, but not necessarily young

As already mentioned, these are rather small perch-like fishes, which are found, sometimes in great numbers, in the older layers in the north-eastern part of the pit (occasionally several specimens even lie together on the surface of the same stratum). By contrast, they appear to be absent from younger strata and from the vicinity of the north-western tributary proposed in the flow model of Franzen *et al.* (1982). The overall length is, even in larger specimens, scarcely more than 8 cm. Mostly, however, only animals of 2–3 cm length are found and tiny specimens of *c.* 1 cm are by no means rare.

All of these small specimens exhibit all characteristics of the perch-like fishes (cf. Fig. 141) and were, for a long time, thought to be specimens of *Amphiperca multiformis*. They are distinguished, however, by several distinct characteristics from *Amphiperca multiformis* as well as from *Palaeoperca proxima*, so that they must, therefore, be placed in a separate genus and species (Gaudant and Micklich 1990). In fact, these are probably not exclusively juveniles. In the larger specimens small ctenoid scales can still be recognized on the front part of the body. In Recent perch, however, the body scales start to develop first in the tail region and progress towards the head (cf. Balon 1959). Still, one must assume that the smallest specimens found are, indeed, juveniles (probably even immediate post-larval stages). Among the distinctive characteristics allowing classification among the new Percoidei of Messel, we shall mention here only the lower number and deviant distribution of the vertebrae, the characteristic pharyngeal teeth (Fig. 143), and the smooth-rimmed preoperculum. In particular, the typical hemispherical or slightly conical teeth found in the pharyngeal region of almost all specimens of this species are a highly reliable diagnostic feature.

Similar pharyngeal teeth are found in extant labrids and cichlids (in these, however, they are borne on medially connected pharyngeal bony plates). They also occur in some other fossil percoids (among others *Dupalis* Gistel 1848, *Dapaloides* Gaudant 1985, *Priscacara* Cope 1877, and *Prolates* Priem 1899). At present there is as little evidence for classification of the small Messel perch with one of these genera as there is for classification with other fossil or Recent percoids. On the other hand, several differences in the skeletal structure of some individual specimens of this new group of Messel perch raise the suspicion that more than one species could be involved (cf. Micklich 1985, p. 126).

The type of pharyngeal tooth arrangement seen in these small Messel perch is, interestingly, found mostly among the 'hard-shell feeders' (for example mollusc-crackers) of Recent Perciformes (cf. Liem 1973) and has also been interpreted in fossil forms as indicating this type of feeding behaviour (e.g. Patterson 1964, pp. 420, 421). Apparently, the Messel Percoidei were not bothered by such considerations: in a large number of the specimens found, smaller fry of the same species can be recognized as prey in the jaws or as yet undigested stomach contents (Fig. 144). On the one hand, this could indicate a lack of the 'right' food (i.e. a food emergency), but, on the other hand, this behaviour might just as well reflect a glut of these very small perch so

Fig. 141: 'Juvenile' form. Length *c.* 7 cm. This is a new genus and species, whose overall length hardly ever exceeds 8 cm. The finds were, therefore, formerly thought to be young specimens of *Amphiperca multiformis*.

Fig. 142: *Amphiperca multiformis*. Length *c.*6.5 cm. Smaller specimens of *Amphiperca* sometimes resemble the new species quite closely. They are, however, easily distinguished from it by their longer fin-spines, the larger number of vertebrae, and the preoperculum, which is strongly toothed even in small specimens. The preoperculum also exhibits a smooth margin in all specimens.

Fig. 143: 'Juvenile' form. Enlargement of X-ray detail from the skull. Width of picture, *c.*20 mm. Among the most marked characteristics are the hemispherical to slightly conical small teeth which can be distinguished in the pharyngeal region. They lie on two upper and lower pharyngeal plates which are not fused with one another. Similar dentition can be found in some Recent perch-like fishes, particularly among the 'hard-shell' feeders such as, for example, some cichlids and their relatives (Labroidei). In these cases, however, the pharyngeal plates are usually fused.

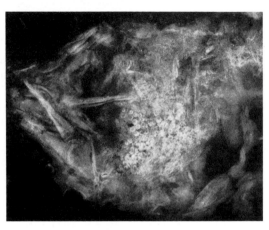

Fig. 144: 'Juvenile' form with prey. Overall length *c.*4.8 cm. In contrast to the adults' apparent specialization for hard-shelled food, the Messel juvenile forms clearly preferred to feed on smaller specimens of their own species.

that the otherwise more specialized larger specimens of the same species turned to them as the most readily available source of food. Cannibalism is not uncommon even in Recent species of perch; among fossils it has been reported in representatives of the genus *Dapalis* among others.

'Salmon' and apparent specialists—an unreliable assemblage

At this juncture some points shall be considered again which, although they have been covered in the descriptions of individual fish, play a significant role in the discussion of possible conditions of life in and round Lake Messel. A warning must be given at the start. Caution is advised, particularly in ecological interpretations: much here is based on analogies with extant forms whose known behavioural habits are transferred, so to speak, to fossil forms.

In some cases this is a procedure which cannot only be defended but is also a sensible and particularly vivid means of 'experiencing' life in the past. This approach is especially successful when applied (as, for example, in this chapter to gars, bowfins, or the eel) to the nearest relatives of an extant group with almost identical skeletal structure, adapted to a very specific way of life; or when special structures are observed in the fossil skeleton, which are, perhaps, in different Recent groups not always identical in every detail, but which, nevertheless, always occur correlated with one and the same ecological peculiarity. But even such seemingly safe applications have their difficulties. Under 'special' conditions (these can even be the mere absence of competitors), such 'specialists' are not concerned that their peculiarities were evolved at such a cost and exhibit as, for example, the small Messel Percoidei, quite different, suspected tendencies. The example of the Thaumaturi has shown what confusion can be caused by hasty—yet constantly demanded—ecological conclusions, drawn when family relations are unclear and when distinct specializations are

lacking. Although classified merely as a separate family in the suborder Salmonoidei (salmon relatives), the putative relationship with salmon or trout has always been enthusiastically discussed by ecologists, in spite of the variations observed in the biology of the individual families and species in the Salmonoidei (cf. Nikolski 1957, pp. 153–216). In view of its relationship with the osteoglossomorphs, which now seems probable, this must at least have led to different (and probably incorrect) conclusions concerning the biosphere of that time. It has to be considered even now, that Recent knife-fish relatives (Notopteroidei) might be quite different biologically and that *Thaumaturus* probably belongs to its own primitive line of evolution, perhaps only sharing the feeding habits of the others.

Only hunters?

It is frequently stated that, astonishingly, there were only predatory fish in Lake Messel and that therefore a complete food chain could hardly have existed. This assumption may even be justified in so far as bacteria were probably the only exploiters of organic detritus on the lake floor; for the rest, there probably were gaps among the consumers at the base of the food chain—thus, according to Lutz (1987), there were no indigenous aquatic insects which could serve as potential 'exploiters' of planktonic organisms and vegetable matter. This assumption does not apply at all to the topic of this discussion, the fish fauna of Messel. On the one hand, one must consider that at the time of the European Middle Eocene, important groups of non-predatory fishes such as carp were still completely absent (cf. Jerzmańska 1977b) and that other forms took their place in the biocoenosis; on the other hand, rather strange cases are known of extant fish whose food chain is formed only from representatives of one and the same species (with plankton- and plant-eating larvae at the start and the mature, largest specimens at the end). Such special cases need not be considered here: to be sure, the primary

consumers appear to be absent (as occurs in quite a number of fish food chains). Nevertheless, secondary consumers, feeding on small animals, including *Thaumaturus*, *Palaeoperca*, and (normal) 'juvenile' forms, can be differentiated from fish-eating consumers of a higher order, such as *Amphiperca*, gars, medium-sized bowfins, or eels. Finally, there exist, in the particularly large specimens of the genera *Cyclurus* and *Atractosteus* found at Messel, carnivores which are almost at the top of the food chain. This gradation is supported by the biological circumstances (in relation to food) of their apparently closest Recent relatives, by their osteological peculiarities, and also by some actual fossil finds (stomach and gut contents, prey in the jaws). Thus, an almost complete food chain could be reconstructed solely by correlation of the fish occurring in Messel. As already partly described when treating the 'juvenile forms', the smaller animals in each case (starting with eggs or larvae even of the very same species) were the potential prey of the next larger specimens. Very young animals of all species probably ate plants or plant detritus at times, just as *Palaeoperca* could have fed occasionally on vegetation (Micklich 1983). Among the secondary consumers, *Thaumaturus* appears to have specialized on insects and their larvae or on small crustaceans, while the pharyngeal teeth of the small-sized new percoid genus and species would 'normally' point to hard-shelled food. As discussed above, this also holds true for *Cyclurus*. By this means, the immediate competition of both forms with other equally ranked species of the food chain was possibly diminished. Recent bowfins are also mainly crepuscular in contrast to gars. The fact that *Cyclurus* and the 'juvenile' Percoidei, which were probably both specialized to feed on snails or crustaceans, also ate fish at Messel deserves attention, especially since molluscs are found throughout Messel. In the case of *Cyclurus* it must be clear that only the most conspicuous prey consisted of fish. Only a few examples are available to support this, while an exact investigation of the gut contents of the majority of specimens still remains to be done. The dentition of the jaw, furthermore, shows

only a moderate specialization for 'hard-shelled' food, which does not exclude fish as an occasional snack. Two explanations for the cannibalism among the 'juvenile' Percoidei have already been offered. An almost arbitrary number of further possibilities could be added. These, however, are based at present, almost exclusively on speculations. Thus, for example, in a biocoenosis which is relatively poor in species we will observe a frequently exponential increase in the population when conditions (a plentiful food supply) are good; the subsequent scarcity of food is followed by a population crash.

Permanent residents or visitors?

The relative paucity of species among the fish fauna of Messel, when compared with other fossil sites or to the lakes of today, suggests, in conjunction with the simultaneously high number of individuals of certain groups, 'unfavourable' living conditions which could be tolerated by only a few species and which were perhaps even advantageous to these. One argument for this is the high proportion of species which, when exposed to de-oxygenated conditions, were able to breathe air through their swim-bladder and so were possibly only in this way adapted for a prolonged stay in Lake Messel (*Atractosteus*, *Cyclurus*, and perhaps also *Thaumaturus*). The striking frequency of damaged and then regenerated scales in representatives of the genus *Amphiperca* probably also points in that direction. It is also striking that some fish species of Messel (*Amphiperca multiformis*, *Anguilla ignota*, *Palaeoperca proxima*, and *Thaumaturus intermedius*) were probably able to cope with euryhaline (i.e. slightly salty) water. Both juvenile and mature specimens of the genera *Amphiperca*, *Atractosteus*, *Cyclurus*, *Thaumaturus*, and of the smaller, new perch genus and species exist. This could mean that their habitat was in the lake itself or in its immediate vicinity. The almost exclusive occurrence of (probably mature) representatives of the genus *Palaeoperca* and the rarity of such finds as well as their even distribution throughout the entire area of the pit suggest that this was only an occasional visitor. The surface structure of its scales in comparison with those of *Amphiperca* shows distinct differences that probably result from the different ecology of both species. There are only a few prominent growth cessation marks on the scales of *Palaeoperca*, which are separated by wide zones of apparently undisturbed growth. Only by very critical examination of these wide zones can we detect signs of occasional slowing down of the growth rate. These structures, to judge by their intensity, could either have been relatively regular yet weak disturbances, or could be so-called spawning-rings, caused by discontinued or diminished food intake during reproduction. The latter may have affected the animals differently according to their bodily constitution. The spurts of growth on *Amphiperca* scales are, by contrast, mostly quite strongly developed and often show an additional resorption of the marginal material. Probably *Amphiperca* was, as a permanent inhabitant of Lake Messel or its immediate vicinity, more strongly subject to unfavourable conditions than the more mobile 'double-finned' Messel perch. Others, also probably constant inhabitants of this region, for example, the gar or the eel, also show distinct growth-markers on their scales. In the case of the eel we can be almost certain that these are not spawning-rings. Eels of the genus *Anguilla*, as is well known, spawn only once, just before their death. It must therefore be assumed that weak seasonal influences predominated in the Middle Eocene of Messel which could, however, have quite different effects on the fish fauna, depending on the special habitat in each case.

If unusual processes like displacement are excluded as reasons for the eel's presence in Messel, the occurrence of the animal there strongly supports the argument that Lake Messel was joined, at least temporarily, to a water system with outlets to the sea. If one takes into account the extraordinarily acute sense of smell and the sensitive reaction of Recent eels to adverse conditions, such a presence is a further sign that conditions in

Figs 145–150: Types of scales from the Messel bony fish. From left above to right below: *Atractosteus strausi* (width of picture *c.*6 cm), *Cyclurus ('Amia') kehreri* (width of picture *c.*3.5 cm), *Thaumaturus intermedius* (back-lit; width of picture, *c.*2 mm), *Anguilla ignota* (back-lit; width of picture, *c.*0.3 mm), *Palaeoperca proxima* (back-lit; photo of a scale surface on acetate film; width of picture *c.*3 mm), *Amphiperca multiformis* (back-lit; photo of a scale surface on acetate film; width of picture, *c.*1.5 mm). The genera and species of the fishes that occur in Messel can be quite well identified by their scales alone. The scales of the gar (*Atractosteus strausi*) are massive, rhomboid, and flexibly articulated. Their surface is covered with a shiny ganoid layer. Proximally to the unpaired fins there are enlarged fulcral-scales, which differ in shape. The scales of the bowfin *Cyclurus (Amia) kehréri* are relatively large and rounded (cycloid) to elongated-oval and lack a ganoin layer. A fine, longitudinal striation can be seen on the surface. The scales of *Thaumaturus intermedius* are also cycloid and lack ganoin, but are much smaller than those of *Cyclurus*. The majority of the surface ridges (circuli) lie almost parallel to the scale margin, but meet at an angle distally. Eel scales are very small and can hardly be seen with the naked eye. In the Messel eel (*Anguilla ignota*) only one scale has been preserved in even half-way complete condition. This,

however, is composed of a circular arrangement of rows of platelets, characteristic of eels. Perch scales are distinguished by having an area in the distal portion which bears small teeth arranged in alternate rows: these are also referred to as ctenoid scales. The tooth-bearing area is quite broad in the double-finned Messel perches (*Palaeoperca proxima*). *Amphiperca multiformis* also normally bears typical ctenoid scales. The toothed area, however, is significantly narrower than in *Palaeoperca* and can at times be entirely absent, for instance in regenerated scales.

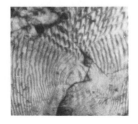

Fig. 151: Overview of fish species in Messel. We have tried to indicate the rank of the various systematic groups by means of different typefaces. The classification, which, in reality, is even more complicated, helps scientists to survey family relationships or the actual organizational or evolutionary levels of individual genera, species, or whole families, quickly. All species of fishes that occur in Messel are so-called Osteichthyes, which means that their skeleton is, at least in part, composed of true bone (the pre-formed 'covering' bone of connective tissue, and 'replacement' bone of cartilaginous tissue). Ray-finned fishes (actinopterygians) are distinguished (among other characteristics) by the fact that the supporting skeleton of their paired fins consists of several parallel fin rays (radialia) of similar size. Typically, these connect directly with the pectoral or pelvic girdle. 'Muscle-finned' fishes (sarcopterygians such as coelacanths and lung-fish) have, instead, an articulated main axis to the fin with skeletal ribbing on either side. The Neopterygii are named after a simplification in the structure of the unpaired fins. This contrasts with the majority of the cartilaginous ganoid-fish (chondrostids, such as sturgeon and paddlefishes), in which every basal support element (radial) still bears only one fin-ray.

Systematic categories
Class: Osteichthyes (bony fish)
Subclass: Actinopterygii (ray-finned fish)
      Neopterygii ('new-finned' fish)

Ginglymodi
    Order: Lepisosteiformes (gar-like fish)
    Family: Lepisosteidae (gar)

Halecostomi
Halecomorphi
    Order: Amiiformes (bowfin-like fish)
    Family: Amiidae (bowfins)

Teleostei
Osteoglossomorpha
    Order: Osteoglossiformes (bony-tongue-like fishes)
    featherbacks, knife-fishes (Notopteroidei)
    Family: Thaumaturidae

Elopocephala
    Order: Anguiliformes (eel-like fish)
    eels (Anguilloidei)
    Family: Anguillidae (fresh-water eels)

Clupeocephala: Eutelostei; Neotelostei
    Order: Perciformes (perch-like fish)
    basal percomorphs (Percoidei)
    Family: Percichthyidae (*sensu lato*)

    Family Moronidae (freshwater and estuarine bass)

Species represented in Messel

Atractosteus strausi (Kinkelin 1884)
Messel gar

Cyclurus ('Amia') kehreri (Andreae 1893)
Messel bowfin

? *Thaumaturus intermedius* Weitzel 1933
'Archaic knife-fish'
'False salmon'

Anguilla ignota Micklich 1985
Messel eel

Amphiperca multiformes Weitzel 1933
High-backed predatory Messel perch

? Juvenile forms
? *Palaeoperca proxima* Micklich 1978
'Double-finned' Messel perch
? *Palaeoperca* sp.

Lake Messel were either not permanently hostile or that these adverse conditions were not predominant over the entire lake. Eels would otherwise hardly have risen in the lake, or, respectively, in its runoff. Speculations about the rarity of eel finds up to now and the possible context have been aired elsewhere.

Also interesting is the occurrence of smaller bowfins and gars, particularly in the north-eastern section. If the physical proportions of their—probably not only systematically closely related but also similar in other ways—Recent relatives are taken into account, then Messel bowfins of up to 10 cm length should still be true larva which possibly still swam about in loose shoals (cf. Reighard 1903). *Amia calva* attains approximately the same maximal length as the larger Messel bowfin finds and in this case the samples up to 10 cm show the behaviour that was just described. In addition, we should then expect extended shallow-water areas in the vicinity of Messel Lake, since Recent bowfins (and also the gars) prefer shallow water for spawning.

7

Amphibia at Lake Messel: salamanders, toads, and frogs

It is striking that, although amphibia are usually dependent on open water, at least for reproduction, only a few species of amphibia have been found at Messel to date. Salamanders, which were strongly adapted to life on land, are rarely found either in Messel or in the Geiseltal. They probably entered the body of water only to breed.

Amphibia at Lake Messel: salamanders, toads, and frogs

MICHAEL WUTTKE

The rare appearance of amphibia at Messel is most striking of all in the case of the batrachosaurids, close relatives of the extant axolotls, which probably spent most of their life in the water. Not a single specimen of these has been found in Messel. Like the extant axolotls, these animals were neotenous, i.e. they spent most of their life in the larval stage during which they even reached sexual maturity. From the brown coal of the Geiseltal, by contrast, more than 300 specimens of the species *Palaeoproteus klatti* have been found to date.

From Messel we have only a single salamander specimen. It was described in 1980 by F. Westphal from Tübingen, as *Chelotriton robustus* (Fig. 153). This genus, of which the Messel specimen is the oldest find, resembles most of all, among the salamandrids, the genus *Tylototriton* (the crocodile salamander). The existence of the latter is documented from the

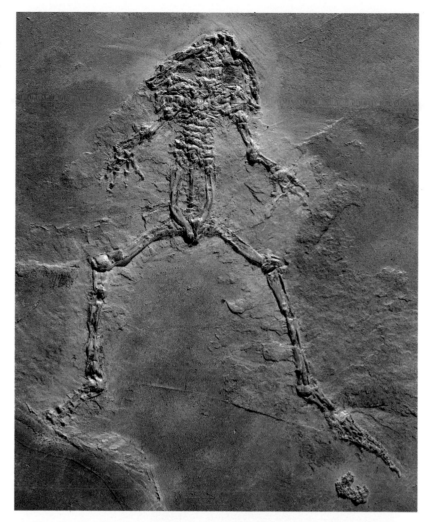

Fig. 152: *Eopelobates wagneri*, a largely terrestrial frog, which only came to Lake Messel to reproduce. Head and body length, *c.* 6 cm.

Fig. 153: *Chelotriton robustus*, a largely terrestrial salamander, the sole specimen found at Messel. Length of animal, *c.* 10 cm.

middle Eocene of the Geiseltal through to the present (distribution: East Asia).

Chelotriton robustus (Latin *robustus*: strong) derives its name from the particularly strong ossification of the skull and vertebral column. The bones of the roof of the skull exhibit a very pronounced puckered and net-like structure; the vertebrae are topped dorsally by broad, puckered bony plates which are in contact with those of the neighbouring vertebra.

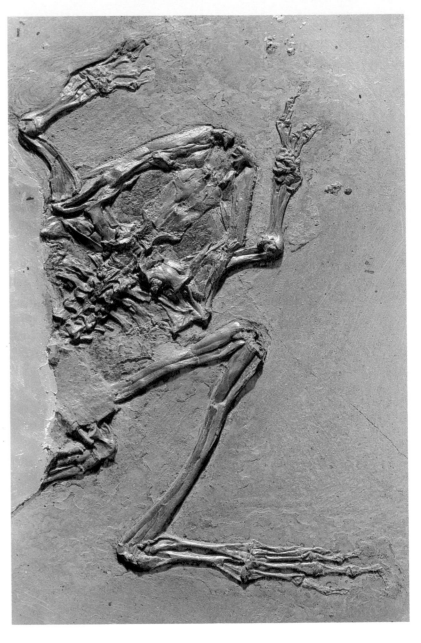

Fig. 155: Among the rarer finds are skeletons of females of the frog *Eopelobates wagneri* with the spawn preserved. Length of skull, *c.*2.5 cm.

Fig. 154: Reverse of find shown in Fig. 155. Spawn is visible as a dark band at the side of the body where it emerged after the skin had decayed. Length of pelvis, *c.*3 cm.

Such extensive ossification is probably primarily associated with the mostly terrestrial life-style of *Chelotriton*, a trend which can be traced in most of the terrestrial amphibia but for which, even today, no one has given a satisfactory explanation as to its function. This mostly terrestrial existence could also be one of the reasons that, to date, only a single specimen of *Chelotriton* has been found in the lake sediments of Messel, even though *Chelotriton* vertebrae are not all that rare in the infills of cracks in the Tertiary karst.

The frog and toad (anuran) fauna, like the salamandrid fauna, is surprisingly poor in species. Only in 1938 was Weitzel able for the first time to demonstrate the existence of

Fig. 156: Tadpole of
Messelobatrachus tobieni.
Parts of the vertebral
column and the skull have
become ossified. Length,
*c.*2 cm.

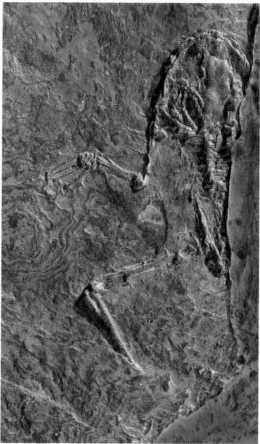

Fig. 157:
Messelobatrachus tobieni,
an aquatic frog. It was
only resident in the lake at
certain periods (see text).
Head and body length,
*c.*3 cm.

Fig. 158:
Messelobatrachus tobieni,
with preservation of soft
parts around the back and
hind legs. Head and body
length, *c.*3 cm.

anurans in Messel. Since that time the number
of species has risen only by three, of which one
is only represented by a single specimen. More
than 100 specimens of the form *Eopelobates
wagneri* studied by Weitzel have already been
found (Figs 152, 154, and 155). This form is
classified in the same family as the Eurasian
spade-footed toads (genus *Pelobates*,
subfamily Pelobatinae). It represents, however,
a separate subfamily: the Eopelobatinae. Fossil
remains of the common ancestors of both of
these subfamilies have still not been found. The
divergence of both lines of evolution should
already have taken place in the Cretaceous.
The Messel *Eopelobates* is the oldest
European example of this subfamily; the oldest
finds of the genus *Pelobates*, however, come
from the Belgian Upper Eocene.

At first glance, the design of the toad appears
to be equally suitable for swimming in water or
for hopping on land. Frogs and toads have,
however, been successful in adapting to the
most diverse habitats, from ponds to puddles in
the leaves of epiphytic bromelians high in
jungle trees. Thus, in the daytime, or when in
flight from predators, the above-mentioned
spade-footed toads bury themselves in the
ground, abdomen first. This behaviour is made
easier for them by corresponding changes in
the foot skeleton. The knowledge of similar
adaptations in the skeleton of extant frogs and
toads permits one to make inferences about the
life-style of the Messel anurans.

Thus, *Eopelobates* probably lived
predominantly on land and only entered Lake
Messel or the mouths of its tributaries for
spawning. Characteristic of such a terrestrial
life-style are the strong ossification of the skull,
the development of large vertebral processes,
and the structure of the pelvis, as well as the
relative lengths of the skeletal elements of the
hind legs. The lower thigh in *Eopelobates* is
considerably longer than the upper thigh (Fig.
152), quite similar to the indigenous green
frogs (Ranidae). *Eopelobates* should therefore
have moved predominantly by hopping. It
would, however, not have been able to jump as
well as the ranids, because the required

Fig. 159: *Lutetiobatrachus gracilis*. Only one specimen of this species has been found to date. Head and body length, *c.*5 cm.

changes in the pelvic vertebrae did not take place. It is astonishing that, although *Eopelobates* entered the lake at the time of procreation, as documented by the preservation of spawn (Figs 154 and 155), no tadpoles of this species have as yet been found.

Far more rare at Messel are finds of an entirely aquatic form classified with the family Palaeobatrachidae. This frog family has, from the Jurassic to the Quaternary, been restricted geographically to the Northern Hemisphere. It derives from terrestrial frogs which returned to the aquatic life in the course of the Jurassic. In its skeletal structure the new genus and species,

Messelobatrachus tobieni (Figs 156–158), forms a link between these Jurassic forms and the palaeobatrachids from the Upper Tertiary, although special skeletal adaptations suggest a separate development. These Messel frogs were the first to accomplish an adaptation to the aquatic life which was as perfect as that of the extant African clawed toads.

One of the reasons that *Messelobatrachus* is found so rarely is due to its occurrence being restricted to the oil-shale facies type 3. The latter amounts to only 10 per cent of the oil shale excavated to date. The reasons for this restriction to a facies may lie primarily in the chemical composition of the lake water of that time which might not have been suitable for *Messelobatrachus* during the remaining sedimentations, in particular those of facies type 1. In contrast to *Eopelobates*, for this species no specimens with spawn exist, but two tadpoles have been preserved. One of these animals died at a very early stage of the development from tadpole to young frog. Although some skeletal elements of the skull and the vertebral column were already ossified, the legs were only just being formed (Fig. 156). In the other animal the development towards the young frog was almost complete and only the tail had not yet been reduced.

The habitat of these animals was probably primarily at the lake borders where they lay in wait for prey hidden under floating plants, or at the bottom of the lake. The considerable elongation of the middle-hand and middle-foot bone (Figs 157 and 158) as well as the relative foreshortening of the lower thigh (ratio upper thigh : lower thigh = 1 : 1) are among the most important of the special adaptations to the aquatic life. In a similar fashion to extant aquatic frogs, *Messelobatrachus* also shows evidence of fusion of some individual vertebrae (7th to 9th). This has a very positive effect on the energy transfer during the push of the hind legs when swimming. It prevents a deflection of the vertebral column from the vertical plane during the transfer of the directional impulse.

Lutetiobatrachus gracilis (Fig. 159), a further new genus and species, is as yet represented by only one specimen. It was also recovered from the oil shale of facies 3 by an

excavation team of the Forschungsinstitut Senckenberg. During excavation and preparation the ventrally exposed parts of skull, the pectoral girdle, and the vertebral column were lost, so that a definitive classification with one of the existing anurid families is, regrettably, not possible. In order to give a more precise systematic identification we must await further finds. Structural characteristics of the skull and the vertebral column suggest a terrestrial ecology. Just as in the case of the salamanders, this mode of life may be the main reason for its rarity as a fossil. This form, like *Messelobatrachus*, obviously came to the lake during the sedimentation of facies type 3.

Although at least two species of Messel frogs are represented by many skeletons, our anatomical knowledge of the skeletons of those species lags far behind that of recent species. This is primarily due to the fact that the skeletons in Messel are as a rule preserved as originally articulated. For an exact reconstruction of the individual skeletal elements, however, it would be far preferable to have disarticulated skeletons. In this way the shape of skeletal structures that, in their original anatomical position, are covered up by other elements, could be recorded and described. Both for the phylogeny of the species and for the reconstruction of the individual ontogeny, finds of all stages from the tadpole to the mature animal would be important. The food spectrum of the Messel frogs is also insufficiently known. In contrast to the mammals (cf. Chapter 26 'Fossilized gut contents: analysis and interpretation') only one specimen has been recovered up until now with its gastrointestinal contents preserved. This frog, *Eopelobates wagneri*, found by the Institut Royal des Sciences Naturelles de Belgique in Brussels in 1987, had eaten a small reptile just before its own demise.

8

Freshwater turtles

By their carapace alone turtles are unmistakable. The earliest turtles are known from the Triassic, more than 200 million years ago, and already carried an arched carapace and a more flattened ventral plate, the plastron. The horny scales in their outer skin, which protect them from desiccation, are enlarged and tightly closed up in close proximity to the underlying bony plates which are fused and form a compact and very tough shell.

Three genera of turtles are found regularly at Messel. Owing to their marvellous state of preservation, the mostly complete specimens can be prepared without embedding and observed from all sides.

Freshwater turtles

THOMAS KELLER AND STEPHAN SCHAAL

The slow walk of tortoises on land, with their limbs strongly splayed sideways which permits only a shuffling gait, is still reminiscent of the early reptiles from the Palaeozoic Era. The characteristic bow-like bend of the upper limb and the thigh is related to the limitations on mobility imposed by the carapace. The limbs of tortoises and turtles (collectively, chelonians) are, however, drastically modified and specialized according to their entirely different terrestrial and aquatic habitats. In the terrestrial tortoises, for example, the digits are very much shortened; in freshwater or oceanic turtles they develop, by elongation of the bony elements, into effective paddles. These characteristics have not changed essentially up to the present day and allow us to draw comparisons with extant chelonians. Genera

Fig. 160: Marsh turtle (*Ocadia* sp.), prepared so as to be free on both sides. Its oval carapace has been considerably deformed by the weight of the rock over it. Length of carapace, 29 cm.

Fig. 161: Ventral shell (plastron) of *Ocadia* sp., Fig. 160. The limbs are only slightly specialized for aquatic life. The five toes are free, which means that the animals could also move freely on land.

Fig. 162: *Allaeochelys crassesculptata* was fully adapted to aquatic life by the modification of its extremities into long flattened paddles. The skull strongly resembles that of the New Guinea soft-shelled turtle; it is unclear whether the nose had a proboscis as does the New Guinea turtle. Length of carapace, 20 cm.

known from the fossil site of Messel are *Ocadia*, *Allaeochelys*, and *Trionyx*.

Small hunters in marsh and stream

The marsh turtle *Ocadia* (synonym: *Palaeochelys*) of the family Emydidae has been known since the Eocene. Representatives of this group, once distributed throughout Europe and Africa, presently still live in Southeast Asia.

In *Ocadia*, the bony, weakly curved dorsal shield and the flat ventral shield are connected to one another and immobile. The carapace, oval in contour, is rather thin-shelled and is therefore often considerably deformed in Messel finds (Figs 160 and 161). In juveniles the fusion of the individual bony plates is not yet complete. Interestingly, the bony surface shows a fine granulation. Weak impressions of bands corresponding to the points of attachment of the overlying horny plates, which are not preserved as fossils, are only occasionally observed.

Carapaces with lengths ranging from 7 to 30 cm have been found at Messel. Whereas formerly the resistant carapaces were principally investigated, we have now, through improved methods of preparation, become better acquainted with the skull and the extremities (Fig. 161).

If we wish to draw conclusions concerning the habitat and life-style of a group of fossil animals, we must, in so far as we have no direct indications, rely on evidence from extant relatives of such a group.

The extant *Ocadia*, for example, prefers quiet water (small ponds, canals, or muddy brooks) in flat, open country (Pritchard 1979). This is also true of many other marsh turtles. One can conclude from this that the animals of Messel also probably preferred similar biotopes; they probably spent the larger part of their time in water. Like many marsh turtles today, *Ocadia* was probably a carnivore. However, the choice of food probably changed according to the particular age of the animal just as can occur among extant turtles.

Among the marsh turtles of the genus

Ocadia, Staesche (1928) established two species in material recovered from Messel; this classification was primarily on differences in size but it is not altogether certain that these are really different species. The species established by Staesche were *Ocadia messeliana* and *Ocadia kehreri*.

Paddlers and swimmers

The turtle *Allaeochelys* (subfamily Anosteirinae) belongs to an extinct group of turtles which existed from the Lower to the Upper Tertiary and was distributed throughout North America, Asia, and Europe (Figs 162 and 163). The Papua soft-shelled turtles are their closest extant relatives. In their structure these turtles represent a transition from bog turtles to the true soft-shelled turtles.

In the carapace of these turtles all bony plates are present, but the horny plates were—most probably—already reduced, since in the Messel carapaces impressions of these horny plates cannot be traced with certainty. Undeformed specimens of *Allaeochelys* from Messel indicate that these turtles, like tortoises, had high-arched carapaces.

The plastron is connected to the carapace only by finger-like processes and not closed as in the marsh turtle *Ocadia*. In the vicinity of the extremities it is reduced to a cruciform shield. This reduction serves to improve the mobility of the swim-paddles which are formed by, in each case, two considerably elongated digits. The sculpture of the dorsal carapace is typical for this genus: short, irregularly undulating ridges alternate with corresponding elongated depressions. The top of the skull is similarly sculpted.

The *Allaeochelys* known from Messel attains maximum carapace lengths of 25–30 cm. The food of these animals can be assumed to have consisted of small, rather slow-moving water-dwellers and also plants.

As for *Ocadia*, two species were described for *Allaeochelys* by Harrassowitz (1922*a*,*b*). The classifications are based mainly on differences in sculpture, but it is not at all clear if these represent real differences at the species level. Harrassowitz named the species

Allaeochelys gracilis and *Allaeochelys crassesculptata*.

Snorkellers, rooters, and omnivores

The genus *Trionyx* represents one of the most unmistakable and, with lengths of up to 60 cm, the largest turtle of Messel (Figs 164–166).

This group (family Trionychidae), which once had an almost world-wide distribution but which is today no longer found in Europe, has been known since the Cretaceous.

The soft-shelled turtle *Trionyx* has a reduced bony shell covered by a thick, leathery skin. The fossil *Trionyx* of Messel already shows a reduction of plates along the margin of the flat carapace and the plastron consists of broad plates which are reduced in extent. The

Fig. 163: *Allaeochelys crassesculptata*. Around the fore and hind legs the ventral shell is reduced to a cruciform shield. This allows greater mobility of the paddle-shaped feet. Length of carapace, 21 cm.

103

Fig. 164: Dorsal view of *Trionyx* sp. This long-necked turtle has free claws, albeit reduced in number. Hence its name in Greek, *tri*: three; *onych*: claw. The large openings on the temple behind the eye sockets provide space for the attachment of the powerful musculature of the jaw. Length of carapace, 30 cm.

Fig. 165: *Trionyx* sp. like *Allaeochelys*, was adapted to its environment by the reduction of its carapace and by the flattening and prolongation of the extremities into paddle-shapes. Recent relatives of *Trionyx* are snorkelers and burrowers. Length of carapace, 28 cm.

carapace and plastron were connected exclusively by skin and bands. One characteristic of extant and fossil representatives of *Trionyx* is the net- or honeycomb-like sculpture of the shell surfaces.

Fig. 166: This tiny turtle belongs to the genus *Trionyx*. It is a rare but not a surprising find. It is known from present-day observations that many young turtles fall victim to predators and hostile environmental conditions.

As well as the sculpture of the carapace, its contour also undergoes changes in the course of the individual's development; although this cannot be demonstrated with any certainty from the fossil material. Differences in the sizes of male and female, which certainly occur in many reptiles, are difficult to substantiate even for extant chelonians. Such difficulties in the interpretation of the fossil skeletons are one more reminder of how very little we know of the appearance of such a prehistoric turtle even if we, as in the case of Messel, find one with an optimal state of fossil preservation.

In 1900 the researcher Reinach named the species *Trionyx messelianus*, and later workers then differentiated between 'variations', i.e. subspecies. The deviations in skeletal structure that gave rise to these divisions cannot today be evaluated as subspecific differences without further evidence. We must here await a re-examination of the extensive material.

The habits and habitat of the Messel *Trionyx* scarcely hold any riddles for us. Most of today's species leave the water only to lay eggs. It is not certain that the fossil species shared this strict preference for the water, but the occurrence of the fossil *Trionyx* is generally taken as a rather clear indication for the existence of constantly flowing water or lakes. The animals are able to protect themselves in dry periods by burying themselves in the mud of water bodies. Of great use here is the long neck which they extend to breathe. Favourite habitats of the extant *Trionyx* are tropical lakes, ponds, and rivers (Pritchard 1979), where it feeds, in part, on fish.

It is not easy to answer the questions as to whether turtles (and, if so, which turtles) lived in Lake Messel or only visited it sporadically. We will observe among the semi-aquatic crocodiles that the death assemblage (thanatocoenosis) gives the false impression of a wealth of species in a limited area, which, in reality, did not exist at such density. *Ocadia*, *Trionyx*, and *Allaeochelys* are inhabitants of running water as well as of the quieter parts of bodies of water. From all of our experiences to date, we must assume that in Messel the transport of dead animals from distant biotopes occurred. From one Upper Tertiary site there is documentation (Schleich 1981) that the crocodile, *Diplocynodon*, and *Ocadia* must have occasionally shared the same biotope (and for *Diplocynodon* we can safely assume that the Eocene Lake Messel formed part of its habitat).

From the needs of today's chelonians we conclude, in general, that closely related fossil representatives shared similar environmental preferences. This is true also for the temperature requirements of chelonians which are, like other reptiles, cold-blooded and cannot sustain an even body temperature. This leads to a restricted distribution for these warmth-loving creatures. According to Brattstrom (1965) we can, as far as Messel is concerned, assume very warm temperatures of 25–30°C for certain periods (reproduction periods).

Among the chelonian fauna of Messel aquatic forms dominate. The marsh turtles as well as the soft-shelled turtles and the terrestrial tortoises (Anosteirinae and Testudinidae) first appear in the European Lower Eocene. *Trionyx* forms are still older and appear in Europe in the Palaeocene.

Zoogeographical relationships are difficult to prove for the turtles of Messel. Distinct faunistic connections should have existed with North America, because we know a turtle fauna from the Eocene of Ellesmere Island in the Arctic which contains numerous 'counterparts' to the European or, respectively, North American Eocene fauna. Even the climatically-demanding giant turtles, which are, surprisingly, still missing in the Messel fossil collection, are represented on Ellesmere Island and an expansion of the ancestors of the marsh- and soft-shelled turtles (*Trionyx*) of Messel into new territory via this Arctic bridge in the early Eocene is still conceivable.

9

Crocodiles: large ancient reptiles

The crocodile order has existed for over 200 million years and is, today, still represented by eight genera. The crocodiles probably derive from bipedal ancestors in the Triassic and have during their further evolution undergone almost no rapid changes in body structure—in distinct contrast to their relatives the dinosaurs. Only a few genera have conquered the seas; the majority have preferred aquatic biotopes on land. Descendants of these old evolutionary lines, once inhabitants of lakes, rivers, swamps, or dry land, are among the typical finds from the Messel oil shales.

Crocodiles: large ancient reptiles

THOMAS KELLER AND STEPHAN SCHAAL

The crocodiles are represented in the Messel oil shales by the genera *Diplocynodon*, *Asiatoschus*, *Pristichampsus*, *Bergisuchus*, *Allognathosuchus*, and *Baryphracta*. Due to its well-preserved skeletons, which are the most frequently found skeletons at Messel, the crocodile genus *Diplocynodon* will be discussed first.

Two species of *Diplocynodon* occur at Messel—*Diplocynodon darwini* and *Diplocynodon ebertsi*. Berg, in 1966, already had fragments of over 30 skulls at his disposal and, since then, the store of finds has increased considerably. It has been possible to recover exquisite complete skeletons of a maximum length of 150 cm: smaller skeletons make up the bulk of the material. The smallest ones have a length of 14 cm and belonged to very young animals.

Diplocynodon is a cayman-like crocodile which is classified with the alligators (subfamily Alligatorinae; family Crocodylidae). Some of the following descriptions have been taken from the work of Berg (1966). The skull of *Diplocynodon* (Fig. 174, middle) shows similarities with the extant genus *Caiman*. The skull as a whole is not broad, the point of the relatively short snout is rounded, and upon it is set the single nostril. The roof of the snout has a scarred to net-like sculpturing. The number of teeth, the arrangement and sequence of the, in part, enlarged teeth, which can exhibit gaps in their placement on the rim of the jaw, as well as the intermeshing of the teeth from the upper and lower jaws, the so-called 'bite', are typical crocodile characters. In *Diplocynodon* the bite is alligator-like because the teeth of the lower

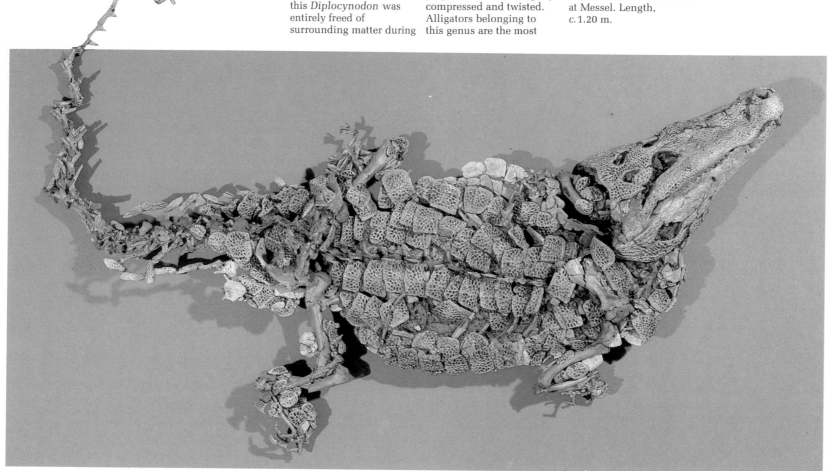

Fig. 167: The skeleton of this *Diplocynodon* was entirely freed of surrounding matter during preparation, and is slightly compressed and twisted. Alligators belonging to this genus are the most common crocodiles found at Messel. Length, *c*.1.20 m.

jaw bite inwards with respect to those of the upper jaw. In the upper and lower jaws two closely juxtaposed, considerably developed, elongated pseudo-canines are conspicuous.

The bony (osteodermal) armour of *Diplocynodon* covers neck, back, belly, and part of the tail (Figs 167–169). The armour plates, which were arranged in rows, were connected so as to be mobile with respect to one another. In some regions of the body, for example, the belly, the plates overlap in a longitudinal direction and abut horizontally. The individual plates show the pitted sculpture which is typical of crocodiles. On the backs of the animals some of the rows of plates are weakly keeled (carinate). These central keels become more prominent on the rows lining the perimeter of the back and form noticeable pointed combs on the tail. Parts of the

extremities and the extremity of the tail are not covered by bony plates.

The two species of the genus can, according to the results of Berg's investigations, be safely distinguished 'for the present only by a characteristic of the teeth'. In *Diplocynodon ebertsi* the crowns of the teeth have ridges of tooth enamel which are missing in *D. darwini*.

Diplocynodon can be regarded as a fish- and flesh-eater which also (especially as a young animal), on occasion, caught different prey, for example, insects. One *Diplocynodon* found at

Fig. 168: Juvenile of the genus *Diplocynodon*. Length, *c.*41 cm. In contrast to *Baryphracta*, only the part of the tail nearest the pelvis is armoured.

Messel exhibited in its body cavity scales from a gar (*Atractosteus*) which it had eaten. The body cavities of almost all adult animals of *Diplocynodon* contain a number of small pebbles (Fig. 169). These were apparently swallowed as diving ballast in order to decrease buoyancy in the water.

Asiatosuchus germanicus was a large and fearsome crocodile which belonged to the true crocodiles (family Crocodylidae: subfamily Crocodylinae). In large individuals from Messel the skull can attain 70 cm in length. The total length of the specimen in Figs 170–172 is almost 4 m. The point of the snout is hemispherically rounded to the front and here too the typical crocodilian constriction between upper-jaw and intermaxillary bone is striking. The robust lower jaw and the bones of the skull display a large-meshed sculpture. The crowns of the pointed conical teeth are equipped with cutting edges. Further back in the rows of teeth, they become blunter (Fig. 172) as, incidentally, for all Messel crocodiles. The bite of *Asiatosuchus* is reminiscent of *Diplocynodon*. The nape of the neck, the back, and the ridge of the tail are covered by osteoderms which consist partly of large,

square, roughly sculpted, and keeled plates and partly of very much smaller cylindrical bones.

The forelegs are short and quite dainty, in contrast to the long, strong hind legs. When swimming, the latter served mainly for steering, while the forelegs were held comparatively motionless at the flanks. Like today's Nile crocodiles, *Asiatosuchus* had a very strong tail which served as a propelling organ. Evidence for this, in addition to the strong tail vertebrae, are the enormous horny scales which have been well preserved in an almost 3 m long specimen (Fig. 173). The speed of propulsion must mainly be attributed to the powerful tail: an advantage which has contributed to the continuing success of these animals during their long history.

Almost all living crocodiles change their diet according to the different stages in their life cycle. This can also be assumed for *Asiatosuchus*; accordingly, its food probably consisted mainly of fish and reptiles or, on the other hand, birds and mammals.

Allognathosuchus haupti is characterized by the particularly striking specialization of its teeth (Fig. 174 below; Fig. 175). This genus, which belongs to the subfamily Alligatorinae (family Crocodylidae), is relatively small; the animals at Messel rarely grew longer than 1 m In the back portion of the upper and lower jaws there are teeth with flat, orbicular, rounded crowns. Towards the front there are one or two teeth which are transitional in shape between the front and back teeth, and among the front

Fig. 169: The belly and throat armour of this skeleton of *Diplocynodon* are preserved undisturbed and complete. Below the two slightly separated longitudinal rows of plates in the middle of the body appear some stones from the stomach which have been swallowed as ballast, to reduce buoyancy when diving. Length, *c.* 1.20 m.

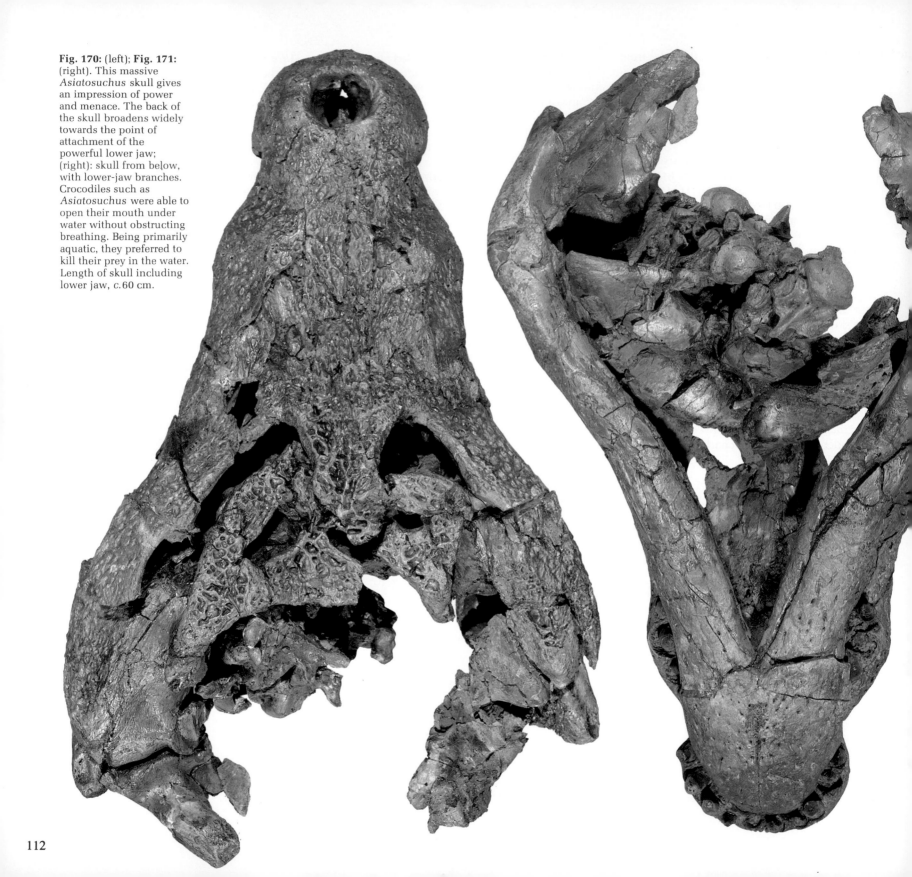

Fig. 170: (left); **Fig. 171:** (right). This massive *Asiatosuchus* skull gives an impression of power and menace. The back of the skull broadens widely towards the point of attachment of the powerful lower jaw; (right): skull from below, with lower-jaw branches. Crocodiles such as *Asiatosuchus* were able to open their mouth under water without obstructing breathing. Being primarily aquatic, they preferred to kill their prey in the water. Length of skull including lower jaw, *c.*60 cm.

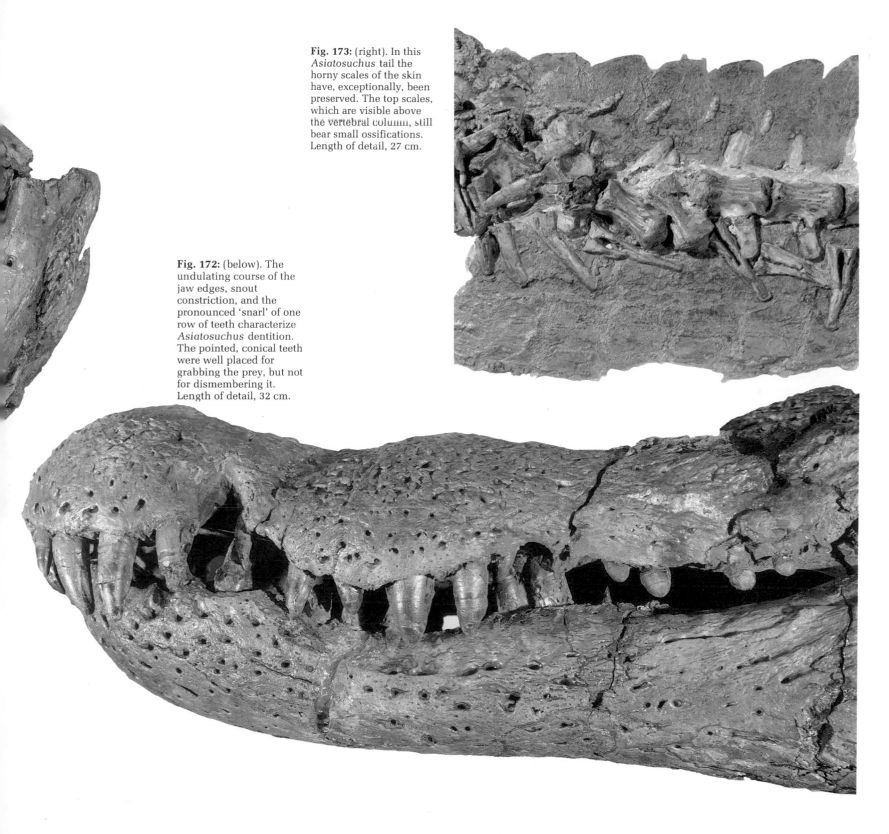

Fig. 173: (right). In this *Asiatosuchus* tail the horny scales of the skin have, exceptionally, been preserved. The top scales, which are visible above the vertebral column, still bear small ossifications. Length of detail, 27 cm.

Fig. 172: (below). The undulating course of the jaw edges, snout constriction, and the pronounced 'snarl' of one row of teeth characterize *Asiatosuchus* dentition. The pointed, conical teeth were well placed for grabbing the prey, but not for dismembering it. Length of detail, 32 cm.

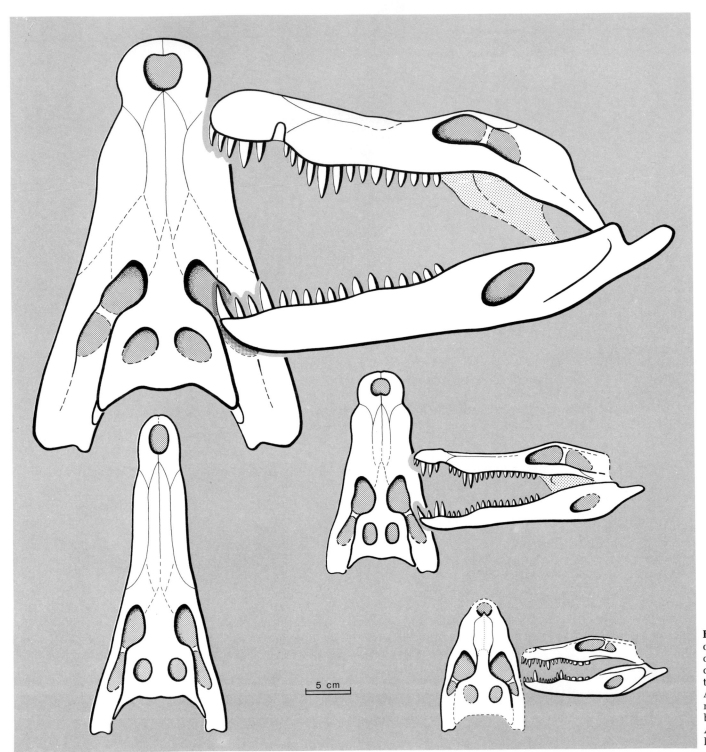

Fig. 174: The Messel crocodile genera are distinguished by distinct differences in the shape of the snout and skull size. Above: *Asiatosuchus*; middle: *Diplocynodon*; below right: *Allognathosuchus*; below left: *Pristichampsus*.

5 cm

114

teeth the conical type dominate. This specialized dentition constitutes an apparatus for catching and breaking up the prey at one and the same time (Berg 1966). However, such teeth must be interpreted as not so much a one-sided adaptation for a particular 'diet' but rather as an adaptation enabling an enrichment of the diet with hard-shelled prey, for example, molluscs. The osteodermal armour is very similar to that of *Diplocynodon*. Here, too, the tail, armoured almost to the tip, on its top side bears two pointed ridges of keel-shaped scales. Just as in *Diplocynodon*, the extremities are almost free of bony plates.

Baryphracta deponiae (Fig. 176) has only recently been described as a new form by Frey *et al.* (1987). The new genus was concealed for a long time among skeletons which were formerly attributed to juveniles of *Diplocynodon*. This is because the finds at Messel attained a length of only *c.* 70 cm. For

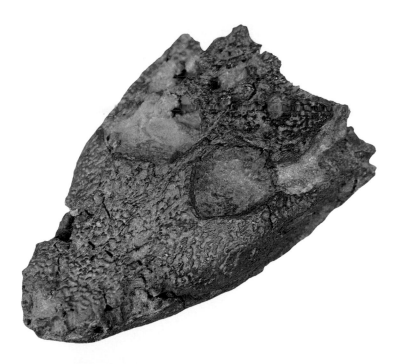

Fig. 175: The small *Allognathosuchus* crocodiles are distinguished by extremely broad snouts. The skull in the illustration is only *c.* 12 cm long.

Fig. 176: This small crocodile is 50 cm long and belongs to the genus *Baryphracta*. The generic name ('heavy armour') refers to the way that the armour extends to the very tip of the tail.

the descriptive diagnosis of genus and species the pattern of skin scales was also taken into consideration, as well as the skull characteristics. The skull of *Baryphracta deponiae* is relatively high with a short snout; its bones are sculpted. As regards the structure of the teeth, it is noticeable that the crowns of the three hindmost teeth in the upper and lower jaws are orbicular and therefore resemble those of *Allognathosuchus*. It is not yet possible to make a reliable statement about the choice of diet of this crocodile because no stomach contents are known. The specializations of the teeth, similar to those of *Allognathosuchus*, cannot be interpreted as a preference for one particular food. In marked contrast to *Diplocynodon*, the tail is also fully armoured and the extremities are more closely covered with bony plates (osteoderms).

Among the rarest finds is *Pristichampsus rollinati*, a species which up to now has been found only at Mèssel (Fig. 177). The meagre fragments of a skull permit the reconstruction of a narrow, if relatively low snout with steep sides; the sides of the upper jaw are fitted over the narrow lower jaw 'so-to-speak, closed by a lid' (Berg 1966). The peg- to blade-shaped, flat and slightly curved crowns of the teeth are characteristic. They are serrated at the edges (Fig. 178). Such teeth are described as dinosaur-like because they resemble the teeth of large, meat-eating dinosaurs. The serrated teeth of the crocodiles evolved then by convergent evolution (Buffetaut 1979). The bite of *Pristichampsus* is alligator-like.

Pristichampsus is classified within the subfamily Pristichampsinae of the family Crocodylidae. We are familiar with the back- and belly-armour of these animals, which attained a length of approximately 1.5 m, from skeletons from the Eocene, in the Geiseltal. The extremities are armoured, down to the middle part of the foot. The tail is enveloped by rings of armour. A peculiarity of this genus is the hoof-like shape of the ends of the toes. But these adaptations should not, without further consideration, be interpreted as evidence for a more terrestrial life-style. The apparatus of *Pristichampsus* for catching prey, with the distinctly elongated snout and blade-shaped

Fig. 177: This fragment of a lower jaw has been attributed to a fish-eater with a narrow snout. The fourth tooth from the front on each side is in the form of a particularly strong canine. Length, *c.* 20 cm.

Fig. 178: This large, 5 cm-long tooth belongs to a fully-grown *Pristichampsus* specimen. It is a replacement tooth; that is, a tooth not yet used, in a row of teeth which will come into use when the current ones fall out.

teeth requiring a scissors-like biting action, indicates that the food consisted predominantly of fish.

Bergisuchus dietrichbergi is founded on fragments of a snout together with fragments of

a lower jaw which belongs to it (Figs 179 and 180). This strange snout fragment from Messel is high and narrow with steep sides. Embedded in the lower jaw is an enormously enlarged fourth tooth. The teeth are approximately elliptical in transverse section; on some, cutting, serrated edges can be recognized, similar to those already noted on the teeth of *Pristichampsus*. The bite is apparently alligator-like and the very much enlarged fang probably fitted into a snout constriction of the upper jaw. *Bergisuchus* has the teeth of a 'tearing' carnivore, that is, of a true meat-eater. As regards the other characteristics of this animal, for example, its armour, we have no information as yet. Based on the unique and incomplete find we may assume that this animal avoided the waters of Lake Messel. This is, without question, the representative of a very ancient group of crocodiles whose roots extend back to the latest Mesozoic of South America or, respectively, of Africa (Buffetaut 1988).

The crocodile fauna of Messel: an overview

Two genera, *Diplocynodon* and *Baryphracta*, predominate among the material excavated at Messel. *Allognathosuchus* and *Asiatosuchus* are rare among the finds and the skeletal remains of the other genera are unique. In the first place, even given the abundance of the Eocene fauna, we should not expect six different genera of crocodiles in such a restricted space. According to Tchernov (1986) different species of crocodiles can coexist, if at all, only in a very extended area and even then only if, by following different adaptation strategies, they can still further distance themselves from one another. For an explanation of this concentration of finds we must assume that the carcasses were transported from different biotopes at varying distances from Messel only to be united in the thanatocoenosis of the lake. A relatively unbroken series of juvenile stages are found at Messel only for a single crocodile genus: *Diplocynodon*, which is also the most frequently found fossil crocodile at Messel. Perhaps the young crocodiles of this genus lived more on land; however, we can assume that in certain age ranges they were clearly inhabitants of the lake. Apart from this the existing evidence of local crocodile presence in or at Lake Messel is very sparse. Typical food remains of potential prey animals have not been found (Koenigswald and Wuttke 1987). Fossil excrement (coprolites) which is attributed to crocodiles are not uncommon at Messel (Fig. 383). According to the communication of Schmitz-Münker, they can contain remains of *Cyclurus* and *Atractosteus*. Food remains vomited up by the marauders, as well as crocodile teeth, which have fallen out or been broken off, are, however, missing. The sunning, nesting, and feeding places of the Messel crocodiles were obviously situated in a more distant area of the ancient lake environment.

Crocodiles like *Diplocynodon* or *Asiatosuchus* were apparently easily able to cross land and water, which accords with the abilities of their extant relatives. The distribution of *Pristichampsus*, which preferred a fish diet, may, by contrast, have tended more towards water habitats and, therefore, have been more restricted.

Diplocynodon is well documented in the Upper Palaeocene and the Eocene of Europe, but most of all in the North American Eocene. It became extinct in the Miocene. *Asiatosuchus* is known from the Palaeocene of Europe and Asia and from the entire Eocene. Crocodiles close to this genus are also known from the Lower Tertiary of North America (Berg 1966). The genus was described first from the Upper Eocene of Mongolia. It is further known (as are *Diplocynodon* and other Messel crocodiles), from the Eocene Geiseltal. Because, according to the information of Berg, *Asiatosuchus* was also able to negotiate ocean straits and was therefore not restricted to land bridges in extending its distribution, an emigration from Asia is conceivable. Even *Allognathosuchus* could, apparently, occasionally invade the marine province. This crocodile is well documented in the European and North American Palaeocene. The genus was, furthermore, also found in the Lower Eocene of Ellesmere Island in the far north which we are assuming was part of an ancient land bridge (see Chapter 27, 'The Messel fauna and flora: a biogeographical puzzle'). This is by no means obvious!

Crocodiles are climatically demanding and, being cold-blooded, prefer warm climates. Probably everyone has seen pictures of crocodiles basking on land, jaws agape, being warmed by the sun. During the day, large reptiles keep an approximately even body temperature by protecting themselves against being overheated by occasional cooling in the water or the shade. During the night, the water offers protection from excessive cooling. Berg (1964) stated minimum temperatures for the biotopes of fossil crocodiles of 10–15°C, this being the average temperature of the coldest month. Palaeobotanical data tell us that during the early Eocene the temperature of Ellesmere Island, which then lay at a latitude of 78° north, was equitable and warm and that freezing temperatures did not occur in winter. There were forests north of the polar circle; the North Pole must have been free of ice. Interestingly, two far more terrestrial reptiles than *Allognathosuchus* have been found on Ellesmere Island, the tortoise *Geochelone* and a monitor-like lizard. These species would not have been able to escape into the water during deviations in temperature.

The additional question arises as to how some of the cold-blooded vertebrates could have coped with the long polar night which must be assumed to have occurred, based on the former geographical latitude calculated for the island. One solution could be the hypothesis that these animals had a longer period of hibernation, which must not, however, be assumed for all the reptiles referred to (Estes and Hutchison 1980).

Pristichampsus appears in the European Palaeocene and in North American, European, and Asiatic Eocene deposits. This crocodile was clearly restricted to fresh water.

Bergisuchus is rather isolated in the European Eocene. According to Buffetaut (1987), there are no closer relationships to the few other European representatives of this ancient group. But the relationships with the South American forms are also minimal. Buffetaut judges *Bergisuchus* to belong to an old 'Gondwana' group. (Gondwana was the huge southern continent which split during the late Mesozoic Era into individual continents— South America, Africa, India, Australia, and Antarctica.)

We find in the Messel crocodiles faunistic relationships with several continents. The climates of the Tertiary which were equable across wide zones favoured a wider distribution of crocodiles than is found today. Only with the constantly worsening climatic conditions of the Upper Tertiary was there a decline in the abundance and dominance of these animals.

10

Lizards: reptiles *en route* to success

The scaly lizards (order Squamata) derive from 250-million-year-old reptiles from the Permian. Among the extant reptiles they form, with 3000 species of lizards (Lacertilia) and more than 2500 species of snakes, the most advanced and numerous group. In the course of their evolution they have pioneered a multitude of aquatic and terrestrial biotopes.

Many fossil finds of rare or hitherto unknown Lacertilia and snakes, often together with remnants of scales or other soft parts, are preserved at Messel in the complete condition for which the site is famous. Without a doubt Messel is one of the most important, if not the most significant, fossil site for Tertiary Squamata.

Lizards: reptiles *en route* to success

THOMAS KELLER AND STEPHAN SCHAAL

In contrast to the mainly aquatic reptiles: the turtles and crocodiles, which are well represented among the excavation finds of Messel, the generally much smaller, predominantly terrestrial groups of lizards (lacertilians) and snakes are far rarer.

Xestops: lizards in armour

A number of good finds are available of *Xestops* (subfamily Glyptosaurinae; family Anguidae (anguids)). A reliable characteristic of these bizarre animals is their dense, nearly unbroken, armour formed from rows of more or less rectangular bony scales. (The bones from which the scales are formed are dermal in origin; as in the corresponding structures of turtles and crocodiles, they are always covered by a horny layer.) Large parts of the skull are also covered by bony scales; sometimes the connection of the actual bones of the skull with the bony scales is quite close. Only a few parts of the body, the feet, for example, are free from this dense protective armour (Fig. 181).

Let us look more closely at these animals which with their resemblance to fully-clad knights are immediately conspicuous among the daintier and slimmer Messel lizards. The body is encased by bony rings of armour in which the individual plates had a certain mobility due to well-developed gliding surfaces. Many of the relatively thick-walled, square to oblong bony scales (osteoderms) bear reinforcing keel-like ridges. The surface of all these bony scales shows a peculiar sculpturing in the form of small, densely juxtaposed knots ('tubercles'), a characteristic only found in members of this subfamily.

The structure of this 'external armour' is

Fig. 181: The lizard *Xestops abderhaldeni* resembles a small crocodile because of its armour plating. The tail of the animal was probably shed during its lifetime. Length, 17 cm.

simple as regards the geometry of its composite elements. It is noteworthy that the dorsal and ventral plates differ slightly but significantly from one another as regards plate type. Where the rows of dorsal and ventral bony plates met on the flanks, an elastic 'seam' may have existed. We recognize such a suture or lateral fold in all four-footed or legless lizards which bear a tightly closed 'armour' of thick-walled bony plates but which need to expand the body for the purposes of breathing or egg development. The armour of *Xestops* enclosed the tail completely. The form and size of the bony osteoderms change according to the varying biomechanical requirements of the different parts of the body.

The vertebral column, pectoral girdle, and limb bones of the armoured lizards like *Xestops* are less well known. This is not surprising when it is considered that these parts are generally covered and are scattered inside the bony armour after death. But one can recognize that the pectoral girdle and pelvis were well developed with strong broad areas of attachment for the musculature. The extremities, in comparison with their proportions in other lizards, are not at all reduced. The hind-limbs are a little stronger and longer than the fore-limbs.

Where and how did these remarkable animals live? Hints about their life-style can be gained mainly from the skull. This is short and broad. Relatively large horn shields on the bony surface, which are not preserved as fossils but whose impressions have survived, made it even more rugged (Fig. 188). The upper jaw is highly triangular, curved, and massive; the lower jaw is similarly robust. The front teeth have pointed crowns and become blunter towards the back, finally becoming almost

resemble the armoured lizards of a North American species, *Xestops vagans*. Deviations in the osteology of the skull in the Messel animals indicate, however, that the geographical separation of the once unified group in the (?) early Eocene was succeeded by a separate European development.

Limbless lizards: snake-like hunters in the undergrowth

A group of extant and also fossil lizards of the family Anguidae have considerably reduced limbs or none at all. The slow-worm, *Anguis*. or the large, Mediterranean glass lizard *Ophisaurus* move over the ground limbless or, respectively, with functionally unimportant appendages formed by the tiny extremities. It is noticeable that some of these limbless forms bear bony scales which are very similar to those of the Middle Eocene *Xestops*. But the scaly armour of these animals is pressed entirely into the service of effective locomotion on the ground. Limbless lizards with this type of bony scale had already appeared in the early Eocene, although the only evidence we have of them from that period is a few of their scales. From the middle Eocene, however, the Geiseltal and Messel sites provide complete skeletons. Let us examine these Messel finds.

Stretched out or bent in a curve, the bodies lie in the rock, shrouded by a dense covering of bony scales. In contrast to the snakes, one can see only very little of the inner skeleton if the resistant, scaly tube-like skin is not split as in the specimen shown in Fig. 182. The animals found at Messel are approximately 20 to 35 cm long; almost half a dozen finds are known. No extremities have yet been found on the skeletons, not even regressed limbs considerably reduced in size such as those known from a contemporary find in the Geiseltal. The thick-walled bony scales are mostly rectangular, arranged in rings, and have a keel-like ridge and a gliding surface. On the underside of the body, and particularly in the tail area, the contour of the osteoderms changes into trapezoidal or rounded polygonal shapes which sometimes have no keel. The surface of the bony scales is lightly sculpted but

peg-like chisels whose striated crowns contrast somewhat with the thicker shaft and which have an indistinct cutting edge that runs from front to back (see Fig. 188). The American palaeontologist Sullivan (1979) is of the opinion that the increased ruggedness of the skull, in conjunction with the blunt-crowned type of the teeth, points to a food specialization. With a less cutting, mainly crushing, tooth function, *Xestops* could indeed have been specialized to take hard-shelled molluscs or, respectively, insects with thick outer skeletons. In finding and recognizing the food the tongue may, perhaps, have played a major role, just as in extant anguids. To date, no stomach contents from the armoured lizards have been found. Also unknown is the more restricted habitat (biotope) in which the armoured lizards preferred to live.

The genus *Xestops* was first found in North America. Here similarly strongly armoured lizards of the same subfamily and of a remarkable number of genera are frequent and widely distributed in Upper Cretaceous and Lower Tertiary deposits. They are, however, much less completely preserved than those

from Messel. In European Lower Tertiary deposits, the variety and state of preservation of this group of lizards are, in comparison, reversed in importance. In Europe we know only two 'safe' genera—*Placosaurus*, a form that is missing at Messel but was phenotypically not unlike the armoured *Xestops*-like lizards of Messel, and the genus *Xestops* itself, which is represented by a surprisingly high number of well-preserved and complete skeletons not only from Messel but also from the Middle-German Geiseltal brown coal. From later deposits of the Swiss Upper Eocene we know a single ancient, unfortunately very fragmentary fossil of an armoured *Xestops*-like lizard. Is it possible to explain such an uneven distribution?

The most ancient anguid lizards had already evolved on the North American continent in the Upper Cretaceous (Estes 1982). They were relatively small animals with a strong armour of osteoderms. Finds from Eocene and, still older, Palaeocene deposits are evidence that descendants of these pioneers migrated to Europe in several waves of expansion. Thus, the armoured lizards from Messel still the tubercles of the armoured lizards (*Xestops*) are not encountered.

The skull of the Messel limbless lizards has not been investigated closely. There are indications that it is relatively long, but this is not clear from the skeletons because the head merges, without a pronounced narrowing of the neck section, into the scaly tube-like body. The animals had slender but strong and pointed teeth.

There is no doubt that these were ground-dwelling animals which were largely incapable of climbing. In the extant glass lizard (*Ophisaurus apodus*), a tendency to dig is noticeable: the outer and inner ear have regressed but separate and freely mobile eyelids are certainly developed, which are common to almost all lizards, but not to snakes. A zoologist from Tübingen, E. Frey (1982), has shown that the extant *Ophisaurus* can move, thrash, and push forward, independently of one another, the individual sections of its dorsal and ventral armour, which are separated from each other by a lateral fold. Only this enables an effective forward locomotion on or through the ground. As an aid to such activity the conical head can, occasionally, serve as a 'drill'. Based on the similarity of their scales to those of the extant *Ophisaurus*, we must therefore assume, for these Messel animals a ground-dwelling, in part burrowing, life-style. According to their dentition and by analogy with their extant relatives, the Messel animals probably ate insects and snails. For the reconstruction of their habitat, we must also look at the extant *Ophisaurus*, even if this implies a certain risk of being wrong. The lizards could have lived in areas densely covered with medium-high vegetation or perhaps even in more open habitats.

Statements as to the closer relationships of the limbless lizards of Messel can only be made when the skull has been investigated. The already very highly evolved adaptations of Messel limbless lizards (which represent a largely 'finished' construction no longer undergoing fundamental development) make the appearance and the evolutionary origin of the ancestors appear all the more mysterious.

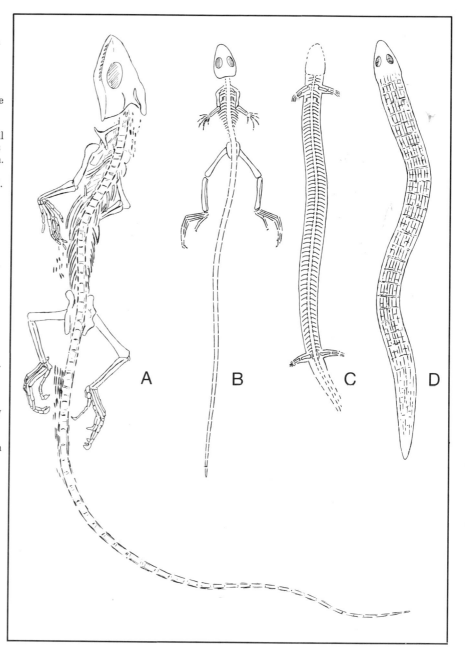

Fig. 183: Structure of Messel lacertilians. (A) (A monitor-like necrossaurian) illustrates the form of one of the larger, terrestrial, predatory lizards. It can be assumed that this animal climbed just as well as it ran. The powerful long tail could probably be used as a weapon. Length *c.* 52 cm. (B) (Iguana), a form with short rump and a long tail. These body proportions identify these reptiles as primarily climbing, arboreal animals. Length, *c.* 30 cm. (C) (Skink-like form). This Messel lizard illustrates an evolutionary trend towards a snake form, widespread among the scaly lizards. The animal moved predominantly by wave-like bends of the body (winding) rather than crawling. Length, *c.* 25 cm. (D) (Snake-like (anguid) limbless lizard, ?*Ophisauriscus*). The limbs of this type probably have completely degenerated; the armour of osteoderms promotes an entirely ground-based, winding locomotion. Length, *c.* 28 cm.

The rich fossil documentation only begins in the Middle Eocene. It follows from matching limb fragments, which turn up occasionally in material from the Geiseltal, that the ancestors were four-footed. Armoured, four-footed anguids such as *Gerrhonotus* (the extant crocodile-lizard), in which a cylindrical rump assists the work of the extremities by lateral undulations, give an impression of what the ancestors of the limbless lizards may have looked like.

Sparse remains from the Upper Cretaceous of North America point to the place of origin of the group. It is also probable that the ancestral forms of the Messel limbless lizards emigrated from North America. Surprisingly, the evolution of the early limbless lizards took place, thereafter, exclusively in the Old World, and these snake-like animals migrated back from Europe to their place of origin, North America, only in the Miocene. However, in the opinion of Gauthier (1982) the relationship of the ophisaurians of both continents is unknown and possibly they may be derived from parallel but unrelated lines of evolution.

The necrosaurians: robbers of the jungle

The appearance in Messel of a group of lizards whose extant representatives belong with the, in some opinions, most modern and advanced lacertilians is remarkable; the group in question is the monitor-like lizards (superfamily Varanoidea). Of this large group, which formerly had many forms, only one family has as yet been found at Messel: these are the Necrosauridae (necrosaurians) which are represented at Messel by three genera and species. Certainly the structure of the skeleton resembles that of the monitors in that it possesses a narrow, relatively high skull, relatively slim body, powerful extremities, and a long tail that is not shed. The animals of Messel, however, bear vestiges of a comparatively ancient osteodermal armour. However, in two of the Messel genera these osteodermal scales are already considerably reduced.

The necrosaurians can, without difficulty, be recognized as carnivores. Their jaws are powerful and long, and the teeth blade-shaped and laterally compressed with cutting edges and thickened bases (Fig. 186). A corresponding thickening of the base of the teeth by folding of the dentine is also found in the true monitors, whose teeth are put through great functional strain (extant monitors shake

and beat the prey in their jaws until it is dead). It has not yet been established whether the lower jaw of the necrosaurians had the greater mobility which characterizes that of true monitors. The carnivore dentition was suitable for overpowering a multitude of small prey animals, including vertebrates. The tongues of extant monitors are well suited to chemoreception which, for example, permits the finding of concealed prey (Pregill *et al.* 1986). Perhaps the necrosaurians, like many extant monitors, were also egg-robbers.

The majority of the Messel necrosaurian forms are not yet known from other localities. Research on and comparison of the finds is not yet concluded and they are therefore given a preliminary designation as 'genera'.

Distinguishing characters of the most primitive genus of necrosaurian (genus 'A') include the osteoderms which are distributed evenly, but with interstitial gaps, over the back, belly, flanks, extremities, and tail. Their basic shape is oval with a high, longitudinal middle keel, but the shape alters, according to the body region, in shape and size. These osteoderms also encroach upon the region of the skull (Fig. 186) but do not fuse with the skull bones. The relatively loose armour still left enough room for the expansion of the hindquarters. These are comparatively squat, with strong limbs, and a rather short tail. The

entire length of the animal was approximately 20 cm. The bones of the narrow roof of the skull bear a sculpture dominated by imprints of the blood vessels and by narrow ridges (Fig. 186). This is reminiscent of primitive stages of evolution of the osteodermal skull covering that has been described for Cretaceous or, respectively, Jurassic forms (Borsuk-Bialynicka 1983, 1984). Similar osteoderms of an obviously closely related necrosaurian are known from the Geiseltal Eocene.

In two further genera of the monitor-like animals represented at Messel, the regression of the osteodermal scales has gone much further. *'Saniwa' feisti*, which was described by Stritzke (1983) as a true monitor (Figs 183A and 184), must be considered first. The Messel finds belong to a distinct genus which is here provisionally called genus 'B'. The teeth correspond with the description given for the family. The narrow roof of the skull is covered by clearly delineated small bony protrusions, but free osteoderms (Fig. 184), enclosed in the skin around the skull, exist as well. The bony protrusions fuse into a massive dorsal armour which extends from the posterior end of the skull to the pectoral girdle. Towards the flanks

Fig. 184: The bone structure of the skull of one of the largest Messel predatory lizards (necrosaurians) is strong and elegant. The skull is covered by numerous small ossifications which were originally embedded in the skin. Behind the head they enlarge to become armour plates, of which only two longitudinal rows are illustrated. Length of skull, 5 cm.

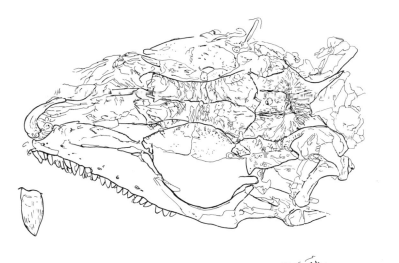

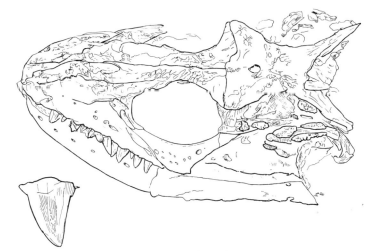

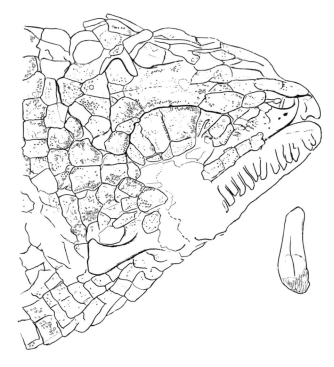

Figs 185–188: The lizards are a group of animals that appear homogeneous but prove on closer inspection to be a collection of families which have similar basic adaptations, but which have been profoundly independent and divergent in their evolution from the Jurassic and Cretaceous onwards. Ancestors of present-day lacertids (Fig. 185, above left) as well as an ancient group of monitor-like lizards (Fig. 186, above right) which are ancestors of present-day iguanas (Fig. 187, centre), and ancestors of extant snake-like lizards (Fig. 188, below) occur at Messel. The individual teeth shown enlarged below their corresponding skulls illustrate, by their specialization into single or multicuspidate forms of different lengths with tearing, squashing, or cutting functions, the different food specializations of these groups.

and the tail the osteodermal scales are soon reduced to narrow, peg-like elements: the thin bony pegs are also found in the skin of the extremities. The shoulder, pelvis, and the extremities are strongly developed. It is of fundamental importance that the rump of this animal is longer and slimmer than in other necrosaurian forms. That is a surprisingly modern adaptation: an increased mobility of the rump, as was obviously achieved in parallel by the true monitors, offered several advantages. With an entire length of *c.* 60 cm, these necrosaurians already belong among the giants of the Messel of lacertilians.

A smaller, graceful lizard (Fig. 189), to date, known from only one specimen, is very reminiscent of the large predators described above. The slenderness of the body and, most of all, the characteristic reduction of the osteoderms from broad, round to oval bones at the front of the skeleton to increasingly elongated, thin pegs towards the end, is noticeable. The sculpture of the roof of the skull and the morphology of the osteoderms exhibit distinct differences from those of the larger forms just discussed, which shows that this small lizard represents a separate genus (genus 'C') and species. The body length of the skeleton is *c.* 20 cm.

The family Necrosauridae is quite old, as recent investigations by the Polish palaeontologist M. Borsuk-Bialynicka (1984) have shown. Together with the ancestors of true monitors, it can be traced back to the late

Fig. 189: This small, slender lizard skeleton has been attributed to an agile, monitor-like predator. The front part of the skull is missing in this find. Entire length, *c.* 20 cm.

Cretaceous. Mongolia has a very rich lizard fauna from this epoch which shows this part of Asia to be an important centre of evolution and distribution of ancient monitor-like forms. It is surprising that one of the necrosaurian genera from Messel shows signs of real family relationships with the more ancient Asiatic forms. Exceptionally in the case of the Messel lacertilians, we can find no close relationships as yet with North American ancestors. This appears contrary to expectations, because one group closely related to the necrosaurians, the true monitors, certainly did use the then-existant land bridges in their migration from North America to Europe. It has been established that the necrosaurian lacertilians had already migrated during the Cretaceous from Asia to North America. The few genera endemic there, however, contrast very strikingly with the Central European material (Geiseltal/Messel). The question remains, therefore, whether, possibly for a very short

time only, there was a migration route from Asia to Europe. The time of the migration of the necrosaurians corresponds, approximately, with the earliest waves of distribution of the North American anguids, which reached Europe during the Upper Palaeocene. The absence of the true monitors at Messel and other Central European sites as well as the obvious dominance of the necrosaurians demands an explanation. On the basis of their well-developed sharp claws, it can be imagined that the necrosaurians were good climbers, that is, inhabitants of the forest biotopes of Messel. The true monitors, which have quite similar adaptations, obviously did not compete with the necrosaurians.

We know now that the true monitors have been more successful in the long term. The necrosaurian lacertilians became extinct during the early Tertiary.

Eolacerta: 'true' lizards at Messel?

Small (*c.* 25–30 cm long) skeletons of a slim, unarmoured lizard (Figs 388 and 390) are remarkable above all for the superlative preservation of their soft parts, of a quality

never before seen at Messel. The relatively long skull shows powerful, long jaw-bones which are densely covered—up to a point below the opening for the eye—with conical, one-pointed teeth (Fig. 185). Small bony plates lie within the eye opening. The relatively narrow bridge of the roof of the skull is sculpted with fine depressions and ridges which can, in a similar diagnostically valuable formation, be found in extant lizards (family Lacertidae; suborder Scincomorpha). Osteodermal scales are missing in individuals from Messel. The upper temporal window in the skull is not closed by a bony plate, a primitive characteristic for the group.

These findings, as well as differences in the shape of teeth, appear at first to argue against a classification with the family of the 'true' lizards. However, close relatives of these lizards are known from the Eocene Geiseltal brown coal and the finds from that site are similarly unspecialized. The genus *Eolacerta*, based on the skeletons of the Geiseltal, is regarded by Estes (1983) as a 'true' lizard. For the animals from Messel, whose skeletons attained sizes ranging from 20 to 60 cm, an exact identification cannot yet be established;

relationships with *Eolacerta* do, however, certainly exist.

The skeletons are quite graceful and slender in appearance; the tail is strongly developed. The posterior extremities are stronger and longer than the anterior ones.

In a small but particularly well-preserved find from Messel, there is a further lizard which has an unusually broad skull roof, with a fine sculpture of small, irregularly shaped tubercles and ridges (Fig. 190). The upper temporal openings are here closed by sculpted bony plates. The entire length of this obviously fully grown animal must be calculated as approximately 11 cm. The find is therefore one of the smallest reptiles found at Messel until now.

The preferred food and habitats of the lacertids are unknown. The specialization of the dentition in the smaller animals points to insects and worms as sources of food; the larger, powerful varieties might even have liked to take vertebrates as prey.

The family Lacertidae is probably one of the few 'indigenous' groups of vertebrates at Messel, but we know hardly anything about the ancestors of these lizards.

Iguanas: reptilian climbers, leapers, and sprinters

A group of lacertilians that was unexpected at Messel was first recognized from its tricuspid teeth. With tricuspid teeth as the only evidence, however, the suspicion that these might be true iguanas had little support: multicuspid teeth occur independently of one another in several groups of lacertilians. However, comparison with Eocene brown-coal finds of the Geiseltal confirmed the initial conjecture.

The skeletons are relatively small when the body sizes are compared with other lacertilian families of Messel, but they are larger than the corresponding skeletons from the Geiseltal (Messel: 20–30 cm body length including the tail). The skull is short and the body distinctly shortened. The posterior limbs are approximately twice as long and much stronger than those at the front (Fig. 191). A relatively

Fig. 190: This skeleton, related to the 'true' lizards, is one of the smallest Messel reptiles recorded. Entire length, *c.* 11 cm.

small fore-leg is accompanied by a strong, long, and broad hind-leg. Like the monitor-like lizards, these animals had particularly well-developed, sharp claws. The strong tails were several times longer than the body (Fig. 183B). In the skull the high-crowned teeth are, at the front, unicuspid with small side cusps which become stronger towards the back and finally form the broad, tricuspid teeth (Fig. 187). Based on preserved stomach contents we know that the food of the Messel animals consisted, at least in part, of plants. Insects were probably taken as additional prey.

Which biotope was settled by these unusual lizards? The question can be answered with reference to the living lacertilians, as the extant basilisks coincide most closely in skeletal structure with the Messel iguanas. The basilisks are bizarre iguanas which move on the ground or climb in bushy or forest areas and are also not averse to water. These excellent climbers and jumpers can, once they have reached a certain speed, stand on their hind legs; they then run upright like small dinosaurs. There is no reason to assume that the Messel iguanas would not have led a similar way of life.

It was doubted for a long time that relatives of New World iguanas (family Iguanidae) could, indeed, appear in the Lower Tertiary of Europe. They are today represented in the Old World and in Australia by the agamids, a group that resembles the iguanas closely in habit. The agamids have evolved, by independent, parallel development, numerous digging, arboreal, or swimming types that resemble iguanas in very close detail. This complementary distribution of both families has now been made relative by the finds of iguanas in Messel and the Geiseltal. Nevertheless, agamids are present in the Lower Eocene of Belgium: their distribution derived from a South-east Asian centre (Estes 1982), while the iguanas are a group whose original centre of distribution was mainly in South America and which migrated into North America only during the Upper Cretaceous. From here iguanas have also moved to Europe.

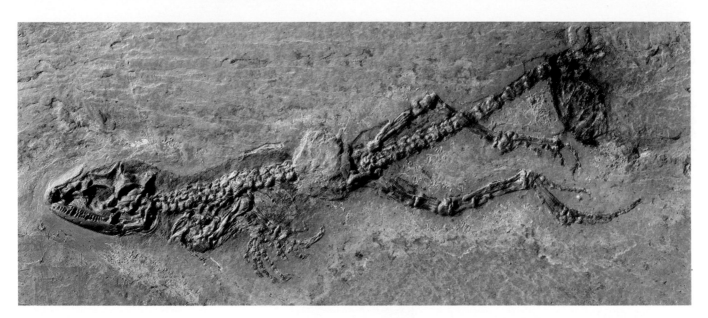

Fig. 191: Fine skin scales form a dark shadow around the front part of this iguana skeleton. The tail, incompletely preserved here, was several times longer than the body. Head–rump length, 9.5 cm.

Finds of uncertain taxonomic position

Groups whose position in the zoological system is not clear are dubbed 'problematical', because they defy a quick classification and act as 'trouble-makers' for a while. They serve to show us how very limited is our knowledge of vertebrate faunas from the early Tertiary. But if identification is successful, they occasionally afford surprising insights.

An unusually long, slim lizard, known only from a single find from Messel, poses such a riddle. Figure 183C shows a structural diagram of the animal which measures approximately 20 cm. There exists only an apparent superficial similarity with the salamanders or the olms; the vertebrae and ribs of the find, which are poor in diagnostically useful characters, point to a reptile. Anguids, gecko-like lizards, and amphisbaenians, in adaptation strategies involving an increase in the number of vertebrae, all produced more or less similar types. Thus, the length of the body, with over 60 vertebrae proximal to the pelvis, along with the relatively small but well differentiated and considerably widely spaced anterior and posterior limbs of approximately even length only allow a classification with the family Scincidae (suborder Scincomorpha).

One lacertilian group is represented in Messel, not by one specimen alone, but with a series of well-preserved skeletons, for which a definite classification among fossil or Recent known forms has up to now not been possible. This is in spite of the notable size of the animals, with lengths of over 50 cm. Limbs and tail are well developed (Fig. 192). The roof of

Fig. 192: The taxonomic position of this lizard is still unknown. The roof of the skull of the animal is noticeably grown over with osteodermal scales. The skeleton, broken in two as it is, represents a rarely found stage of decomposition. Entire length, *c.* 26 cm.

the skull is striking because of an extraordinarily pronounced sculpted pattern on the bony surface. Similar to the xenosaurians, a group of North American anguimorph lacertilians, the sculpture on the parietal bone is oriented towards the centre of the bone. At the distal rim it forms a regular scalloped ridge. The parietal eye (a rudimentary, unpaired, light-sensing organ) which has developed in almost all the Messel lacertilians as a small, round hole, is missing here.

The forms that have been introduced in this section are not known from other comparable fossil sites.

The Messel lacertilian fauna: an overview

The middle Eocene scene was characterized by the occurrence of a number of 'new' groups, whose origin must be looked for on other continents. The dominance of anguid-like (anguimorph) lizards is notable and is also a characteristic of the Geiseltal fauna. The significant ecological niches were already occupied in the Eocene by those groups of lacertilians that still occupy them today; to that extent the lizard fauna is 'modern'. However, ancient groups (necrosaurians; 'true' lizards) maintain their position.

Giant snakes and pipe snakes

Snakes arose very late in the evolutionary history of reptiles. The earliest documentation stems from the late Cretaceous, 70 million years ago. The evolutionary pace quickened in the Tertiary: *Boa* and *Python* appear in the early Tertiary while there is no evidence of the highly evolved poisonous snakes before the Miocene. More than 2500 extant species are known today, most of which live in warm countries.

Owing to the lack of fossil material, it has not been possible to elucidate the derivation of snakes and this leaves room for speculation. There is a consensus at the moment that snakes do not originate from relatives of any extant

Fig. 193: This small representative of the family Boidae is barely more than 50 cm long. The skull bones, thin in snakes, are preserved remarkably well in this fossil.

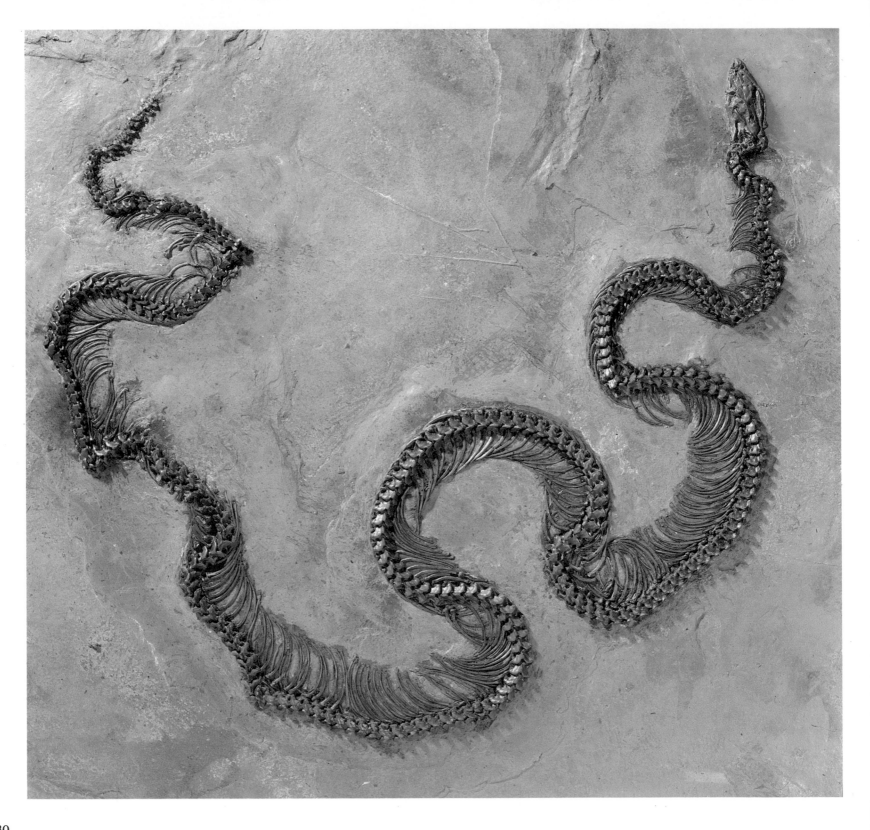

Fig. 194: (left). This snake, the largest yet found in Messel, belongs to the genus *Palaeopython*. The grey skull was added on from the cast of a second complete specimen of the same genus. The arrow marked on the upper part of the synthetic base plate points North. Together with other measurement data (e.g. excavation site, layer) it records the exact position of the find. Length, *c.*2.01 m.

Fig. 195: This snake, only 55 cm long, has well over 400 vertebrae. It belongs, as do most of the Messel snakes, in the family Boidae.

group of reptiles. Possible ancestors may be animals which resembled the monitor lizards or some digging forms of early lizards.

The scientific evaluation of the snakes of Messel has begun only recently so that special publications on this topic are not yet available. The finds of snakes known of to date are, therefore, introduced here only briefly; the exact identifications must await later publications. The Messel oil shales promise rich results for the research on fossil snakes. Although snakes are among the rarer finds of each excavation season, those that are found make available, thanks to the complete preservation (especially of the skull), a quantity of information which far exceeds the usual amount of such documentation. The systematics of fossil snakes are otherwise primarily based on finds of the far more massive and therefore more frequently preserved individual vertebrae.

Among the recoveries of Messel snakes, it is possible at present to recognize at least six species. A statement as to the frequency of the different species, based on the material as yet seen, can only be provisional. Of the already prepared specimens, most belong to the Booidea, only a few to the Anilioidea. In the presence of very meagre finds of stomach

contents only comparisons with close living relatives can give information about the choice of food. All snakes produce soft excrement; with a view to fossil material no further results can be expected here.

At Messel we find that nearly all representatives of the giant snakes belong to the family Boidae. At least four forms can be documented. The largest specimens belong to the genus *Palaeopython* (subfamily Boinae) and are over 2 m long (Fig. 194). The tail is short and accounts, as in the giant snakes alive today and in the vipers, for less than 10 per cent of the entire length. The base of the tail can be recognized in all snakes by the lack of free ribs. The illustrated *Palaeopython* is 201 cm long. Its skull was completed by a skull-cast taken from a second specimen of almost the same length and of the same species. The lower jaw with the many, unspecialized teeth is protruded, so that the skull, rounded towards the snout, appears pointed. The snake in the illustration possesses *c.* 280 vertebrae. These allow conclusions as to the mode of life and are an important aid for classification at species level. All boas are snakes which strangle their prey by constriction and their vertebrae and ribs, put under great strain during the catching of the

prey, are, therefore, strongly formed. The vertebrae of the body exhibit a considerable correspondence with those of *Palaeopython ceciliensis*, a species which has been known for a long time from a different Middle Eocene fossil site, the Geiseltal near Halle (Barnes 1927; Kuhn 1939). It is well known that giant snakes live in the warm areas of the earth. They are ground- and tree-dwellers; many prefer the vicinity of water. Among the food taken are reptiles, birds, and mammals. In Messel a specimen was found, in whose stomach a small alligator had been preserved (Greene 1983).

A second, somewhat smaller representative of the Boidae that occurs frequently in Messel is closely related to the genus *Palaeopython*. Its 200-plus vertebrae can, by their shape, hardly be distinguished from those of the larger form. The structure of the skull distinguishes the approximately 50-cm-long snake (Fig. 193) from *Palaeopython*. The illustrated specimen is very well preserved; the skull, which consists of thin bones, is, in contrast to almost all other snake skulls from Messel, not crushed. On the left lower jaw numerous evenly formed teeth are visible. Habitat and nutrition should be almost identical with the preferences of *Palaeopython*. As all snakes swallow their prey

131

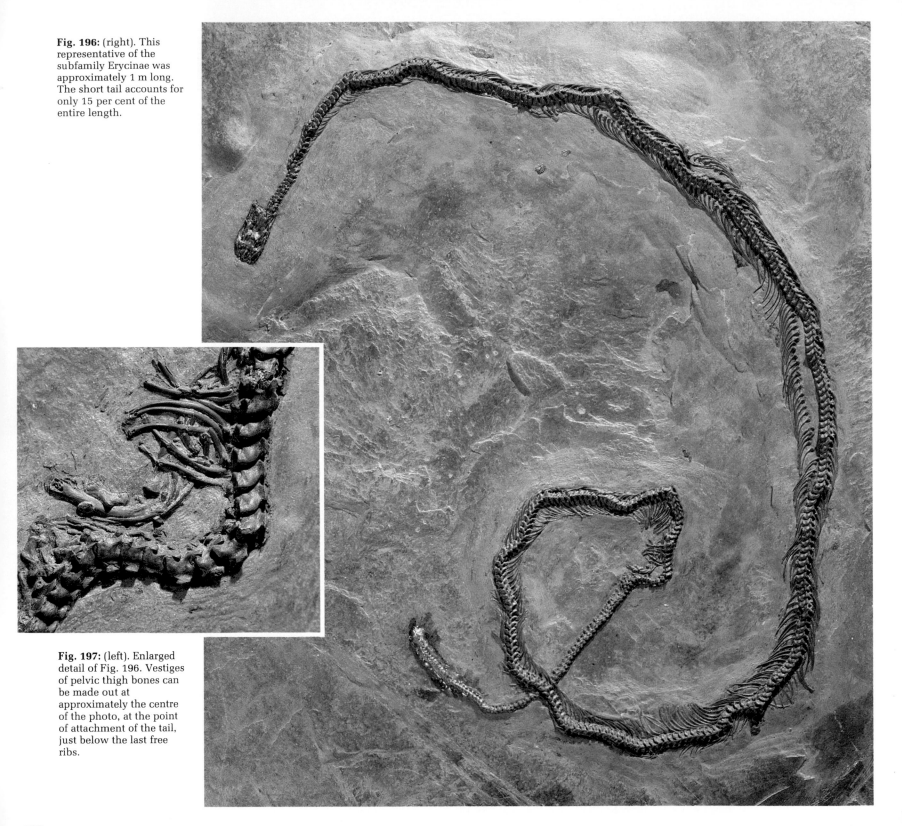

Fig. 196: (right). This representative of the subfamily Erycinae was approximately 1 m long. The short tail accounts for only 15 per cent of the entire length.

Fig. 197: (left). Enlarged detail of Fig. 196. Vestiges of pelvic thigh bones can be made out at approximately the centre of the photo, at the point of attachment of the tail, just below the last free ribs.

whole, there are natural limits to the size of the victim.

Among the rare finds is a third form which can probably be classified with the ground-dwelling boas (subfamily Tropidopheinae). The representatives from Messel have a slender build and attain a length of 55 cm. Their skull ends in a blunt snout. The specimen in Fig. 195 is remarkable for the preservation of its soft parts and the partial preservation of its windpipe (trachea) which can, however, only be observed under the microscope. The habitats of extant forms are soft forest floors in tropical woods.

Some snakes can be classified with the subfamily Erycinae (family Boidae) (Fig. 196). They are barely 1 m long and their tail makes up 15 per cent of the entire length. The skull ends at the blunt snout, just like the skulls of the Tropidopheinae. It is especially interesting that in some specimens fragments of the pelvis and extremities have been preserved (Fig. 197). In most extant snakes the skeleton consists only of skull and vertebral column. The loss of legs and pelvis can be traced back to a formerly subterranean life-style. The (vestigial) remains of the pelvic girdle, therefore, indicate that these fossils are of 'primitive' snakes. These pelvic and limb remains are no longer connected with the vertebral column and are at the base of the tail, close to the last free ribs. In the enlarged detail (Fig. 197) we can recognize, in addition to the thigh (femur), remains of the pelvis which are identified as pubis, ilium, and ischium.

Close family relationships exist between the small Messel pipe-snakes (superfamily: Anilioidea; family: Aniliidae) and the giant snakes. Fossil pipe-snakes are known from the Upper Cretaceous and the Eocene. The approximately 50-cm-long specimens from Messel have a head which is rounded at the snout, the small vertebrae are strongly developed, and the tail amounts to only 15 per cent of the entire length. The biotope of the extant pipe-snakes consists of soft forest floors in tropical woods where they search for their favourite prey, small reptiles (Bellairs 1971). No exact statement can as yet be made about the choice of food of the fossil aniliids. One specimen contains the remains of one or several tiny snakes in the posterior part of its body. The small lower jaw of one of these snakes measures just 4 mm. It cannot be ascertained if these are vestiges of food or of young that did not hatch. According to the position of the remains, they should lie in the region of the gut. Snakes produce very strong digestive enzymes which speed the decomposition of the prey, including hair, feathers, and bones. The preservation of a small snake in the region of the gut suggests, therefore, a young animal which did not emerge.

We also must place some finds, hardly longer than 25 cm, among the digging snakes. These are strongly built snakes with a short tail. The 270-plus vertebrae of body and tail are very short. Due to the bad preservation of the bones no definite statement can be made as yet regarding the systematic position.

Is it possible to use snakes as indicators of climate? The closest extant relatives of the Eocene snakes of Messel live in tropical and subtropical forests. Snakes are cold-blooded; this means that they lack the ability to regulate their body temperature. The temperatures of air and ground are, therefore, of immense importance. Moisture and equable high temperatures influence and support procreation and the development of the young (Trutnau 1981). In South America, for example, the numbers of species of snakes increases towards the warmer latitudes. While at the southern latitude of 50° only one species is known, this had already increased at the southern latitude of 35° to over 50 species (Schwarzbach 1974).

It would, however, certainly be wrong to draw conclusions from the species-poor Messel snake fauna concerning inclement climatic conditions during that time. It must rather be assumed that, under the conditions of deposition of the Messel oil shale, a large number of the animals once present were not preserved and that others have not yet been found. The low number of species may also be connected with the fact that the snakes had not yet pioneered all biotopes, so that the radiation of snakes took place only later. In any case, there is no contradiction between the subtropical climatic conditions postulated for Messel on other grounds and the snake fauna discovered to date at Messel.

11

Messel birds: a land-based assemblage

In contrast to the mammals and crocodiles, the fossil birds of Messel were neglected for a long time. Nevertheless, the first species, namely, the ibis *Rhynchaeites messelensis*, was described in 1898 by Wittich. After that, however, nothing happened for a long time. It was not until 1965 that Berg recorded another bird that was, however, quite spectacular: a giant, flightless species of the genus *Diatryma*.

The lack of attention to the birds was by no means due to a dearth of finds. Quite the opposite—birds were and are recovered very frequently from the Messel pit. Many specimens are available as complete skeletons and show remains of feathers, stomach contents, and other soft body parts. The intensive investigation of the birds, started in recent years, no longer leaves any doubt that the birds of Messel are just as significant for our knowledge of the evolution of their extant relatives as are the Messel mammals. Some examples illustrating this will be introduced here.

Messel birds: a land-based assemblage

DIETER STEFAN PETERS

Recent researches show that the birds during the Cretaceous were highly varied in form and of world-wide distribution. They belonged, however, in part to systematic groups that appear very strange to us now and that obviously did not survive the end of the Cretaceous. In any case, in the Tertiary we find exclusively birds whose general appearance does not differ from that of extant birds. They can also be separated into two systematic supergroups into which the Recent birds also fit. One can give these groups the rank of superorders. Both are represented at Messel, although by very unequal numbers of species.

Superorder: Ancient palate birds (Palaeognathae)
Order: Ostrich-like birds (Struthioniformes)
Family: early Ostriches (Palaeotididae)

Among birds living today, besides the ostrich-like birds ('ratites'), only the tinamous (Tinamiformes) from the New World tropics are classified with the Palaeognathae. Among other characters this group is distinguished from the new palate birds (Neognathae), as the names suggest, by the construction of the mobile jaw–palate apparatus. We must remember on this occasion that birds, unlike mammals, can move not only their lower but also their upper jaw. This alone makes the beak an apparatus capable of so many functions. In the Palaeognathae the movable bony palate consists of stiff elements and is in this similar to the palates of many reptiles, while the Neognathae possess an additional joint in the construction of their palate. This difference

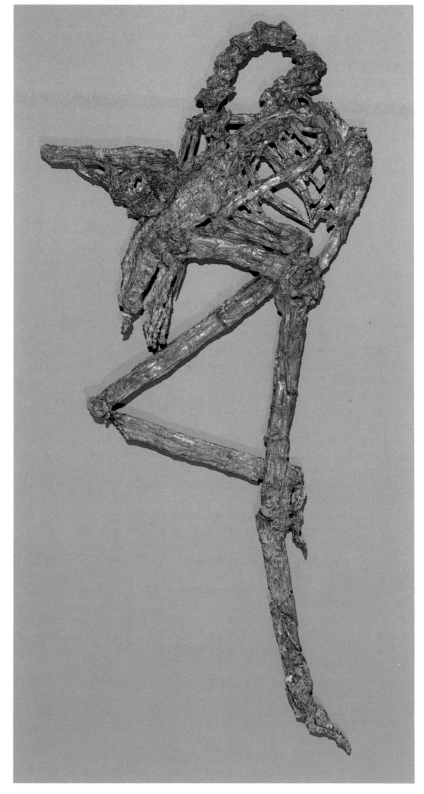

Fig. 198: *Palaeotis weigelti* Lambrecht 1928. Known first from the Geiseltal, this palaeognathous bird species has also been found at Messel. The illustration shows the most complete specimen discovered so far. In contrast to most Messel fossils, its robustness meant that it could be freed entirely from the surrounding matrix. Note the strong legs (due to compression, they appear somewhat thick) and the weak development of the sternum, pectoral girdle, and wings. Reduction, 0.26×.

leads to further differences in the construction of the bird skulls which cannot here be given in greater detail. Only so much must be said, that palaeognathy and neognathy can be understood as two ways of solving the mechanical problems of the bird palate and that in this case neognathy appears to be the variation more suitable for further development (Peters 1987a).

Palaeognathous and neognathous birds are sister groups, genealogically speaking. Among these, the latter have been more successful and the larger number, by far, of extant species are recruited from them. Fossil finds are evidence, however, that in the Lower Tertiary the Palaeognathae were also richer in taxa, so that the few extant representatives must be regarded as relics (Houde and Olson 1981; Houde 1986). It is of biogeographical importance that some of the representatives from the Tertiary were found in the Northern Hemisphere. This removed the most important basis for the argument that the Palaeognathae, or at least their flightless, 'ostrich-like' representatives, evolved exclusively in Gondwanaland (the southern ancient continent), a thesis based on their present distribution.

Palaeotis weigelti plays a role here. Lambrecht (1928) described this bird from material in the Geiseltal as a putative fossil bustard (*Palaeotis*: ancient bustard). Based on more recent finds from the Geiseltal and the Messel pit, Houde (1986) recognized that this was no bustard, but a flightless, palaeognathous bird. Houde believed that this bird could be recognized as an ancestor of the African ostrich. A further, very complete specimen of this species was recovered in 1986 at Messel (Figs 198 and 199). The findings of Houde were largely confirmed by this, but the immediate genealogical connection of *Palaeotis* and the African ostrich is questionable because, strictly speaking, no unequivocal commonly derived characters (synapomorphies) can be documented (Peters 1989a). On the contrary, in the new specimen, as in the South American rhea, only three pairs of ribs are attached to the sternum. The ilium and ischium were most probably connected by

a small bone bridge at the distal end, again in the same form as that found in the South American rhea, but differing from those of the emu and cassowaries which also have a connection between the above-mentioned pelvic bones. There is no such bridge in the ostrich.

From the zoogeographical point of view a close relationship between *Palaeotis* and the South American rheas appears, at first sight, strange, but, for example, in the ant-eaters and *Eurotamandua* we have a similar relationship. A phylogenetic relationship between Europe and South America is, therefore, not impossible. We can, however, exercise caution and assume that the congruence in the number of ribs and in the pelvic structure in *Palaeotis* and the rhea was brought about by convergent evolution. Under such circumstances *Palaeotis* could still belong to the group of the 'early ratites' which gave rise to the ostriches, rheas, and perhaps also emus and cassowaries. A closer connection between the different ratites has, moreover, long been expected.

In any case, we have with *Palaeotis weigelti* a ratite, that is, a palaeognathous running (cursorial) bird with strong legs whose small wings, flat, keel-less sternum, and stiff unit of scapula, clavicle bone, and coracoid show it to be flightless. The fossil evidence reveals distinct differences in size which may, perhaps, be interpreted as sexual dimorphism. The larger specimens, when alive, had a maximum height of 90–95 cm.

The term 'ostrich-like bird' (ratite) evokes spontaneously the image of an inhabitant of open steppe-like landscapes. The cassowaries, however, prove to us that large cursorial birds can also live in dense forests. The presence of *Palaeotis* does not, therefore, militate against the frequently supported impression that the Eocene Lake Messel was surrounded by forest.

Still immature and therefore hard to identify leg bones found in Messel, which in size correspond with *Palaeotis*, could have belonged to a young bird of this species; this would support the hypothesis that *Palaeotis* bred in the vicinity of the lake.

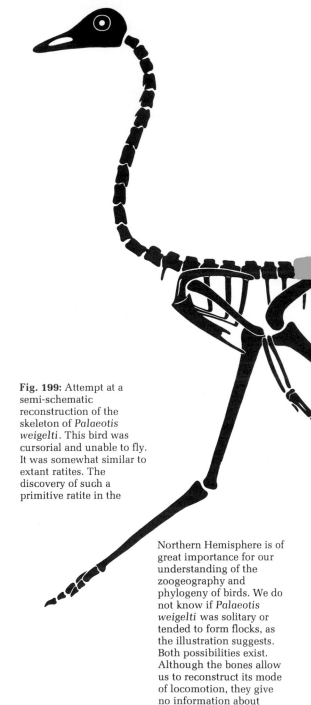

Fig. 199: Attempt at a semi-schematic reconstruction of the skeleton of *Palaeotis weigelti*. This bird was cursorial and unable to fly. It was somewhat similar to extant ratites. The discovery of such a primitive ratite in the Northern Hemisphere is of great importance for our understanding of the zoogeography and phylogeny of birds. We do not know if *Palaeotis weigelti* was solitary or tended to form flocks, as the illustration suggests. Both possibilities exist. Although the bones allow us to reconstruct its mode of locomotion, they give no information about behaviour.

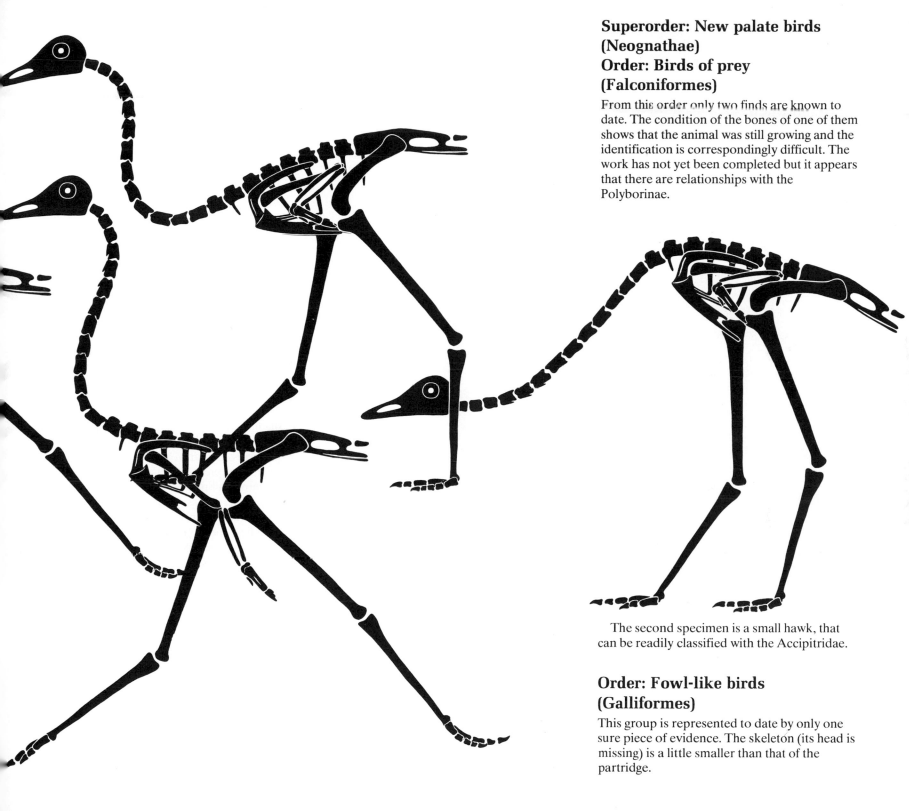

Superorder: New palate birds (Neognathae)
Order: Birds of prey (Falconiformes)

From this order only two finds are known to date. The condition of the bones of one of them shows that the animal was still growing and the identification is correspondingly difficult. The work has not yet been completed but it appears that there are relationships with the Polyborinae.

The second specimen is a small hawk, that can be readily classified with the Accipitridae.

Order: Fowl-like birds (Galliformes)

This group is represented to date by only one sure piece of evidence. The skeleton (its head is missing) is a little smaller than that of the partridge.

Wittich (1989) as *Rhynchaeites messelensis*. In his opinion the skeletal structure of this bird appeared to 'occupy an intermediate position between Limicoles [charadrians] and rallines'. This reminded Wittich of painted snipes (Rostratulidae). The same species was described once more under the name *Plumumida lutetialis* Hoch 1980 and again similarities between this fossil and some plovers were emphasized.

A renewed investigation of the fossil material was able to confirm the results of earlier authors but also showed that *Rhynchaeites messelensis* quite clearly belongs, on the basis of its beak structure, to the ibises (Peters 1983). This was not a contradiction because repeated comparative studies on extant ibises have shown that these species too have many connections with cranes and plovers. In the Messel ibis these similarities are still more distinct. This species therefore offers good grounds for the assumption that ibises form a group that is close to those birds from which gruiform and charadriiform birds have evolved. What effect this insight will have on the classification of the birds concerned depends on a scrupulous phylogenetic analysis of the three orders, Ciconiiformes, Gruiformes, and Charadriiformes, which, however, still requires protracted work. For this reason we relinquish, at present, a classification of the ibises into a definite order.

Rhynchaeitis messelensis is, however, one of the oldest known ibises and at the same time the smallest species of the family and the most completely preserved among the fossils. Its skeleton is evidence that it was a good flier and runner, if only on relatively short legs (Fig. 200).

Order: Uncertain
Family: Ibises (Plataleidae)

The ibises form a group which is characterized by the singular construction of its beak (a combination of schizorhinal nostril and desmognathous palate). As a family they are usually classified within the order of the storks (Ciconiiformes), although Garrod (1876) had already alluded to consistencies in the bone structure of ibises and the cranes (Gruiformes). Olson (1979) took up this argument once again and saw in the ibises a connection between cranes and plovers.

One bird from the Messel pit supports this view very convincingly. It was described by

Order: Crane-like birds (Gruiformes)
Family: Diatrymas (Diatrymidae)

These large birds, which, in the Eocene, were widely distributed in North America and in Europe, also lived in the vicinity of Lake Messel. The evidence for this is, however, only a cast of a natural imprint of the bone of a thigh (Berg 1965).

The species described for Messel could be identical with that described from the Geiseltal, *Diatryma geiselensis* Fischer 1978. The diatrymas;, without a doubt, were then among the most impressive animals in Europe. They were very robust, flightless birds with powerful beaks. Their strong, but relatively short legs probably made them look rather plump. Fischer (1978), on the basis of finds from the Geiseltal, states a maximum height of 175 cm. The diatrymas are usually classified among the gruiforms, close to the rails, but recently Andors (1991) argued convincingly that *Diatryma* and *Gastornis* are the sister-group of the Anseriformes (waterfowl).

Family: Phorusrhacidae

Representatives of this family are also large flightless birds that were very probably predators and that have been documented mainly in South America. But there are also a few finds from North America, which were joined quite unexpectedly by some further evidence from France (Quercy) and from Messel (Mourer-Chauviré 1981; Peters 1987*b*).

To date, the remains of three specimens are known from Messel, one of which has been preserved relatively complete. They belong to a species that attained the size, approximately, of a large rooster and that was described as *Aenigmavis sapea*. Its small wings were, perhaps, just capable of short, clumsy flight, but the hind limbs were very strong and armed with long claws (Fig. 201). Perhaps these birds had even entirely lost the ability to fly.

The Messel species is far smaller than the often man-sized American species, but this does not negate their belonging to one and the same family. A giant size must undoubtedly be regarded as a derived character that presupposes smaller original forms. In good agreement with this supposition is the fact that *Aenigmavis sapea* is the oldest-known representative of the family; the American evidence stretches from the early Oligocene to the late Pliocene.

Based on the evidence available we must assume that the phorusrhacids belonged to the early forms of the Gruiformes and were more widely distributed than was thought for a long time. It is not possible to decide at present if, as one might expect, the species from Quercy and Messel were particularly close to one another, as the material available does not permit the necessary comparisons.

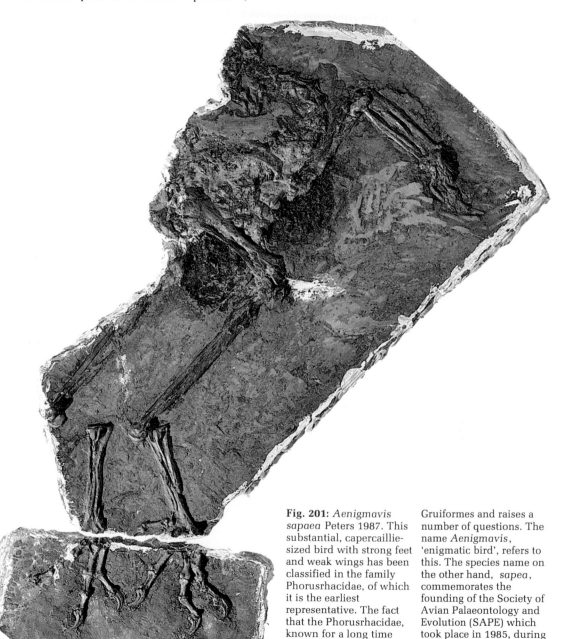

Fig. 201: *Aenigmavis sapaea* Peters 1987. This substantial, capercaillie-sized bird with strong feet and weak wings has been classified in the family Phorusrhacidae, of which it is the earliest representative. The fact that the Phorusrhacidae, known for a long time only from America, also occurred in Europe throws new light on the early phylogeny of the Gruiformes and raises a number of questions. The name *Aenigmavis*, 'enigmatic bird', refers to this. The species name on the other hand, *sapea*, commemorates the founding of the Society of Avian Palaeontology and Evolution (SAPE) which took place in 1985, during an international palaeo-ornithological congress in Lyons. Reduction 0.37×.

Family: Seriemas (Cariamidae)

Only two South American species are counted today among the seriemas, namely the red-legged seriema (*Cariama cristata*) and the black-legged seriema (*Chunga burmeisteri*). They are long-legged, mainly terrestrial birds, approximately as large as a stork; they fly badly but still nest in trees. They have relatively short necks and hooked beaks. One must regard both species as being remnants of a group that was widespread and much more varied in the Tertiary.

Mourer-Chauviré (1983) reported evidence from the phosphorites of Quercy (France) of 12 species in five genera. All European fossil Cariamidae are united in the subfamily Idiornithinae. From North America, in addition, many fossil species are known which belong to a different subfamily (Cracraft 1968, 1971). It is, therefore, not surprising that these obviously once so frequent birds have also been found in Messel. Here, however, we have, to date, only three species.

Evidence for one of these species is the lower end of a tibiotarsus, the tarsometatarsus, and parts of the toes. The pieces can be classified with the genus *Idiornis* Oberholser 1899, known at present from the Middle and Upper Oligocene, but they probably belong to a new species which still awaits description. The bird was dainty and could, most probably, run well, perhaps similarly to the extant seriemas, although, as far as one can conclude from the available leg bones, it only attained half their size.

The classification of the second species is more difficult. Here we have a relatively well-preserved skull together with some neck vertebrae (Fig. 202). As far as can be seen, this fossil agrees well with the corresponding parts of extant Cariamidae. It is therefore natural to suspect that it is a representative of the Idiornithinae. Unfortunately, skulls of Idiornithinae are unknown at present. It is, therefore, hardly possible to state whether the Messel skull belongs to a known species or not; even the affinity with a known genus cannot be decided. For the above-mentioned *Idiornis* species, at least, the skull appears to be too large. Although the similarities with the Recent seriemas leave no doubt as to the family relationship, the find cannot be named at present. There is the hope, of course, that future discoveries will shed more light.

The third species, represented by a nearly complete skeleton, is even more enigmatic. To judge from most of its bones it clearly belongs to the Cariamidae but some features of the pelvis are unusual for a neognathous bird, and the structure of its cervical vertebrae is unique among birds.

Family: Messel rails (Messelornithidae)

By far the most frequent birds of the Messel oil shale are the so-called Messel rails. Extensive investigations, performed by Frau Angelika Hesse at the Forschungsinstitut Senckenberg (Hesse 1988), showed that the temporary classification with the rails was incorrect. Although the moorhen-sized birds (Fig. 204) are to some extent reminiscent of rails, their relationships to sun bitterns (Eurypygidae) are much stronger; however, they are again so different from these that they deserve to be placed in a separate family.

Among the several hundred specimens recovered, there are some whose soft parts were preserved. From these we know, for example, that the head supported a helmet-like ornament (Fig. 205) consisting of a fleshy or horny but boneless formation. The tail feathers were quite long and this also does not fit in with the rails.

Strangely, only adult birds have been found. Regarding the frequency of the finds this permits only two interpretations. Either the Messel rails did not breed in the vicinity of the lake but only visited it outside of the breeding season (during migration?) or they did breed here, only not on the shore, but at some distance on trees, so that the chicks remained in the nest. In this context it is significant that the only extant species of sun bittern very frequently breeds in trees.

To the new family belong, besides *Messelornis cristata* Hesse 1988, at least one species from North America and some further ones from Europe that must, in part, be placed into different genera (Fig. 203).

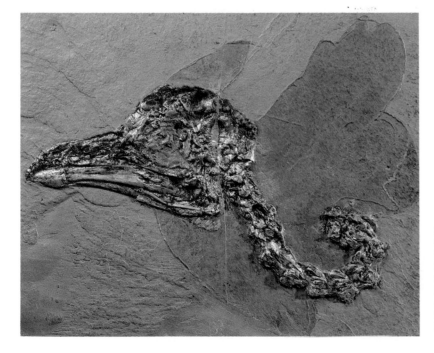

Fig. 202: Skull and neck vertebrae of a species belonging to the Idiornithinae, related to the seriemas, which are today extant only in South America. Identification is not yet possible (see text). Reduction 0.67×.

Fig. 205: *Messelornis cristata* Hesse 1988. Head with vestiges of a fleshy or perhaps horny crest which has contributed to its name (*cristata*: crest-bearing). As has occurred frequently in Messel when the soft parts of a specimen have been preserved, the bones have only been incompletely preserved. But, because a large number of finds in all states of preservation are available, there is no doubt about the taxonomic status of this find. Magnification, 1.8×.

Fig. 203: This bird from the Green River Formation in Wyoming is a current subject of research. It belongs to the Messelornithidae, and demonstrates that this family was widely distributed in Europe during the Eocene. Reduction 0.51×.

Fig. 204: *Messelornis cristata* Hesse 1988. The 'Messel rails' are the most common Messel fossil birds. Their vernacular name is so well established that it is worth retaining, even though these are not true rails, but a separate family. Reduction, 0.67×.

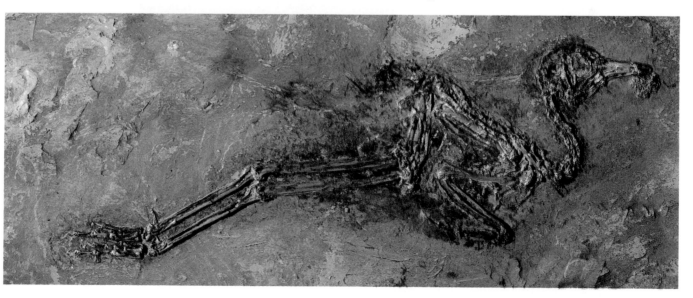

Order: Plover-like birds (Charadriiformes)
Family: Flamingos (Phoenicopteridae)

A very well-preserved, almost complete skeleton of a long-legged bird, approximately as large as a demoiselle crane, for a long time was held by the specialists at Messel to be a 'heron' or a 'crane' (Fig. 206). A first impression could, indeed, lead to this opinion. A thorough study, however, confirmed that at most the skull was superficially crane-like; the rest of the skeleton, however, did not fit. Many more similarities were observed with extant flamingos and particularly with the fossil genus *Juncitarsus* Olson and Feduccia 1980, so that the Messel bird was described as *Juncitarsus merkeli* Peters 1987.

To appreciate the real value of this result one should realize that the phylogenetic relationship of the flamingos was always controversial. Most specialists saw a connection of these highly specialized birds either with the storks (Ciconiiformes) or with the waterfowl (Anseriformes). More recently, Feduccia and Olson (Olson and Feduccia 1980) have produced evidence that flamingos are much more likely to be connected with the Charadriiformes, particularly with the family Recurvirostridae, which includes the avocets and stilts. The two authors base their conclusions on numerous results from the skeleton, system of muscles, feathers, and breeding habits. The afore-mentioned genus *Juncitarsus*, whose bones combine the characteristics of stilts and flamingos, served as palaeontological evidence. The skull of *Juncitarsus* remained unknown. All the same, such a fossil skull would offer particularly informative insights, because the structure of the beak is very different in storks, waterfowl, and plover-like birds.

The Messel specimen has a pointed, moderately long beak, the upper jaw of which is schizorhine, that is, due to the slit-like expansion of the nasal opening, the skeleton of the upper jaw has become a framework of slender bony bars. Such a construction does not occur in Ciconiiformes or Anseriformes,

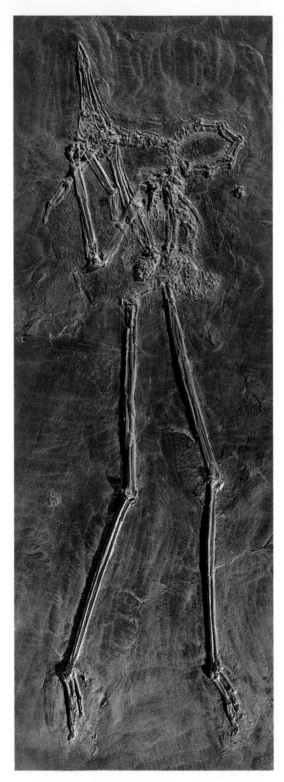

Fig. 206: *Juncitarsus merkeli* Peters 1987. This splendid find has helped to resolve a long controversy over the phylogeny of flamingos— neither storks nor ducks, but rather stilts and avocets (Recurvirostridae), which belong to the Charadriiformes should be considered their closest relatives. This is a prime example of the scientific significance of the Messel fossils. The photograph shows a copy of the holotype from a private collection. Reduction, 0.33×.

but is very characteristic for Charadriiformes and some Gruiformes. Because nothing in the remainder of the skeleton of our fossil reminds us of gruiform birds, we can omit these birds from further consideration. Thus, the plover-like birds remain.

How is this identification to be arrived at? We start from the established fact that the beak of the extant flamingo is extremely specialized, namely as a filtering apparatus which can be used to greatest advantage in water or thin mud. Such a beak can only have evolved under the selection forces to which a wading or swimming bird, whose food consists of small particles or organisms retrieved from the water, is exposed. The sequence of adaptations is therefore established. The ancestors of the flamingos must have already been waders before they evolved the specialized beak.

Juncitarsus correlates entirely with this stage of evolution. The body proportions and, most of all, the legs, are already those of a flamingo-like wader; the beak, however, is still not flamingo-like. But it is also not stork- or goose-like, and suggests a charadriiform relationship. The opinion of Olson and Feduccia is therefore supported by virtually irrefutable evidence (Peters 1987*c*).

This example elucidates the particular value of the Messel fossils in that they frequently consist of entire skeletons or of large parts of skeletons. If, in the present case, skull and limbs had been found separately, they would probably not have been thought of as parts of the same bird; it would, at least, have been very difficult to justify such a combination.

The interpretation of fossils is also always dependent on the opinions and the concepts of the investigator. If, for example, someone is convinced that flamingos are related to storks he or she will be 'justified' in supposing that the isolated skull of *Juncitarsus* is not the skull of an early flamingo since, in his or her opinion, early flamingos should be more stork-like and have, therefore, a beak of completely different construction. Nevertheless, the skull of *Juncitarsus merkeli* was not isolated and the findings from the rest of the skeleton could be reasonably combined only with relationships with flamingos and plovers. Most of all, the findings fitted the evolutionary model outlined above. Only then did the fossil become evidence for the phylogenetic derivation of the flamingos.

Our model can even be expanded and supported by further evidence which mediates between *Juncitarsus* and the recent flamingos. We must here mention first of all *Phoenicopterus* (*Harrisonavis*) *croizeti* from the Upper Oligocene and the Lower Miocene of France, which has an only weakly bent beak. The extant flamingos, however, must be regarded as relics of an earlier relationship that was once far more diverse in forms. If one includes all the fossil representatives, then the family Phoenicopteridae can be separated with ease into three subfamilies, namely Phoenicopterinae, Palaelodinae, and Juncitarsinae, of which only the first is still extant.

We can conclude that, due to the lucky find from Messel, there are now few doubts that flamingos must be linked not with the storks or geese but with the stilts. *Juncitarsus merkeli*, however, is the only species among the birds of Messel to date whose life-style is associated, with a fair degree of certainty, strictly with the water. As only one specimen is known at present—although the large bones should be easily found if they are present—one may suspect that this bird did not live entirely at the lake, but visited it only sporadically.

Order: Owls (Strigiformes)

The oldest known owls are known from the Palaeocene. From the Eocene several families are already known, of which the barn owls (Tytonidae) are still represented today by a dozen species. According to the most recent information (Mourer-Chauviré 1987), most Eocene owls of Europe belong to this family.

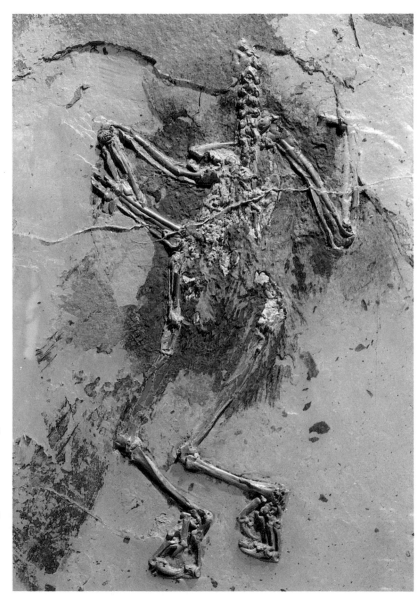

Fig. 207: *Palaeoglaux* sp., the only owl known from Messel up to the present time. Although the feet are strongly reminiscent of barn owls, other differences of detail have led to the erection of a new family, the Palaeoglaucidae. The pinions of the left wing are on the left. At higher magnification, the body feathers reveal structures which are unknown in Recent owls. Reduction 0.93×.

The Messel owl is approximately the size of a little owl and is documented by a rather well-preserved specimen (Fig. 207). It belongs to the genus *Palaeoglaux* Mourer-Chauviré 1987 which has to be placed in a separate family Palaeoglaucidae. The strange plumage of this owl is worth mentioning. Relevant imprints are evidence of the narrow, long, almost ribbon-

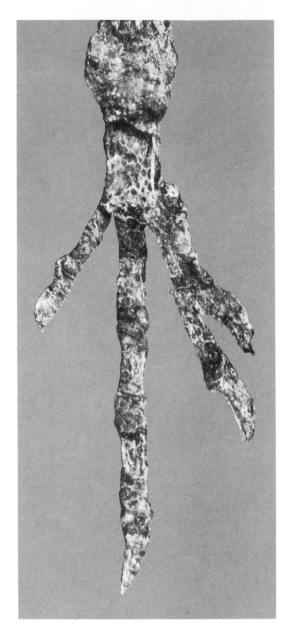

shaped feathers—probably on the lower back—carried by the bird. They are preserved mainly on the opposing face to the fossil, which is not illustrated here. Similar feathers are not known in extant owls. They could have been ornamental feathers, which are, of course, unexpected in a nocturnal bird. But were the 'early owls' already distinctly nocturnal? It is a fact that skeletons cannot provide information on the pattern of activity of a bird particularly when the heads, which could afford information about ears and eyes, are missing (Peters, in press).

Order: Nightjar-like birds (Caprimulgiformes)

Caprimulgiformes are poorly documented as fossils. All the more interesting are two species which were recently recovered from the Messel pit. One of them is strongly reminiscent—mainly because of its large and massive beak—of the frogmouths (Podargidae), which are distributed nowadays from South-east Asia to Australia. The other resembles the nightjars (Caprimulgidae) more closely (Fig. 208). But neither correlates in all aspects with the criteria of the families mentioned, and further investigations are required.

Order: Swifts (Apodiformes) Family: Aegialornithidae

The Aegialornithidae are a family of swifts that are only known from fossils and of which several species were discovered at different European sites. The description of *Aegialornis szarskii* Peters 1985 from the Messel pit is based on four individuals. According to our present knowledge this is the smallest species of the family (Fig. 209–211) and in size resembles a medium-sized hummingbird rather than any extant swift.

With the four specimens from Messel, there were available, for the first time, substantial parts of the skeleton of an *Aegialornis* species in an articulated condition, so that the correlation of the individual bones was beyond doubt. With this, and using the method of

analogy, the placement of individually discovered bones from other sites could be placed on a more secure basis than hitherto. More than that, it was finally proved that the Aegialornithidae were swifts. On the exclusive basis of the comparison of upper arm bones,

Figs 209–211: *Aegialornis szarskii* Peters 1985. X-rays of three specimens, original size. Above (Fig. 209), the holotype. X-ray photographs are extremely helpful. In this case, details of the vertebral column and the bones of the wing could only be interpreted by this means. In the lowermost specimen the bones of the wing are partly broken.

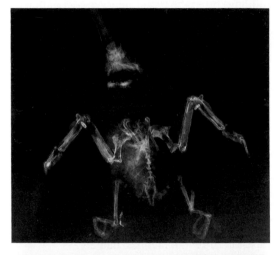

Fig. 208: Lower side of the left foot of a nightjar, which is as yet undescribed. On the tarsus and on the third (longest) toe in particular, horny scales are visible as small, polygonal shapes. They are quite similar in extant nightjars. The fossilization of such scales is very rare indeed, even at Messel. Magnification, 4.86×.

146

without considering their function, some authors had believed previously that it was necessary to classify this family with the nightjars.

Aegialornis szarskii is not only documented by well-preserved skeletons; in one case, for the first time in this family, the imprints of pinions were even preserved (Fig. 212). The

Fig. 212: *Aegialornis szarskii* Peters 1985. Only single bones of members of the swift family were known hitherto. The Messel pit revealed not only the first complete skeletons, but also vestiges of feathers. In this picture wing bones and wing feathers are recognizable, especially on the left side where a few full length primaries are visible. Their enormous development relative to body size indicates that their owner was a high-performance flier, comparable with extant swifts. The same inference can be made from the proportions of the skeleton of the wing (a very short upper limb, and a very long 'hand'; see also Figs 209–211). Magnification, 1.73×.

feathers show, as was expected given the bone structure of the fore-limbs, that this bird had wings constructed for high-performance flight. Obviously, the Aegialornithidae were on a path towards specialization for sustained flight similar to that followed by the true swifts (Apodidae) whose oldest representatives also stem from the Eocene. However, the Aegialornithidae never attained the extreme specialization of the Apodidae. One can speculate, that this is one of the reasons why the Aegialornithidae became extinct, leaving the air space to their competitors.

Order: Roller-like birds (Coraciiformes)

The recent roller-like birds appear to be a very disjunct group as they include, among others, such diverse forms as kingfishers, hoopoes, hornbills, and bee-eaters. Such groups that give the impression of being jumbled together always lead one to suspect that they are either polyphyletic groups or old, vestigial groups whose ranks have been thinned by the extinction of connecting forms. Indeed, the status and delineation of the rollers, especially from the woodpeckers, is under dispute.

Palaeontological finds provide evidence that the major part of the tree-living birds of the early Tertiary, at least in North America and Europe, were recruited from the rollers. The great number of finds, however, has not really contributed to clarifying the systematics of this order and has in part even led to confusing publications.

It is apparent, and this is also applicable to the finds of Messel, that some fossils agree quite well with Recent groups, for example, the true rollers (Coraciidae) or the ground rollers (Atelornithidae = Brachypteraciidae) (Fig. 214), while other finds represent obviously extinct families. Particularly remarkable are some species whose feet are reminiscent of birds of prey or owls, although the rest of the skeleton points unequivocally to the rollers (Figs 213 and 215). The habit of these birds was obviously similar to that of birds of prey.

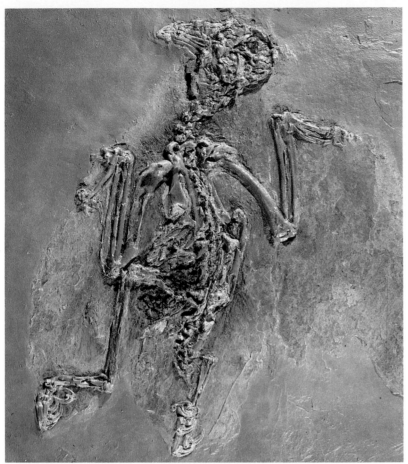

Fig. 213: Numerous roller-like birds lived in Messel at the time of the Eocene lake. Among these were specimens whose toes were strongly reminiscent of those of falcons or owls, just as in the species illustrated here, which has not yet been described. Reduction 0.7×.

Fig. 214: A bird which is reminiscent of ground-rollers in some osteological respects. The specimen is exceptionally well preserved, and includes the pinions. Unfortunately, the hind part of the body is missing. Reduction, 0.75×.

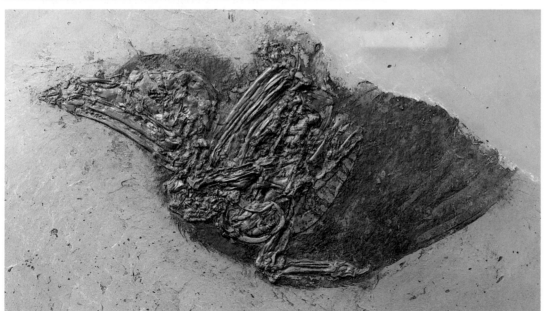

The structure of the toes with their partly shortened digits and the long, pointed claws hardly allow the supposition of another function. One could regard the shrikes (Laniidae) as a recent case of a similar convergence towards the appearance of a bird of prey. Thus, one can assume that in the Eocene of Europe and North America, predatory rollers had partly taken over the ecological niche of birds of prey. Besides this there were, of course, still many other representatives of the rollers. For some we can even picture their attitude in flight because pinions and tail-feathers have been preserved. A long tail and relatively short, rounded wings enable the conclusion that the flight was agile but not long-sustained.

A most interesting species is under study now. It has the appearance of a todi bird, but some of its features can be found also in passeriform birds. Since the origin of the Passeriformes is still obscure, this little bird may reveal exciting information.

Order: Woodpecker-like birds (Piciformes)

As in the case of the rollers there are differing opinions concerning the size and delimitation of this order. A distinction between woodpeckers and rollers appears to be even more difficult in the representatives of the Lower Tertiary than in Recent species. This can be understood when it is supposed that both groups are closely related and could, perhaps, at that time not be separated at all. At least some North American fossil species, described as primitive woodpeckers, appear rather to belong to the rollers (Olson 1985).

There are several bird remains at Messel which one would like to attribute to the Piciformes. Figures 216 and 217 show a specimen for which this classification even appears to be safe. It has distinctly zygodactylous feet (toes I and IV point backward, II and III forward). This feature alone is, to be sure, still not definite evidence, since, apart from the woodpeckers, cuckoos and parrots, for example, share the zygodactylous arrangement of toes. However, the proportions of the feet as well as the development of the remainder of the skeleton, argue for placement in the Piciformes, in particular for barbets (Capitonidae), even if the beak appears to be too thin for this family.

The research on the Piciformes and Coraciiformes of Messel is very time-consuming, due to the countless necessary comparative investigations on Recent and fossil material. In any case, it promises significant new knowledge about these disputed groups that lived, obviously with a great number of species, in the Eocene forests of Messel before the passerines, dominant today, appeared on the scene.

Fig. 215: This particularly fine specimen shows a further species of roller-like bird with feet like those of birds of prey. The bird had obviously moulted, as is seen in one of the tail feathers whose tip has just emerged from its quill. Reduction, 0.78×.

Outlook

The preceding survey of the birds of Messel is by no means complete because only those finds were used whose classification appears approximately safe, at least in the larger systematic framework. Besides these there are, however, still numerous other remains of birds whose preparation is not yet complete or which have, up to now, defied classification. Publicizing uncertain speculation hardly assists in the acquisition of knowledge and we have refrained from doing so here. Unfortunately, some specimens, however promising they may appear at first glance, will, probably, always remain undetermined, because their features are not sufficiently preserved (Fig. 218). A different problem is posed by the young birds which, although found rarely, would be the more interesting scientifically. Because their bones are not fully grown and, in particular, do not show the joints, it is, as a rule, only possible to indicate those groups to which the young birds most probably do not belong. An exact, positive classification cannot be achieved except under special circumstances. The latter is true for the above-mentioned young *Palaeotis weigelti*, whose size corresponds with

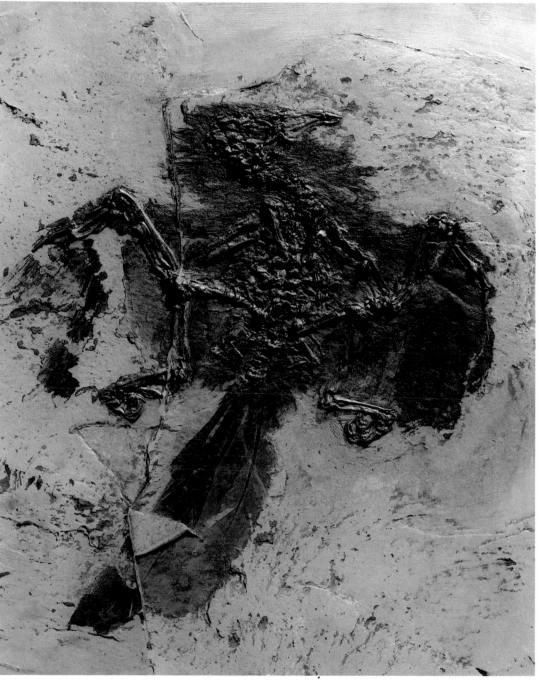

that of no other bird of Messel which therefore makes the classification of this species plausible.

In spite of these difficulties the Messel birds have helped us to achieve a series of important discoveries which can be briefly summarized as follows.

1. **Phylogeny and systematics.** In the case of ostriches, ibises, cranes, flamingos, and

149

Fig. 216: With the rollers (Oraciiformes), the woodpeckers (Piciformes) represent tree-living small birds at Messel. Their typical zygodactylous arrangement of toes is particularly clear in this specimen. The (relatively short) second and third toes point to the front, the first and fourth to the back. The latter two partly cover one another, but the first is the very much shorter of the two. Magnification, 1.35×.

2. **Animal geography.** The birds of Messel exhibit, according to each period in time studied, different zoogeographical connections. When they are compared with faunas of approximately similar age, then there are, exclusively, European and North American similarities (e.g. Messelornithidae, Cariamidae, *Junciarsus*, Coraciiformes), but if they are compared with younger or even Recent faunas, relationships to South America (e.g. Phorusrhacidae, Rheidae) and South Asia (e.g. Podargidae) are added. This means that the Messel birds are always older than their southern relatives. The above-listed southern areas are, palaeo-ornithologically, in part insufficiently researched. It cannot, therefore, be stated with certainty whether groups from Messel, which occurred later and in Recent times only in the tropics and subtropics, were at that time already distributed beyond the northern areas; we cannot be sure if the present-day limits of distribution are the result of a regional extinction or if later migrations led to the displacement (Peters 1991).

3. **Concerning the ecology of Lake Messel and its environs**

a. With the exception of *Juncitarsus merkeli*, birds that could be regarded as typical water birds are absent. And even this species is presently known from only one specimen; it was, therefore, not frequent on the lake which it perhaps only visited sporadically, without breeding there. This statement agrees well with other findings which support the view that the lake was not a very welcoming body of water. Nevertheless, the Messel rails might have preferred to stay close to the shore which might explain why they are so frequently found.

b. Like the other vertebrates, the birds are also found frequently as complete skeletons or even with parts of their plumage intact. This is also true for species which have undoubtedly not lived exclusively at the water's edge (e.g. swifts, nightjars). This leads us to suspect that they were transported over short distances

swifts, investigations of the Messel birds have contributed decisively to the clarification of phylogenetic and systematic questions and, as shown in the above discussion, partly forced a revision of opinions held up to now. Similar results must also be expected in future. Particularly promising here is the research into the Coraciiformes and Piciformes, which contains most of the arboreal species of Messel. The order of perching birds (Passeriformes) is, by contrast, entirely absent. Although some small Messel birds have occasionally been called 'sparrow-like birds', I could not substantiate this classification in any of the cases that were checked. This finding agrees with the discoveries that have been made at other sites in the Northern Hemisphere. In the Eocene no passerines occurred anywhere here. It is proposed that this later very successful group evolved somewhere in the Southern Hemisphere and did not reach the more northern areas in sizeable numbers before the Miocene. However, the todi-like bird mentioned above may yield some surprises in this connection.

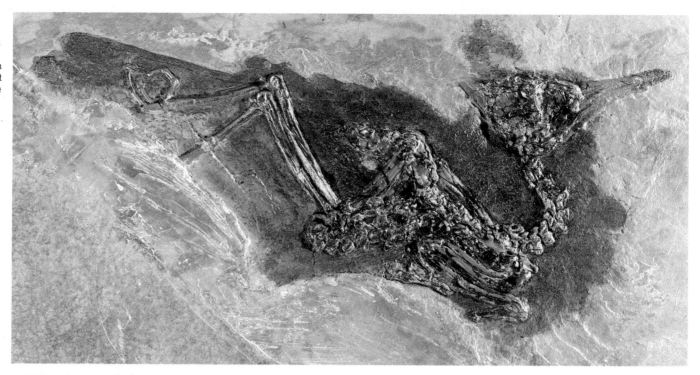

Fig. 217: A further piciform bird, whose stomach (roughly between the elbow and the knee in the picture) was filled with many seeds (see Fig. 82). It should be stressed that the Messel Piciformes are not woodpeckers in the true sense. Reduction, 0.86×.

Fig. 218: Finds like this are also recovered from the Messel pit. This is clearly a small bird with a feathered crest, but it will probably never be identified because no details can be discerned from the mushy fused skeleton. A touching but unidentifiable testimonial of life in the distant past. Reduction, 0.61×.

and that the burial of their corpses by sediment was undisturbed; like the preceding argument, this indicates at least temporary hostile conditions on and in the lake.

c. There are many species whose proportions and anatomical characters indicate a life in bushes or in trees. Here the ornithological results agree with other indications that argue for a forested environment around the lake.

d. Several species were primarily or exclusively ground-dwelling cursors; two or three of them had even lost their capacity for flight. This, however, must not be taken as evidence of a life spent in steppe-like ecosystems; the body structure allows us to postulate a cursorial habit, but not where the birds lived, whether on the forest floor or in the desert. Among extant birds too, there are many that are predominantly 'pedestrians' and, nevertheless, live in the forest (e.g. fowl-like birds, tinamous, trumpeter birds, etc.).

151

12

The marsupials: inconspicuous opossums

Marsupials and 'higher' or placental mammals are sister groups. They separated from common ancestors in the Lower Cretaceous more than 100 million years ago. Extant marsupials give birth to immature young after a very short gestation period. This premature birth takes place before the immune system of the mother becomes fully active with respect to the embryo, because—in contrast to the placental mammals—the body of the mother has no safety barrier.

The marsupials: inconspicuous opossums

WIGHART VON KOENIGSWALD AND GERHARD STORCH

The skeleton and dentition of marsupials show many characteristics that distinguish them from placental mammals. Owing to the name, one is tempted to regard the marsupial bones, with a bony spine on each pubis, as of particular importance (Figs 223 and 224). But these bones actually do not have anything to do with the marsupium and the means of procreation. They are present in both sexes as supporting bones to the musculature of the belly. They are formed in many primitive mammals but are retained among the modern mammalian fauna only by the marsupials. For this reason the marsupial bones, present in all marsupial skeletons at Messel, can be used only conditionally as features for allocation to taxonomic groups.

The tooth formula, however, is more characteristic. While no Eocene placental mammal has more than three incisors, four premolars, and three molars, the opossums (Didelphidae)—to which the marsupials of Messel belong—possess five incisors in the upper jaw and four in the lower (Fig. 222). It was directly from the number of incisors that the first marsupial skeleton could be identified with certainty as an opossum (Koenigswald 1982); as a juvenile animal it had few other distinguishing characteristics. In contrast to the placental mammals, the opossums have only three premolars, but possess four molars (Fig. 221). This character cannot always be checked, particularly in the young animals that are not infrequently found at Messel.

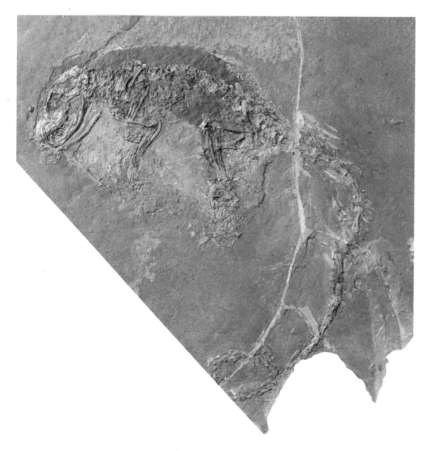

Fig. 220: Small undetermined opossum from Messel. Total length, 82 mm.

Fig. 219: *Peradectes* sp. Small opossum with a long prehensile tail. Similarly adapted marsupial climbers of the family Didelphidae live in South America today. Head and body length, 92 mm; length of tail, 165 mm.

Marsupials live today in Australia and its adjacent islands, extending to Sulawesi and Timor, and also in South and North America. The opossum, the only marsupial of North America, had migrated approximately 3 million years ago from South America and spread at first through the eastern half of the USA and south-eastern Canada. Their introduction into the western USA led to a rapid dispersal extending to British Columbia—some opossums are very well able to overcome the competition of the 'higher' mammals!

If we go back in time, all continents were been inhabited at some time by marsupials. In

Europe they lasted for the considerable time-span of at least 45 million years before they died out, approximately 14 million years ago, in the middle Miocene. Africa and Asia were assumed, until a short while ago, not to have been settled at all by marsupials. Only a few years ago they were discovered in early Tertiary sediments in North Africa (Crochet 1986) and in Kazakhstan (Gabunia *et al.* 1985). The first record for the Antarctic too is of a recent date: in 1984 it was possible to describe scientifically *c.* 40 million-year-old finds from the Antarctic Peninsula (Woodburne and Zinsmeister 1984). This

Fig. 221: The upper four cheek teeth of *Amphiperatherium* sp. The three main cusps and

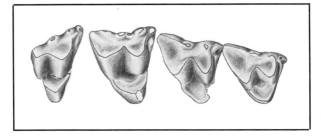

broad outer shelf have small additional and characteristic cusps.

marsupial is the very first land mammal of the Antarctic! At present the marsupials are most diverse in Australia and are represented here with almost all types of adaptation known for placental mammals. Their documentation as fossils, however, begun just recently through the efforts of M. Archer and collaborators, is scarce—with early Tertiary finds at the Tingamarra local fauna. From the middle Miocene onwards, there is more information about the history of Australian marsupials available.

The quest for the cradle of the marsupials and their migration routes across the globe has occupied scientists since the founder of scientific vertebrate palaeontology, Georges Cuvier (1769–1832), described the first fossil marsupial in 1804. It originated from the gypsums of Montmartre in Paris and therefore from a continent where marsupials are no longer extant. The 'when' and 'where' of marsupial occurrences is, however, not only of interest for palaeontology. As a geologically ancient group with a once global distribution, they are very well situated to help reconstruct the former distribution of land and water on earth. Thus, if at a given time, the same or at least similar forms inhabited two continents, which today are separated by deep oceans, then we must conclude that there was an

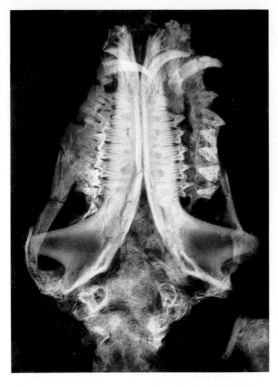

Fig. 222: X-ray photograph of the skull of *Amphiperatherium* sp.

Note the inwardly bent angular processes of the lower jaw.

intercontinental connection at one time. This implies that the present distribution, as well as the fossil history of the marsupials, can be explained reasonably only on the basis of continental drift and the geologically late formation of today's oceans (see Chapter 27, 'The Messel fauna and flora—a biogeographical puzzle').

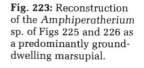

Fig. 223: Reconstruction of the *Amphiperatherium* sp. of Figs 225 and 226 as a predominantly ground-dwelling marsupial.

In fossil sites of the Cretaceous/Palaeocene of Peru and Bolivia a great diversity of marsupials is already evident. This is an indication that their ancestry stretches much further back in time here and also that they underwent their initial radiation in South America. But, because South America and Africa had only finally separated in the late Cretaceous, approximately 90 million years ago, Africa must also have belonged to this area of early radiations (Storch 1986). This may also be true for parts of the Antarctic. Although the Antarctic had already begun previously, together with Australia, to split from a southern land-mass, the Antarctic Peninsula was for a long time still connected to the southern tip of South America by chains of islands. Australia probably obtained its basic stock of marsupials only after the South American/African 'initial spark' and then, like a kind of Noah's Ark, carried the marsupials on its long northward drift to the neighbourhood of southern Asia. From South America the marsupials were also able to spread during the late Cretaceous to North America.

For our species at Messel this means, therefore, that their ancestors were able to migrate on the direct route from North America, as also via the Tethys Sea, the forerunner of our Mediterranean, from Africa (Crochet and Sigé 1983). The last continent, Asia, was finally reached by the marsupials either from Europe and/or from North America via the Bering Strait.

If one speaks of an almost global distribution of the marsupials, then this concerns only one type among them, namely, the little-specialized opossums (Didelphidae). Highly specialized forms, as are known in South America and Australia, were not able to evolve in areas inhabited by placental mammals, or they disappeared when placental mammals encroached in larger numbers, as in South America.

The opossums (Didelphidae) as a rule are small in size. The widely known North American opossum (*Didelphis virginiana*), is one of the larger forms and attains, approximately the size of a domestic cat.

However well the opossums can be separated by their various characters from the placentals, it is difficult to distinguish between their main fossil forms and evolutionary lineages. The dentition is rather uniform in all species and often only minute differences and measurements can be used for the separation of one form from another. The finds documented from Messel open new avenues of investigation, because the skeletons belong to two different adaptive types.

One is a large form (Figs 223, 225, and 226) with a head and body length of approximately 15.5 cm and a tail of only just 17 cm, which is relatively short for opossums. According to measurements of the teeth, the species is a member of the largest lineage within the genus *Amphiperatherium* documented by Crochet (1980). The arrangement of small additional cusps on the outer shelf of the upper molars fits well into this picture (Fig. 221).

The small form is clearly distinct from the larger form (Figs. 219 and 224). The head and body length was here measured as 8.5 cm. The tail, by contrast, is, at 16.5 cm, very long. The measured skeleton is not fully adult but, as also confirmed by the teeth measurements, is much smaller than the large species. Based on the teeth, an assignment to the opossum genus *Peradectes* is probable. The species determination cannot yet be given. The difference in the relationship of the body to the

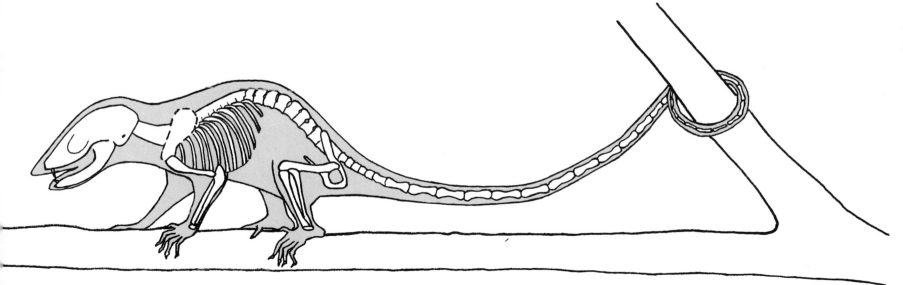

Fig. 224: Reconstruction of the *Peradectes* sp. of Fig. 219. Note the long prehensile tail and the marsupial bones.

Fig. 225: (left); **Fig. 226:** (right). *Amphiperatherium* sp., a large, short-tailed opossum from Messel. When shale splits apart in the field, fossils may tear open and be subsequently prepared in two halves (Fig. 225 (left) and the opposite plate Fig. 226). Total length, 32 cm.

length of the tail, indicates the occupation of a completely different ecological niche.

A very young opossum, described by Koenigswald (1982), also probably belongs to this small species. Opossums are mostly crepuscular, often climbing, omnivores. To judge by the short tail of *Amphiperatherium* (Fig. 223), this genus should have lived mainly on the ground. For *Peradectes* (Fig. 224) we must, however, assume a climbing habit among thin branches. The long tail may have served as an additional prehensile organ for climbing, just as it is used by many opossums today.

The two ecologically so clearly differentiated genera *Peradectes* and *Amphiperatherium* may possibly have reached Europe along different routes. Crochet and Sigé (1983) assume that *Peradectes* migrated via North America in the Palaeocene, while they suppose that *Amphiperatherium* arrived in Europe via a different route from South America via Africa.

13

Primitive insectivores, extraordinary hedgehogs, and long-fingers

The insectivores have been assigned an important role as ancestral group of the higher (placental) mammals. The delimitation of the insectivore group, however, and its systematic subdivision are problematic. According to earlier, broadly based concepts, groups which were primitive and had already mostly appeared at an early date in the phylogeny of mammals were placed in the insectivores. But similarities based on shared primitive characters provide no evidence for relatedness and, of late, the insectivore systematic unit has been dismantled step by step. The insectivores, in the more restricted sense, which remained after that reorganization (order Lipotyphla) are represented among our extant animals by hedgehogs, moles, and shrews. The hedgehog group (suborder Erinaceomorpha) is the most primitive and is known from the uppermost Cretaceous of North America.

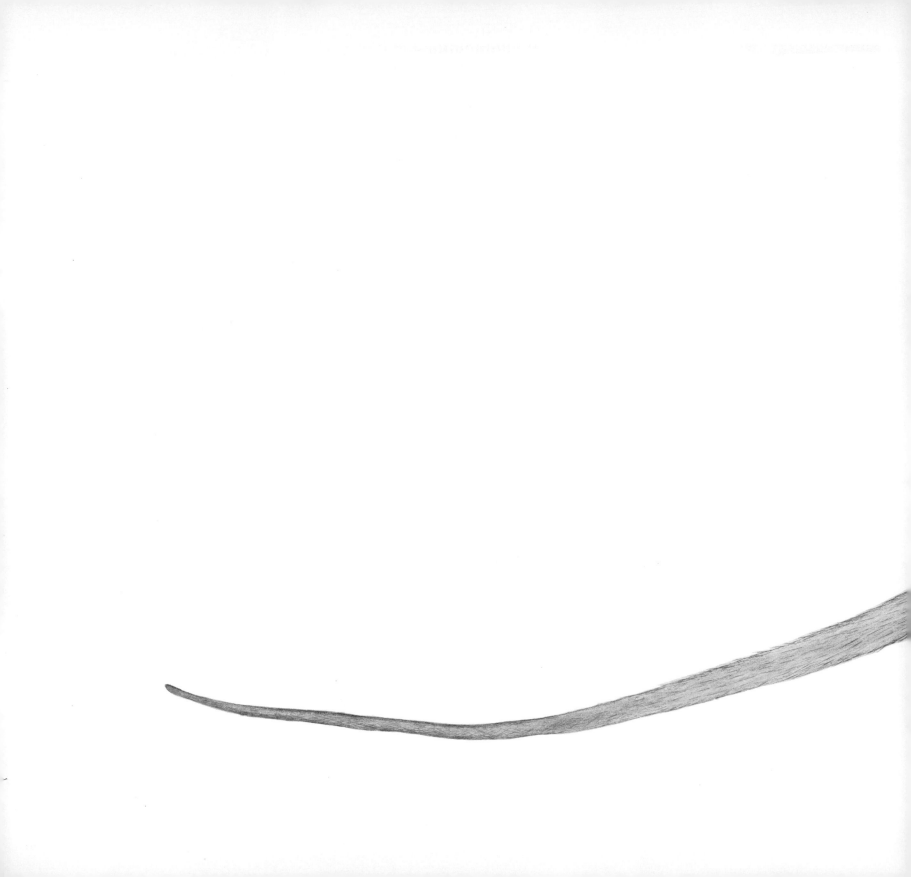

Primitive insectivores, extraordinary hedgehogs, and long-fingers

WIGHART VON KOENIGSWALD,
GERHARD STORCH, AND
GOTTHARD RICHTER

Three Messel species related to the hedgehog have been scientifically described: *Macrocranion tupaiodon* Weitzel 1949, *M. tenerum* (Tobien 1962), and *Pholidocercus hassiacus* Koenigswald and Storch 1983. In addition, two genera of primitive mammals occur at Messel that were originally assigned to the insectivores, namely, *Leptictidium* and *Buxolestes*. These, however, as discussed by Storch and Lister (1985), must be regarded as members of the order Proteutheria—an order that includes primitive families (all extinct) from the late Cretaceous and the early Tertiary of North America, Eurasia, and Africa. These Proteutheria should hardly be regarded as an important ancestral group for other eutherian mammals; rather, they form a side-branch which split off very early.

Heterohyus represents the family Apatemyidae at Messel. Apatemyids are considered to belong either to the Proteutheria or are recognized, due to highly specialized features, as a separate order: the Apatotheria. Relationships with the lipotyphlan insectivores are also under discussion.

The swift hunter *Leptictidium*

This genus is among the most striking and remarkable of the Messel mammals. On the one hand, as far the morphology of the molars or of the pelvis is concerned, the animals are very primitive and still resemble late Cretaceous mammals (Storch and Lister 1985). This does not at all exclude, on the other hand, the unusual and even bizarre specializations of the locomotion and feeding system (Maier *et al.* 1986). *Leptictidium* belongs to a family with the complicated name Pseudorhyncocyonidae, which is known to us only from the early Tertiary of Central and Western Europe. At the end of the Eocene it became extinct (Storch and Lister 1985).

To date no fewer than three *Leptictidium* species have been described scientifically from Messel. They differ in body size and in features

Fig. 227: Reconstruction of *Leptictidium nasutum* at rest.

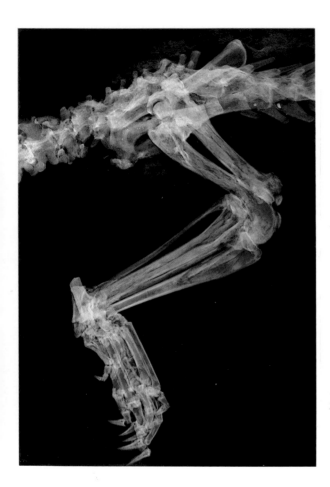

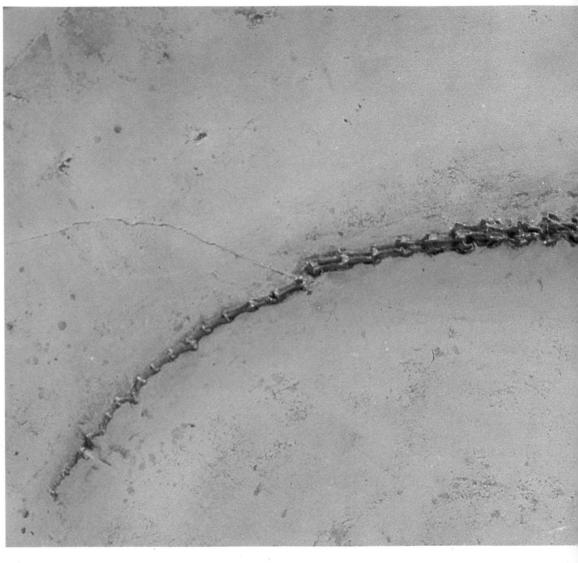

Fig. 228: *Leptictidium nasutum*, X-ray photograph of the skeleton. Note that the tibia and fibula are not firmly fused, and the weak articulation of the pelvis.

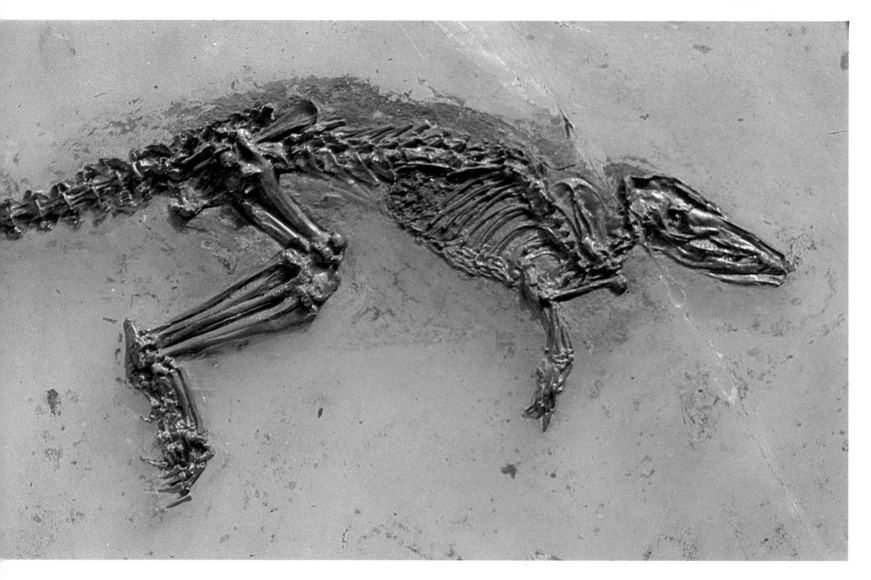

Fig. 230: *Leptictidium nasutum*. Reconstruction of this swift bipedal hunter which did not jump, but ran on two legs.

of their dentition, but agree in the skeletal morphology, including the striking adaptations to a particular mode of locomotion. It appears remarkable to us that three mammalian species of such a homogeneous and highly specialized type are able to share a habitat. The smallest species *L. auderiense* Tobien 1962 has an overall length of scarcely 60 cm, the medium-sized *L. nasutum* Storch and Lister 1985 (Figs 227–230) measures approximately 75 cm, and *L. tobieni* Koenigswald and Storch 1987 (Fig. 231) is almost 90 cm long. For each species only a few specimens are known.

Although our biological/ecological interpretation of morphological characters is based primarily on skeletons of *L. nasutum* (Maier *et al.* 1986), it can certainly be applied to the other two species (Koenigswald and Storch 1987).

The unusual body proportions are immediately apparent. The fore-legs are extremely shortened, and the hind-legs elongated. The tail is extraordinarily long without, however, having evolved prehensile or supporting functions. The vertebral column of the tail consists of over 40 vertebrae, a number which is not nearly approached by extant mammals, even by the long-tailed kangaroos or primates. The thorax is rather weakly developed. The centre of gravity of the body was therefore balanced above the hind-legs and, even at low speed, locomotion should have proceeded on two legs. One immediately thinks of kangaroos or jerboas and, therefore, of jumping on two legs. Peculiarities in the structure of the skeleton, however, contradict this assumption (Fig. 228). The weak fixation of the ankle-joint—tibia and fibula are not firmly fused to one another—and of the sacroiliac joint—only one vertebra was connected to the slender ilia of the pelvis—suggest manoeuvrable bipedal running. Body and tail were extended horizontally over the hind-legs (Fig. 230). The reduced inner and

grabbing, the next area for holding and piercing, and the rear area for a stabbing/cutting and mincing action. The dentition suggests predominantly animal prey and—in connection with the skeletal construction—a fast, agile predator on the forest floor. However, quite certainly, only relatively small prey animals could be killed and eaten, because

Of the three species of *Lepticidium*, which probably had similar feeding habits, three gut contents from *L. nasutum* could be investigated. Each one is distinctly different from the others, but all three substantiate and complete the picture, which is derived from the anatomical/morphological investigations of the skeleton

Fig. 231: *Leptictidium tobieni*. Skull with long snout and muscle depression in front of the orbit. Length of skull, 10 cm.

outer toes show a specialization for running on the ground. Extant mammals no longer exhibit a corresponding mode of locomotion (Maier *et al.* 1986).

The skull is characterized by a very long, slender snout, and muscle depressions in front of the orbits suggest a short trunk. Long gaps between the teeth partition the dentition into areas of different functions: the frontal area for

the dentition is weak in relation to the size of the animal, the musculature for chewing is weakly developed, and the poorly developed fore-legs are little suited to holding down larger prey. The prey was, presumably, tracked by hearing as well as by smell and tactile senses. The rather large, well-delimited orbits suggest that vision was an important factor in pursuit during the hunt (Maier *et al.* 1986).

and certainly can be related to the entire genus.

In the very rich contents of one gut there are bones and bone fragments of a reptile: bones of the lower leg, ribs, several toe phalanges, and fragments of numerous vertebrae (Figs 232 and 233). The prey animal may have been a small, lizard-like reptile which was, according to cautious appraisal and size comparisons of

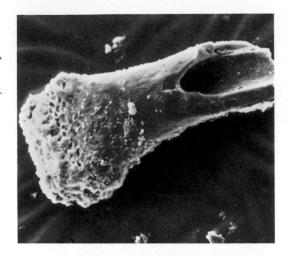

Fig. 232: Stomach contents of *Leptictidium nasutum*. Bone fragment of a reptile, magnification, 48×; fragment of a finger bone (phalanx) with a clean (sharp-edged) break. Scanning electron micrograph (SEM): magnification, 200×.

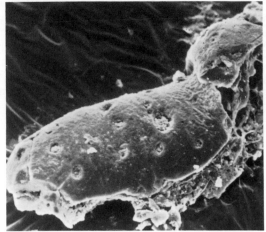

Fig. 234: Stomach contents of *Lepticidium nasutum* (different animal). Two segments (split longitudinally) of the antenna of a large insect. Each segment of the antenna is convex at its distal end and encloses the proximal end of the segment in front. The cuticle is covered with bristle sockets. SEM: magnification, 130×.

Fig. 233: Stomach contents as in Fig. 232. Fragment of vertebra with a strongly spherical joint typical of reptiles. SEM: magnification, 100×.

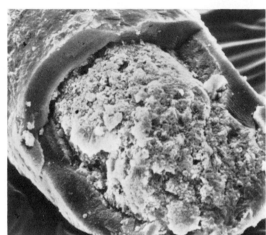

Fig. 235: Stomach contents of *Leptictidium nasutum* (animal of Fig. 234). Insect antenna, single segment, broken transversely. This is one of a few cases in which a hollow chitinous form has been preserved without deformation. SEM: magnification, 300×.

individual bones, perhaps the size of a Recent European mountain lizard.

A second specimen had only meagre gut contents, of which the only identifiable part is a fragment of a lower leg bone. It is the bone of a mammal, approximately the size of an extant house mouse. The third *Leptictidium* had eaten an insect. The remains of the food consist of noticeably thick fragments of cuticle with hair pores and a very uniform patterning, as well as of segments of an antenna, the latter partly preserved as complete, not compressed, hollow forms (Figs 234 and 235). From the size of antenna segments alone one can hardly estimate the size of the prey animal. Together with the thick fragments of the cuticle (thicker than 120 μm), however, they indicate a large insect. In form, size, and surface sculpture the antenna segments resemble those of large Orthoptera, such as mantids or grasshoppers.

Small reptiles and mammals, as well as large insects, like the prey species hunted by *Leptictidium*, usually have fast reactions and are agile. *L. nasutum* with its weak frontal extremities was neither able to dig out earth mounds, nor to topple heavier stones; it either had to discover and grab its prey before it reached a safe haven or it had to lie in wait for it. In both cases, however, a predator must initiate its decisive attack either with a jump or (as is more likely with *Leptictidium*) with a sudden quick onrush. Accordingly,

Leptictidium lived on a high-quality animal diet and hunted relatively small and presumably very agile prey animals.

The fish predator *Buxolestes piscator*

Buxolestes, which belongs to the family Pantolestidae is, in addition to *Leptictidium*, a further representative of the Proteutheria, the group of 'primitive insectivores'.

In overall body size, *Buxolestes*, with a head and body length of 46 cm and a tail length of 35 cm, corresponds approximately with *Leptictidium nasutum*, but has an entirely different appearance. The body of *Buxolestes* (Fig. 236) is much more compact, there are no significant differences in length between the strong fore- and hind-legs, and the rather strong tail is short. Before discussing the ecological role of *Buxolestes*, we must look at the family Pantolestidae, whose finds are by no means restricted to the Messel oil shale.

The pantolestids first appeared in North America in the middle Palaeocene and in Europe from the late Palaeocene onwards. Accordingly the pantolestids were at home on both sides of the Atlantic before the great faunal interchange between both continents at the start of the Eocene, and they, therefore, certainly hail back to a very early origin.

The pantolestids are documented in Europe with several evolutionary lineages until the middle Oligocene, but never played a significant role within the mammal fauna. The skeletons from Messel, which derive from the Lower Mid-Eocene, are the most complete finds of the pantolestids anywhere. Only a badly crushed skull with a few tooth remains is documented from the Upper Middle Coal of the Geiseltal (Koenigswald 1983). Better-preserved teeth were recovered in the limestone of the Upper Mid-Eocene of Bouxwiller in Alsace and the genus *Buxolestes* is based on these (Jaeger 1970). While only teeth or skulls are known from many fossil animals, some long bones were already described very early on from pantolestids of North America (Matthew 1909). Matthew concluded then, due to the similarities of the bones with *Lutra*, the otter, that the pantolestids must have had a semi-aquatic life-style. When it then became possible to allocate an almost complete skeleton from Messel to the pantolestids (Koenigswald 1980), it was of great interest whether the ecological assignment, based only on a few bones, could be confirmed.

The approximately 9-cm-long skull (Fig. 239) is strongly built and shows the attachments of a strong jaw musculature. The canines are moderately large and the multicusped teeth can cut. This is certainly not a highly specialized meat-eater, but a predatory life-style can certainly be assumed. The fore-arms are very strong and allow vigorous

rotation, because the ulna and radius are not fused with one another. The fingers terminate in large bony claws, the horny sheaths of which may have been very much larger still. They are not claws as possessed by specialized tree-climbers, e.g. *Kopidodon* (Fig. 343), but, taking into consideration the large muscle attachments, one could envisage their use for the building of underground dens. Signs of a

Fig. 236: *Buxolestes piscator*, skeleton of an otter-like fish predator. Note the strong vertebral column of the tail, used as an organ of propulsion. Total length, 80 cm.

mostly swimming life style cannot be observed on the frontal extremities. That does not, however, in any way exclude the possibility of a semi-aquatic life-style, because in most swimming mammals the main propulsion comes from the rear part of the body.

The hind-legs are also strongly built, but cannot be rotated so far because the fibia and tibia are fused firmly to one another. The hind-feet are not obviously specialized and bear

large claws, similar to those on the fore-feet. As a particular adaptation to swimming, the toes of the South American giant otter *Pteronura brasiliensis* are considerably elongated, so that a large paddle has been formed. In *Buxolestes* one searches in vain for such adaptations. The extremities of the fore- and hind-limbs do not differ from those of a cursorial land animal walking on all fours. But one must keep in

mind the fact that this is also true for the otter (*Lutra*) and that swimming is one of the general abilities of mammals. Many mammals which are specially adapted for swimming use (apart from the hind-legs with which simple running locomotion is performed underwater) the tail for forward propulsion. In an extreme case the latter can, as in the beaver (*Castor*), be reshaped into a flat paddle, but other rodents, for example, the coypu (*Myocastor*), swim

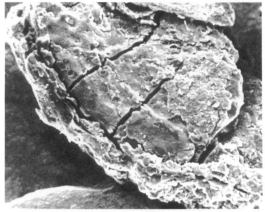

Fig. 237: (above). *Buxolestes piscator*, gut contents; includes a flat oval seed in its seed coat. SEM: magnification, 50×.

Fig. 238: (below). *Buxolestes piscator*, gut contents with remnants of fishes. Width of picture, 4.5 mm.

Fig. 239: *Buxolestes piscator*, crushed skull with large crest above the occipital condyles (treated with NH₄Cl vapour for the photograph).

excellently with a completely round tail. A prerequisite is for the tail musculature to be so arranged, so that force can be transferred to the tail. In this context it is very important that in *Buxolestes* the first vertebrae of the relatively short and strong tail carry particularly widened, transverse processes. These serve as areas of attachment for a strong musculature. It can be proven, therefore, from the bone morphology, that the tail was moved powerfully, which really makes sense only for movement in water. To judge from the bone morphology, it is therefore probable that *Buxolestes* did, indeed, have a semi-aquatic life-style.

In Messel, however, more elegant proof can be obtained from the very frequently preserved gut contents. According to the dentition, *Buxolestes* should be a predator rather than a plant-eater. It was not surprising, therefore, to find a mass of small bone splinters in the gut contents of the first skeleton, among which fish scales and even a fragment of a fish jaw with three teeth (Fig. 238) could be observed. Due to this, the species was named *piscator* (fisher) (Koenigswald 1980).

Even if everything here appears to fit seamlessly and *Buxolestes* must be regarded as a frequent swimmer and predator on fish, which also moved quickly on land and possibly even built underground dens, one must be very cautious with such reconstructions and must not force these animals into too narrow a scheme. A warning in this regard came from

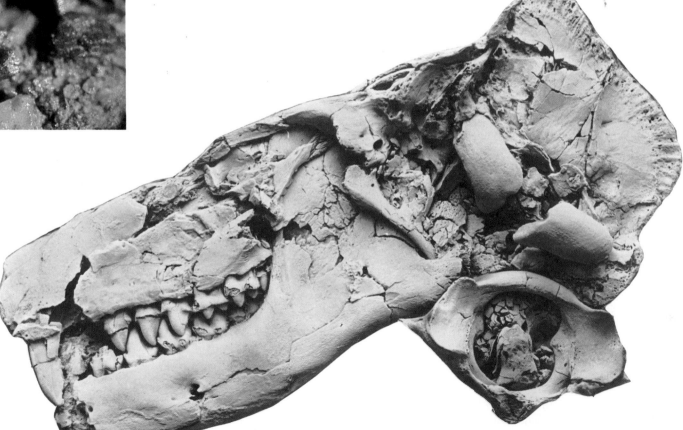

Fig. 240: *Macrocranion tupaiodon*, a jumping relative of the hedgehog with a particularly well-preserved outline of the soft tissue. Total length, *c.* 30 cm.

the gut contents of the second specimen of *Buxolestes*, which consisted exclusively of plant material. Particularly frequent are fruits and seeds of an unknown plant species (see Fig. 237 and Richter 1987).

Macrocranion and *Pholidocercus*, highly specialized relatives of the hedgehog

Both genera *Macrocranion* and *Pholidocercus*, each of which were first discovered in Messel, belong among the relatives of the hedgehog. They differ from one another only in details of the morphology of their molars. In the frontal dentition, which is preserved sufficiently in only a few fossil insectivores, the differences become somewhat more distinct. According to the rather unspecialized dentition one would hardly expect that both genera belong to entirely different adaptive types. This is exactly what the skeletons show.

Macrocranion (Figs 240 and 241) is the genus with the lighter body construction. Two species were described—a larger one, *M. tupaiodon*, and a significantly smaller one, *M.*

tenerum (Tobien 1962; Russell *et al.* 1975). If one disregards the bats, *M. tupaiodon* is one of the most frequently found mammals in the Messel oil shales. The skull of this insectivore is approximately 47 mm long, the head and body length is 16 cm, and the tail adds a further 12 cm. The numerous skeletons permit certain conclusions regarding the life-style of this animal (Maier 1977, 1979). The slender and pointed head shows muscle attachments in front of the orbits, from which one may conclude that the tip of the snout was highly mobile. The orbits are astonishingly small, so that the sense of sight probably played no significant role for the animal. In addition to using its sense of smell, the animal probably assessed its orientation acoustically. The soft-body outlines, preserved in some skeletons, show relatively large ears and in one case even long tactile hairs on the muzzle. This could be evidence of nocturnal activity or an adaptation to the darkness on the jungle floor. From its pointed teeth, the animal could have coped with quite a variety of food. A particular specialization cannot be deduced here and thus omnivory is assumed. The remainder of the skeleton gives information on the mode of locomotion. The fore-legs are slender, and

noticeably short in relation to the hind-legs. The last phalanges of the fore-limbs are short and do not look as if they once bore sharp or large horny claws as would be demanded for a digging or climbing animal. It is much more likely that these last phalanges could have borne nails or even hoof-like structures which would indicate fast running. The hind-legs are long and powerful and show *Macrocranion* to have been a runner on the ground (Figs 240 and 241). Large muscle attachments and the elongation of the metatarsals lead us to conclude that they were fast runners, and the last phalanges of the toes are, again, formed like small hooves. Possibly *Macrocranion* could also have jumped using the two hind-legs. The approximately 12-cm-long tail can have served, to a certain degree, as an aid to balancing during running. Since none of the skeletons, which often show outlines of the fur, have hairs visible on the tail, it was probably naked. Therefore *Macrocranion* was probably an agile forest floor-dweller in the gloom of the jungle. The preservation of relatively large numbers of skeletons in the oil shale might be explained by the animals being drowned during flooding of the forest floor and their bodies then being washed into the lake.

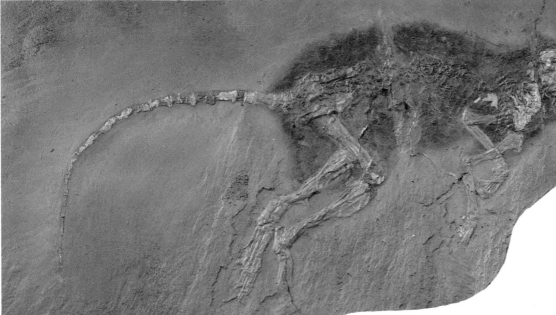

Fig. 241: *Macrocranion tupaiodon*. Note the elongated hind legs and the stiff, balancing tail. Total length, *c.*24 cm.

structured hairs or narrow scales (Figs 243 and 244); these are very similar to the hairs of the caddis-fly found in bats of the genus *Palaeochiropteryx*.

For an animal of the size of *Macrocranion*, caddis-flies (Trichoptera)—if the hairs or scales did indeed originate from these insects—are, in spite of their occasional occurrences in large numbers, only a very small individual prey item which surely was taken only *en passant*. It would also be wrong to assess *M. tupaiodon* as a pure carnivore (fish and insects) because plant residues play an important role in the gut contents of all the preserved specimens. The plant residues consist of leaf fragments—mostly with cell preservation—and somewhat thicker tissues with very corky, thick epidermis (bark or skin of fruit?).

According to its diet, pieced together from gut content analysis, *M. tupaiodon* lived in close vicinity to the border of Lake Messel, perhaps even making occasional trips into the lake itself. Two of the seven investigated specimens have remains of fish in their guts, consisting of simple bones or fin-rays with joints (Fig. 242). All other specimens contain varying amounts of insect residues in the gastrointestinal area. Unidentifiable cuticles with varying surface sculpture predominate, along with rare occurrences of very finely

Fig. 242: Stomach contents of *Macrocranion tupaiodon*. Fish bones and fin ray with basal joint (species unknown). SEM: magnification, 75×.

Fig. 243: Stomach contents of *Macrocranion tupaiodon*. Collection of scales and hairs, possibly from Trichoptera. SEM: magnification, 1400×.

Fig. 244: Stomach contents of *Macrocranion tupaiodon*. Single scales (or scale hair) with characteristic fine structure. SEM: magnification 7000×.

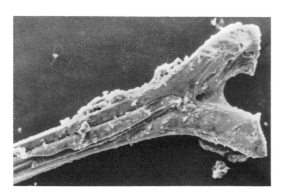

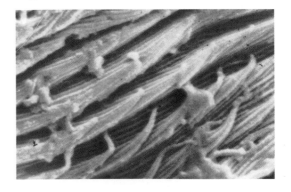

Macrocranion tupaiodon was an omnivore. In relative 'weighting' of the food residues, one may well assume that—as presumably applies to all Messel omnivores—the emphasis was on animal food. Plant residues are, as a rule, better and more completely preserved and are correspondingly recognized more easily than animal tissues. It is possible that *M. tupaiodon* occasionally caught small fish or water insects when swimming or diving (comparable perhaps with the European Recent water-shrew), but it is more likely that the animals searched for food along the lake shore and occasionally in the course of this took a half-dead or dead fish.

After the dentition of *M. tupaiodon* had become completely known from Messel, the genus could also be recognized from the Geiseltal (Storch and Haubold 1985) and in North America (Krishtalka and Setoguchi 1977).

Only a single skeleton of the smaller species

M. tenerum is known (Tobien 1962; Russell *et al.* 1975). Because it was found before 1960, it was not possible to transfer it to artificial resin and it has been preserved very badly in the oil shale. With a skull of *c.* 27 mm in length and a head and body length of scarcely 10 cm, it attained only little more than half the size of *M. tupaiodon*. The tail has not been preserved. Although the specimen is a young animal whose dental replacement is not yet complete, the last molars have already formed, and it should, therefore, be not far short of its full body size.

The life-style that has been traced for *M. tupaiodon* cannot be applied easily to *M. tenerum*, since the small, partial skeleton gives at least the impression that the front legs of *M. tenerum* were not so considerably shortened as in the larger species. Further finds are necessary to also gain insight into the biology of this species.

Pholidocercus hassiacus, the 'scaly-tail'

(Koenigswald and Storch 1983; Figs 245 and 246) is, in comparison with *Macrocranion*, much more robust. The head is large and has, as a peculiarity, delicate pits and runnels on the bones of the nose and forehead (nasalia and frontalia) (Fig. 248). They originate from veins impressed into the bones of the skull. One can find such structures when a horny plate covers the bone, but also when a thick 'bolster' is formed. Some animals that burrow using nose and forehead have such a structure.

Above the vertebrae of the back several specimens show a distinct outline of fur, which exhibits relatively long bristles. The tail, finally, is surrounded by a tube of small bony scales which lie above one another like tiles (Fig. 247). Such bony scales in the skin of the tail are otherwise developed by only one rodent, the spiny mouse *Acomys*. The legs of *Pholidocercus* are, in relation to the body length, shorter than in *Macrocranion*. The hind legs are, to be sure, somewhat longer than the

Fig. 245: *Pholidocercus hassiacus*, a well-defended relative of the hedgehog. Note the bristly hair of the back and the tube of bony scales around the tail. Head and body length, 19 cm.

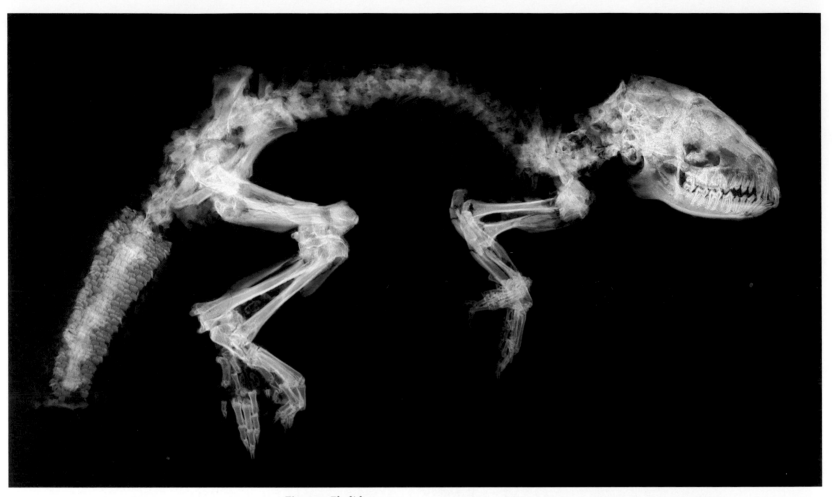

Fig. 246: *Pholidocercus hassiacus*, X-ray of the skeleton. Note the cleft terminal phalanges as well as the X-ray shadow of the bony scales. Head and body length, 19.4 cm.

fore-legs, but not to such a degree as in *Macrocranion*. The animal must have moved exclusively on all fours. The difference to *Macrocranion* is particularly marked in the terminal phalanges of fingers and toes; they are developed as large, deeply cleft bony claws. They were sheathed in still larger horny claws.

If one tries to combine these different peculiarities into an ecological picture, one obtains, first of all, a ground-dweller running on all fours, which, however, does not exclude some facility for climbing and swimming. The feet, armed with claws, suggest a scratching digger which searched through the carpet of leaves for varied food. Thus far, *Pholidocercus*

Fig. 247: Detail of the scaly tube of the tail. The scales, which normally overlap like tiles, here lie one behind another. Detail from Fig. 245.

172

does not offer anything of exceptional interest. Nevertheless, the protective adaptations are unusual (Fig. 249). Bristly fur on the back, a horny plate or leathery callus on the forehead, and especially a tail protected by bony scales turn *Pholidocercus* into an animal that was probably attacked by few enemies. In their lifestyle, the extant hairy hedgehogs (Echinosoricinae) might, perhaps, be compared with *Pholidocercus*. It is known that they keep their enemies at bay by means of a foul smell. In spite of the excellent preservation of the Messel fossils, such a strategy cannot be proven for them, but fits in well with the very defensive character revealed in the skeleton of *Pholidocercus*.

Before the complete skeletons of *Pholidocercus* were discovered at Messel (Koenigswald and Storch 1983), the amphilemurids, which include *Pholidocercus*, were only known from parts of their dentition; these, due to their moderate specialization, were transferred back and forth several times between primates and insectivores. The temporary assumption that these were primates, can be judged from the names given: *Alsaticopithecus* (ape from the Alsace; Hürzeler 1947) for an almost complete dentition from Bouxwiller, and *Amphilemur* (almost a lemur) for a lower jaw from the Geiseltal (Heller 1935).

The entire body morphology and the appearance of *Pholidocercus* clearly place all the genera, which are included in the amphilemurids, with the true insectivores.

This relative of the hedgehogs probably lived on scavenged morsels of food, in a similar fashion to the extant hedgehog, that is, it was an omnivore (Richter 1987). In the gut of one of the three specimens investigated, plant remains predominate: leaf epithelia preserved as cells, or lumps of cells with large lumens and very thin cell walls (soft fruit pulp?). Only the imprint of a scaly insect cuticle on the sediment is evidence that this specimen, too, had ingested animal food. By contrast, among the food remains from two further specimens, animal residues in the form of insect chitin cuticles predominate. Which insect groups

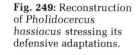

Fig. 249: Reconstruction of *Pholidocercus hassiacus* stressing its defensive adaptations.

Fig. 250: Stomach contents of *Pholidocercus hassiacus*. Thick chitin fragment with strong longitudinal (?) bulges. Possibly part of a beetle elytron (wing-case). SEM: magnification 325×.

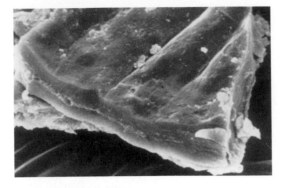

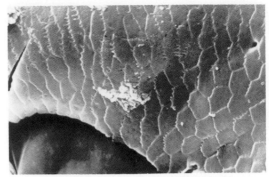

Fig. 253: Stomach contents of *Pholidocercus hassiacus*. Fragment of an insect cuticle with variable scale sculpturing. SEM: magnification, 800×.

Fig. 251: Stomach contents of *Pholidocercus hassiacus*. Insect cuticle with scaly sculpturing. Free scale margins are serrated.

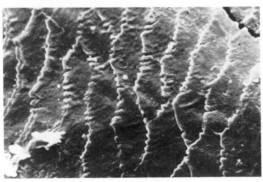

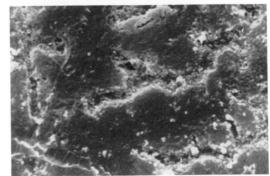

Fig. 254: Stomach contents of *Pholidocercus hassiacus*. Chitin fragment, severely eroded by bacterial action. Fossilized bacteria are scattered individually on the surface and collected in corroded indentations. SEM: magnification, 700×.

Fig. 252: Stomach contents, as in Fig. 251. Insect cuticle with scaly sculpture. Free scale margins are smooth. SEM: magnification, 1600×.

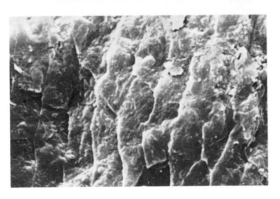

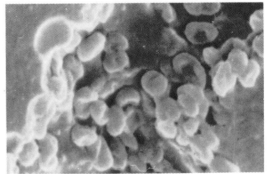

Fig. 255 Greatly enlarged detail of the chitin fragment of Fig. 254. Depression with many (slightly collapsed) fossilized bacteria and traces of dissolved article caused by individual bacteria on the surface of the cuticle (below). SEM: magnification, 8750×.

these belong to cannot even be speculated upon because, as usual, neither mandibles nor extremities can, unfortunately, be recognized. Only one thick fragment of chitin is reminiscent of the wing-case of a beetle, with strong, longitudinal ribbing (Fig. 250). Most cuticle fragments, however, are relatively thin, packed in thick layers, and frequently bear a surface sculpture of fine scales with serrated or smooth margins (Figs 251 and 252). How strongly the scale shape can vary on one and the same part of the body of an insect is shown

in Fig. 253. Any attempt at a taxonomic interpretation of such sculpture is hopeless.

One of these chitin fragments is, upon closer inspection, particularly interesting. As part of a certainly dead insect it was eaten by *Pholidocercus*—presumably together with many living ones—so to speak, by mistake. Its strongly corroded surface shows not only traces of the activity of chitin-decomposing bacteria, but the fossilized bacteria themselves, physically preserved (Figs 254 and 255).

The long-fingered *Heterohyus*

Heterohyus (Fig. 257) is a small animal that is hitherto documented in Messel by only three skeletons. According to its specialized frontal dentition it belongs to the family Apatemyidae. Because this dentition is markedly modified, due to the two large incisors, this family cannot be classified easily with the archaic insectivores and the rank of this group is still uncertain. There are authors who judge the Apatotheria to be an independent order of mammals

174

Fig. 256: Reconstruction of *Heterohyus nanus* searching for food.

edge of the teeth. Behind this first incisor follows a serrated scissor-like formation, which consists of a further upper incisor and the premolars in the upper and lower jaw. The molars are relatively small.

The body is relatively small in relation to the skull. The tail, composed of long, equal-sized vertebrae, is strikingly long and is certainly a balancing aid rather than a prehensile tail.

The hind-legs are strong, and the feet are large. At the end of each toe there is a sideways flattened claw which is suitable for anchoring the foot, when climbing, in the bark of trees and branches.

The arms appear to be more delicate than the legs. The lower arm bones (ulna and radius)

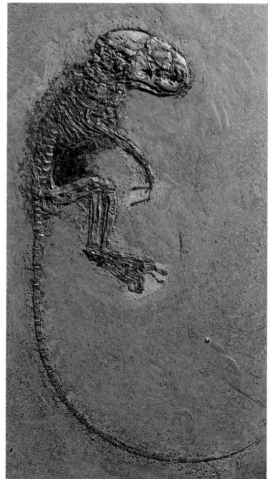

(Russell *et al.* 1979). The significance of the three skeletal finds of Messel lies in the fact that here, for the first time, apart from remains of the teeth and skull, the entire skeleton has been preserved, and permits a glimpse of the possible life-style of these animals (Koenigswald and Schierning 1987; Koenigswald 1990).

First of all, the dentition. In the large skull, strongly curved incisors are formed in the upper and lower jaws (Fig. 253), which can be protruded from the base for an extended period, to compensate for the wear of the front

Fig. 257: *Heterohyus nanus* a small arboreal animal which has specialized adaptations for preying on wood-boring insects. Head and body length, 12.5 cm, length of tail, 18 cm.

175

are not fused and permitted an excellent mobility of the hand. The outstanding peculiarity is the unusual elongation of the fingers (Fig. 258). Not all five fingers, but only the second and third ones, are extremely prolonged and bear, like the others—claws at the end that are laterally flattened.

In the first skeleton, found at Messel in 1973 by a private collector, one could recognize elongated finger bones but could not correlate them to a particular finger. The second skeleton was recovered in 1984 by workers of the Landessammlungen in Karlsruhe; it shows that the second and third fingers are the elongated ones (Koenigswald 1987a, b), but the last digits of these fingers were lost. A third specimen, found at Messel in 1986 by the Hessisches Landesmuseum (Koenigswald 1990), already revealed the elongated fingers on the X-ray picture before preparation. Here even the small terminal claws could be substantiated (Fig. 258).

How can this strange picture now be interpreted in relation to the life-style of this extinct group of animals? We find the enlarged incisors together with the selective elongation of fingers in two genera among modern mammals, namely, in the lemur *Daubentonia* and in the marsupial *Dactylopsila*. Both genera have a very similar way of life: they tear the bark open with their teeth and with their fingers winkle out the insect larvae from their burrows. The combination of these two characters is so marked, that one may assume a similar life style for the 'long-finger' from Messel (Fig. 256). The long fingers can, in addition, be used to anchor the animal: *Dactylopsila* holds fast with the prolonged finger while the prey is moved to and fro between the remaining fingers (Möller 1976).

Due to similarities in the dentition, *Daubentonia* and *Dactylopsila* had already been considered earlier (Stehlin 1916; McKenna 1963; West 1973), but a similarity in life-style can only be proven with the find of the elongated fingers.

In nature it can happen that certain combinations of characters develop independently several times, so-called convergent evolution, which simulates a relationship only at the first, superficial glance. But just as the wings of the winged saurians, of the birds, and of the bats are differently constructed and thus record a different evolutionary history so, in the three specialized feeders on woodworm the dentition and the hands are constructed differently in detail.

Whereas *Heterohyus* has a serrated scissor-like formation between the enlarged incisors and the small molars, *Daubentonia* has a wide gap (diastema) between the teeth and *Dactylopsila* has only peg-teeth. While the second and third fingers are elongated but equally strong in *Heterohyus*, *Daubentonia* has

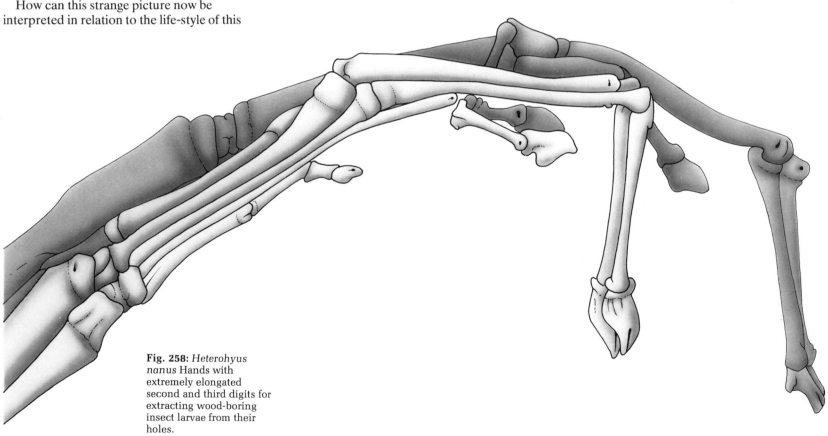

Fig. 258: *Heterohyus nanus* Hands with extremely elongated second and third digits for extracting wood-boring insect larvae from their holes.

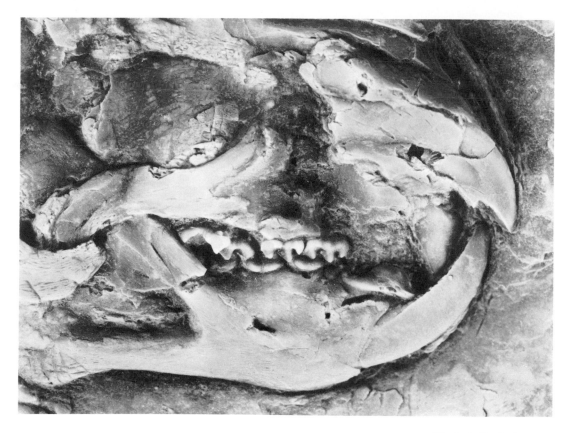

Fig. 259: Skull of *Heterohyus nanus* with enlarged incisors.

a very thin, elongated third finger for winkling out morsels, but which, when at rest, is laid across the even longer, but stronger fourth finger. In *Dactylopsila* only the fourth finger is elongated. This shows that all three forms are independent of one another and are therefore adapted somewhat differently to the special requirements of this hunt for food.

It is usually the niche of the woodpeckers to dig out wood insects from infested tree trunks. Madagascar and New Guinea, the homes of *Daubentonia* and *Dactylopsila*, respectively, have never been settled by woodpeckers, and mammals can, therefore, keep this niche (Cartmill 1974). Is it possible to conclude from this that there were (as yet) no woodpeckers in Messel? This would, perhaps, be somewhat too rash, even though remains of fossil woodpeckers extend only into the late Tertiary.

Recently, however, a woodpecker nest-hole has been described in a silicified piece of wood from the Eocene of Arizona (Buchholz 1986). Thus the life-style of the extant woodpeckers may possibly reach back much further than was previously believed. Accordingly, the least that can be said is that the contest for this food source had not yet been won by the woodpeckers.

14

Bats: already highly specialized insect predators

Bats are among the most successful mammals living. The conquest of the air at night has opened a niche where there is almost no competition. The almost 900 bat species are exceeded in number only by rodents; their feeding habits are more varied than those of other groups of mammals; and only polar areas and some oceanic islands are excluded from their world-wide distribution.

Bats: already highly specialized insect predators

JÖRG HABERSETZER, GOTTHARD RICHTER, AND GERHARD STORCH

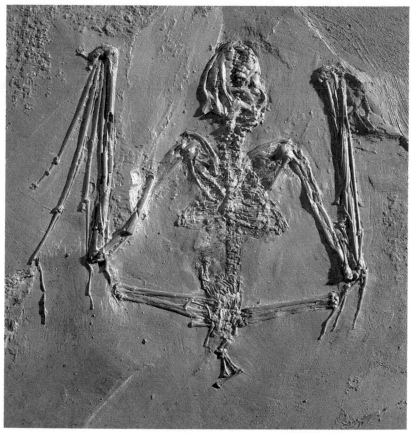

Fig. 260: *Archaeonycteris trigonodon* is a primitive bat species. Note the short, straight forelimb. Length of forearm, 52.5 mm.

The known fossil documentation of bats is in marked contrast to their evolutionary success. In fossil sites bats are usually rare and this is surely a consequence of their life-style. From hidden refuges such as caves, holes, or crevices in trees, the bodies of these animals only rarely end up in fossil deposits. Furthermore, animals that can fly only rarely have accidents in natural traps on the ground and do not become victims of catastrophes (for example, floods). Deposits in karst fissures and caves, in which former bat roosts have produced an increased concentration of remains, are an exception to this paucity of fossils.

This situation is, however, completely different at Messel: here bats are by far the most frequent mammals. So far, there are several hundred finds. Although this fact is still puzzling, one possible explanation may be as follows. Most animals have been preserved with their gastrointestinal contents. This implies that they must have died during their nocturnal hunt or shortly afterwards, since, in extant species, the indigestible parts of the food are voided very quickly. To judge from the wear on their teeth and from marks on their skeletons, the animals were mainly healthy adults. Unusually numerous finds of bats in the prime of life that were active shortly before their demise lead us to infer unnatural death. Predators can largely be excluded from consideration because there are no traces of their activities. We therefore regard it as probable that clouds of poisonous gas, emitted from Lake Messel, led to the fall and the drowning of animals that hunted over the surface of the water.

Such an overwhelming number of completely preserved bat fossils exists only at Messel. They are, furthermore, among the oldest. They are only preceded by early Eocene finds from North America and France that are perhaps up to 4 million years older. But these are either veritable rarities (very few specimens from Wyoming, USA) or only fragmentary pieces (isolated teeth from the Paris Basin).

Even though the Messel bats cannot contribute to the understanding of the evolution of active flight, because their flight apparatus was already perfected, they are still important for the understanding of early phylogeny. As regards phylogeny the opinions of van Valen (1979) have been widely accepted. He groups all geologically old bats, including those of Messel, in a single group having family status. This group is also

supposed to represent, as suborder Eochiroptera, the group from which both extant bat suborders, namely, the Microchiroptera (bats) and the Megachiroptera (flying foxes, fruit bats) originate. In this context the Eochiroptera would be a group of primitive and, moreover, unspecialized species. We do not concur fully with this view. The oldest fossil finds are of 'fully evolved' bats, which allows us to postulate an origin that reaches far back prior to the Eocene and gives sufficient time for adaptive radiations up to the period represented by Messel. We have, furthermore, quite reliable concepts of a tropical/subtropical Messel habitat with a multitude of ecological niches. These could in no way have been partitioned and used by a pool of species that had remained unspecialized. We should rather expect a biologically well-balanced bat

Fig. 261: A newly described species of *Archaeonycteris* is *A. pollex*. The large, robust skull is remarkable compared with *A. trigonodon*. Length of forearm, 61.1 mm.

Fig. 262: In contrast to the other Messel species and to extant microchiroptera, *Archaeonycteris* has a primitive feature, a claw on the index finger. X-ray

detail from specimen in Fig. 260.

fauna which was taxonomically diverse and included already highly specialized species.

Seven species from Messel have been described scientifically: we classify these within three Microchiroptera families (Habersetzer and Storch 1987).

Archaeonycteris trigonodon Revilliod 1917 (family Archaeonycteridae) is a primitive species (Fig. 260). It still has a claw on its index finger which is no longer present in extant Microchiroptera (Fig. 262). The fore arm (radius) is short and straight. The last upper molar remains very primitive with a complete W-shaped outer border (ectoloph). Lower molars, if found isolated, could be taken, at first glance, for marsupial molars (Fig. 267); the last lower premolar is very similar to the

molars. A second species, *A. pollex* (Fig. 261) was described recently.

Palaeochiropteryx tupaiodon Revilliod 1917 and *P. spiegeli* Revilliod 1917 (family, Palaeochiropterygdae) are small, specialized species (Figs 263 and 267); the former is by far the most frequently found bat in Messel. The fore-arm is relatively long and bent in the proximal section. The complete tooth formula for bats is present (incisors (upper jaw) 2/ (lower jaw) 3; canines 1/1; premolars 3/3; molars 3/3).

Hassianycteris messelensis Smith and Storch 1981, *H. magna* Smith and Storch 1981, and *H. revilliodi* (Russell and Sigé 1970), which belong to the family Hassianycterididae, are by far the most advanced and the most specialized species (Figs 264–267). The animals are large. Their fore-arm is very long and bowed, the fifth finger is markedly shortened and the second finger is reduced to one phalanx. The dentition is strong and high. The anterior upper premolar is a tiny peg or completely absent. The posterior lower premolar is simple with a high main cusp.

The bats of the Messel pit are often preserved with a soft-body outline which permits the complete reconstruction of the wing skeleton and the flight membranes. With their aid all parameters that are important in describing the flight characteristics can be determined. The Messel species can be compared most closely with extant tropical bats, for example, with 31 species of the Indian subcontinent (Habersetzer and Storch 1987). Their hunting biotopes are strongly species-specific and this is reflected in the adaptations of the flight apparatus as well as in the echolocation sounds. The adaptation of the wings to particular ecological niches is indicated, among other parameters, by the relationship of flight membrane area to body weight. Species that fly close to the ground or in the foliage have a relatively large wing area. The resulting low wing-loading enables slow flight in an area rich in obstacles. Another functionally important ratio is the so-called 'aspect ratio' as an expression of the wing shape. High aspect ratio values are found in narrow wing forms that are typical for many species that fly fast and high.

Like all other hitherto investigated bats from the early Tertiary, *Archaeonycteris trigonodon* has a fairly small finger flight-membrane (dactylopatagium). In relation to the wing (Fig. 268), however, the long body is remarkable. *A. trigonodon* conforms in its morphology most closely with the ordinary vespertilionid bats, for instance, with the large mouse-eared bat *Myotis myotis* or also with species of large horseshoe bats (rhinolophids). It can therefore

Fig. 263: *Palaeochiropteryx tupaiodon*, with its rather delicate body, is the most common and the smallest Messel bat. Length of forearm, 36.7 mm.

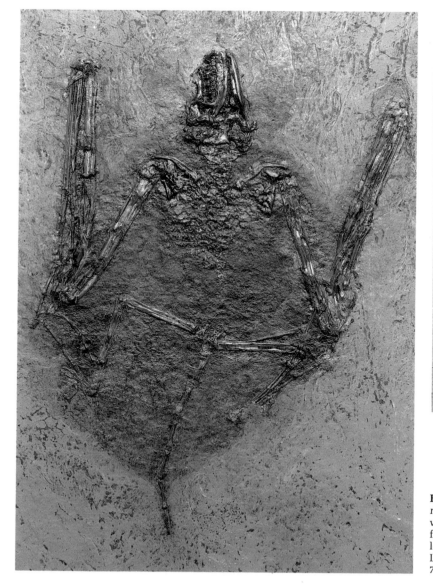

Fig. 264: *Hassianycteris messelensis* is a species with many advanced features. Note the very long and bowed forearms. Length of forearm, 71.6 mm.

be regarded as relatively unspecialized in flight biology.

The wings of *Palaeochiropteryx spiegeli* and *P. tupaiodon* are of very similar proportions. When compared with *Archaeonycteris trigonodon*, both bats share a common characteristic: the rather delicate body. With very broad wings (very low aspect ratio) and

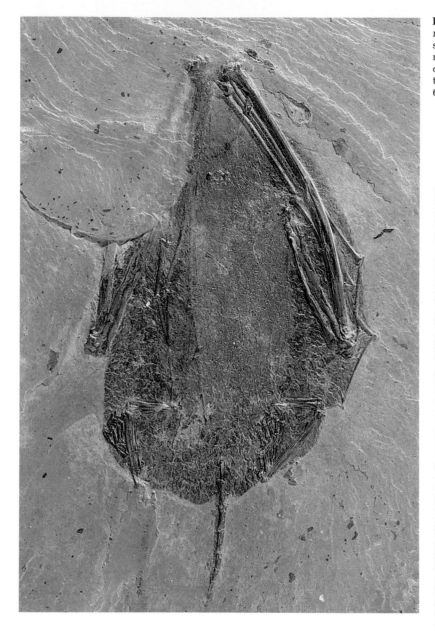

Fig. 265: *Hassianycteris messelensis*. Note the superbly preserved flight membrane, enclosing half of the vertebral column of the tail. Length of forearm, 64.5 mm.

foliage. *P. tupaidon* was even more capable of a sustained hover-flight than *P. spiegeli*. It can be affirmed with certainty that *P. tupaiodon*, among all the Messel bats, offered the best morphological prerequisites for this type of flight.

Hassiancyteris magna, with a wing-span of 47.5 cm, is the largest of the Messel bats. *H. magna* and the smaller *H. messelensis* have a

with very reduced wing-loading, *P. tupaiodon* represents an extreme form of specialization. On the basis of similar wing parameters in extant, extremely specialized, horseshoe bats (Rhinolophidae) and Old World leaf-nosed bats (Hipposideridae), one can assume for them a type of flight that was slow, agile, and occurred close to the ground or in dense

Fig. 266: Skeleton of *Hassianycteris messelensis*, with full articulation preserved. The dark area marks the fossilized gastrointestinal contents. Length of forearm, 64.5 mm.

significantly narrower wing (high aspect ratio) than the other Messel species (Fig. 268) and are very similar to the extant fast-flying bats that hunt high in the air. This characterization is confirmed by a second parameter, the high wing-loading, because a high body weight coupled with a relatively small wing area demands a high flight speed as well. Both *Hassianycteris* species show even higher parameter values than the high-flying tropical smooth-nosed bats (Vespertilionidae) and mouse-tailed bats (Rhinopomatidae). Because of their similarity to extreme specialists, the mastiff bats (Molossidae) and sheath-tailed bats (Emballonuridae), *H. magna* and *H. messelensis* can be assessed as very well adapted to flight in the open air. Interestingly, in certain wing proportions (humerus/radius), they show a closer relationship to the opposingly specialized *Palaeochiropteryx* species than to the primitive *Archaeonycteris trigonodon*.

Finally, two peculiarities regarding flight biology remain to be mentioned: the important parameters, aspect ratio (an expression of wing shape) and wing loading, vary in the Messel bats quantitatively to such an extent that they embrace practically all the variability of the extant species of the Indian subcontinent. The specializations of the Messel species traced above are also confirmed by comparisons with the extant bats of different continents. A second peculiarity concerns the narrow-winged *Hassianycteris* species. In contrast to Recent narrow-winged bats they exhibit a proportion-based type (wing-tip/radius) similar to that of the broad-winged horseshoe bats. This explains why the wing shapes of all Messel species are very similar at first glance. Thus, the five species of bats have managed to gain niches in the most important flight and hunting biotopes of Messel by means of a modification of wing shape similar to that of extant horseshoe bats. In this way they represent a biologically balanced fauna which probably already corresponds with a present-day tropical bat fauna.

Thanks to the good preservation of the bat finds, it was possible to investigate all species except *Hassianycteris revilloldi* for remains of food in the gastrointestinal area. Most species,

however, are represented only by one or a few specimens (Richter and Storch 1980). The only exception here is *Palaeochiropteryx tupaiodon*, the most frequent species among fossil finds. With 16 specimens, of which two are immature, *P. tupaiodon* is the Messel species that has been most completely investigated for its gut contents.

P. tupaiodon was a specialist, feeding almost exclusively on Lepidoptera. In all cases investigated, the majority of the gut contents consist of Lepidoptera scales in thick layers with an admixture of thin particles of cuticle. As is usual with gut contents, the chitin has been preserved superbly and without any surface corrosion, so that even the faintest surface structure of the scales can be discerned. However, they give, on their own account, still no hints as to the family relationships of the Lepidoptera consumed, because the same and similar types of scale occur in completely different families; on the other hand, the scale cover of an individual insect can consist of different scale types. The shape of the scales is somewhat more instructive. Thus, in the gut contents of the Messel bats, one scale type, the so-called 'sinus scales' of butterflies, is completely missing. All scales are narrow, spatula-shaped (without a distinct stalk), and terminate in three to five, approximately equally large, points (Fig. 269). Although scales of this kind occur in all families of Lepidoptera, they are the only type of scale in most primitive Microlepidoptera. This applies similarly to the most frequently found scale structure, a very regular cross-ribbing between the embossed longitudinal ribs of the scale surfaces (Fig. 270).

The fragments of cuticle from the gut contents furnish an important indication as to the taxonomy and also the ecology of *Palaeochiropteryx tupaiodon*. Between the rows of scale pores a significant number of cuticle fragments bear a cover of so-called aculeae, i.e. simple, trabeculate or bristle-like processes of the cuticle (Fig. 271). Among the Recent Lepidoptera aculeae are known, however, only from a few families of very primitive Microlepidoptera. These are all small moths and weak night fliers. If we also assume such characteristics in the prey animals of *P.*

Fig. 267: Molars of the Messel bats. Left: first upper molar in occlusal view; right: first lower molar, lingual view. Above: *Archaeonycteris trigonodon*. Note the deeply notched labial margin of the crown (left) and the low, posterior inner cusp (right) at the back. Centre: *Hassianycteris messelensis*. Note the small additional cusp on the labial margin of the crown. Below: *Palaeochiropteryx spiegeli*. Note the notched outline of the labial crown margin (left) and the low posterior cusp (right).

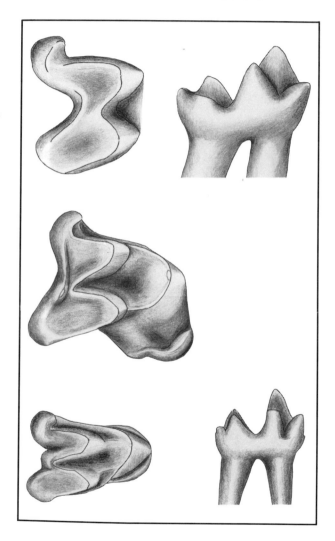

tupaiodon, then it is established, independently of wing morphological data, that this species was a nocturnal hunter and that it hunted presumably low-flying, mainly small and slow-

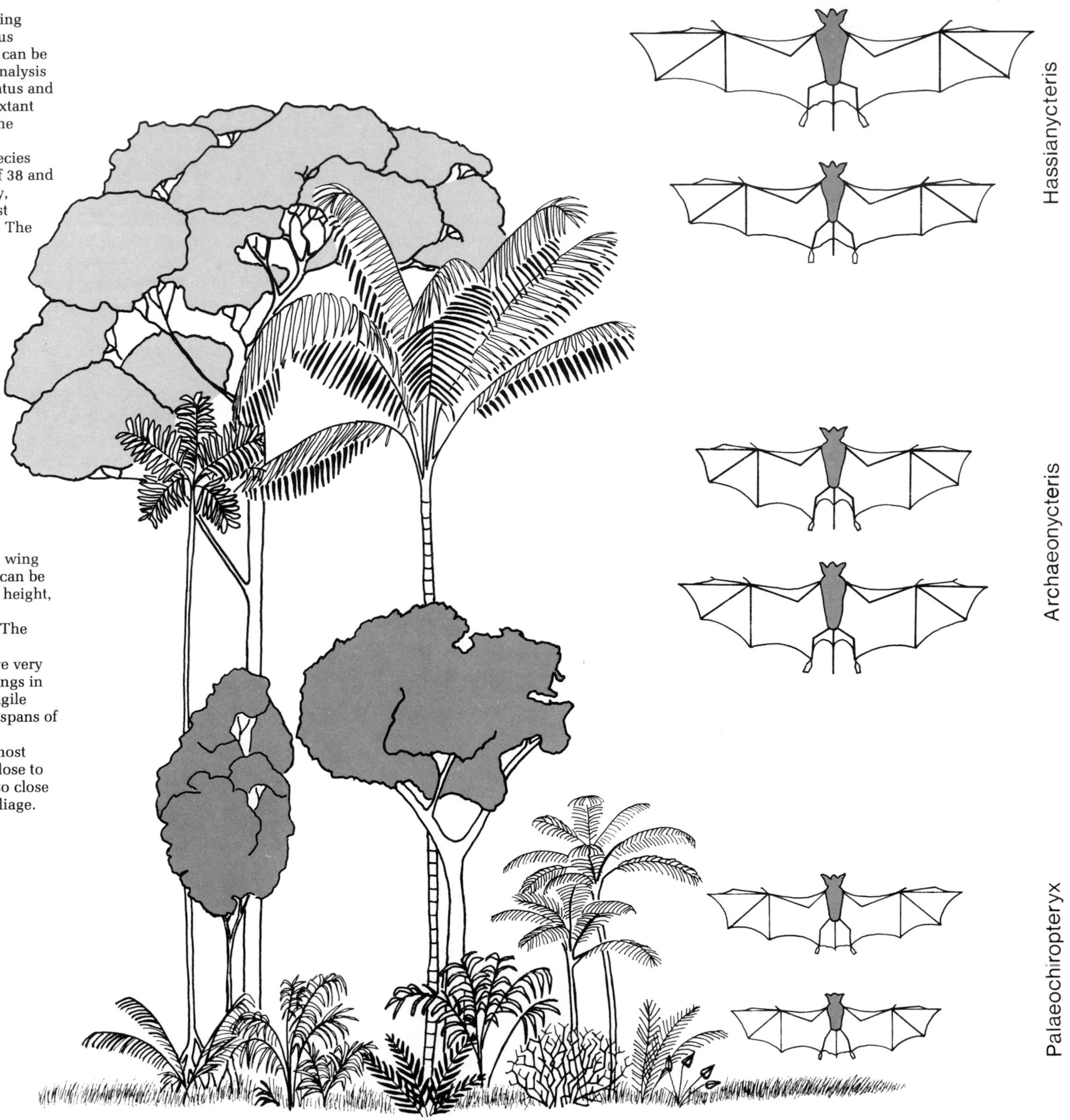

Fig. 268: The hunting arenas of the various Messel bat species can be discovered by an analysis of the flight apparatus and comparison with extant tropical species. The narrow-winged *Hassianycteris* species with wing spans of 38 and 48 cm, respectively, hunted in the forest canopy and above. The rather compact *Archaeonycteris trigonodon*, with a wing span of 32–37 cm, can be located at medium height, in the open spaces between the trees. The *Palaeochiropteryx* species, which have very broad and large wings in relation to their fragile bodies, have wing spans of 25 and 30 cm, respectively, and most probably hunted close to the ground and also close to or among the foliage.

Hassianycteris

Archaeonycteris

Palaeochiropteryx

Fig. 269: Gut contents of *Palaeochiropteryx tupaiodon*. A simple lepidopteran scale. The upper scale membrane has only been retained at the scale tips, the rest of the membrane is detached. The micrograph shows the framework of longitudinal ribs and trabeculae which stiffen the body of the scale. SEM: magnification, *c.*1700×.

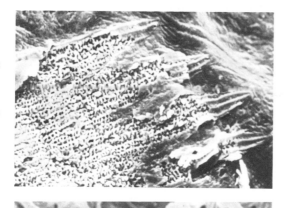

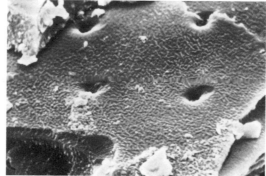

Fig. 273: Gut contents of *Palaeochiropteryx messelensis*. Thick cuticle with bristle sockets and wrinkled surface sculpture. Cuticles of similar thickness and surface sculpturing have been found in Messel coleopteran fossils. SEM: magnification, 1600×.

Fig. 270: Gut contents of *Palaeochiropteryx tupaiodon*. Fine structure of the surface of a lepidopteran scale. SEM: magnification, *c.*9000×.

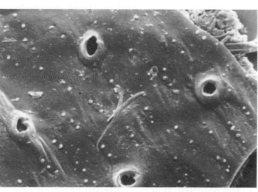

Fig. 274: Gut contents of *Palaeochiropteryx spiegeli*. Hair with strong longitudinal ribs, with star-shaped cross-section, and scaly fine structure. Hairs of similar form and structure occur commonly in extant caddis flies (Trichoptera). SEM: magnification, 7250×.

Fig. 271: Gut contents of *Palaeochiropteryx tupaiodon*. Moth cuticle with scale insertions (the scales have been lost) and with irregular rows of hair-like structures that are generally only present in the form of pale hair bases. At the top right of the micrograph densely packed layers of scales and bristles can be seen. SEM: magnification, *c.*950×.

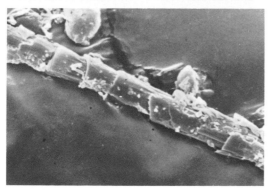

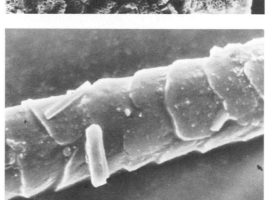

Fig. 275: Agglomerations of pollen in the gut content of *Palaeochiropteryx tupaiodon*. In two cases large amounts of pollen were found in the gut contents of the bats studied. The pollen was probably dusted on the scaly surfaces of prey animals and hence simultaneously ingested. SEM: magnification, 425×.

Fig. 272: Gut contents of *Hassianycteris magna*. Bat's hair with characteristic 'equisetum' structure. Comparisons with extant bats indicate that these are the animal's own hairs, swallowed during grooming. SEM: magnification, 1600×.

Fig. 276: Gut contents of a Recent, African bat (*Eptescius guineensis*). Hair from its own fur, swallowed during grooming. SEM: magnification, 1800×.

flying Lepidoptera. The same is probably true of *P. spiegeli*, the second species of the genus (two samples investigated). In this species also the gut contents consist of remains Lepidoptera (scales) and of ribbed hairs (Fig. 274) that are characteristic of the imagos of Trichoptera. In size and flight habit the caddis-flies (Trichoptera) correspond closely with the Microlepidoptera above. They are active at night and at the time of hatching they can swarm in myriads low above the water surface at the border of the lake. Larva tubes of caddis-flies, formed from grains of sand, are frequent in the Messel oil shale. Presumably the larvae lived in the aerobic shallow area of the lake border or in tributaries.

The remaining finds of bats with gut contents are rather scarce and biological conclusions on feeding habits are correspondingly uncertain. *Hassianycteris magna* and *H. messelensis*, at least, were not specialist feeders. The gut contents of the animals investigated consist of a mixture of Lepidoptera remains (scales) and of very thick, cuticle fragments; the latter often have a structure that occurs also in beetle remains from Messel (Fig. 273). In contrast, vestiges of Lepidoptera are completely absent in *Archaeonycteris trigonodon* whose gut contents consist predominantly of thick chitin fragments, among which are some with iridescent structural colours. On the basis of the investigation of only two animals, it cannot be decided if *A. trigonodon* was a specialist feeder (Coleoptera) as well.

Bats mostly hunt in flight but some species also take prey from the surface of leaves or from the ground. Some extant bat species, for example, *Hipposideros fulvus*, feed seasonally on larger quantities of locusts/grasshoppers, spiders, and caterpillars (M. Eckrich, oral communication). Corresponding evidence in the food analyses of *Palaeochiropteryx tupaiodon* is lacking at present. The gut contents investigated mostly lack strange admixtures. They only occasionally contain particles which do not form part of the actual food. Twice these were lumps of flower pollen which were probably ingested while sticking to the scales or hair of prey animals (Fig. 275). In three further cases bat's hairs (easily

recognized from the *Equisetum*-like structure) were swallowed during the important grooming of fur, as indicated by comparisons with Recent bats (Figs. 272 and 276).

During the nocturnal hunting flights the acoustic orientation of bats is of great importance. Obstacles are actively located, i.e. the bat itself produces acoustic signals and recognizes the obstacle due to the reflected echoes. Prey objects that themselves emit sounds (flight, landing, or crawling noises as well as communication signals) can be passively located by bats. The prey can also be detected, according to the given situation, either with the help of or exclusively by actively acoustic means, i.e. with the aid of the echolocation system.

In the open air there is no disturbing background and during a flight in this environment a relatively simple location of prey insects by ultrasound is possible. In this echolocating situation bats often use rather low-frequency sounds between 17 and 45 kHz. In this frequency band, sounds (and echoes) are only very slightly weakened over larger distances. Hence the great range of such sounds is particularly advantageous for high- and fast-flying species because they have to initiate a successful catching manoeuvre early on. For the auditory system, i.e. for the receiver of such ultrasound signals, no particular specialization is required because the frequencies lie in a band which is also typical for many nocturnal mammals. The inner ear of extant representatives of fast- and high-flying species (Figs 277 and 278) shows, just as in *Hassianycteris messelensis* (Figs 282 and 283), no morphologically remarkable features. It is remarkable, however, that Recent as well as fossil bats have a relatively larger cochlea (inner ear) (Novacek 1985) when compared with other mammals and with the flying foxes which also use echolocation.

Echolocation in the foliage or above ground vegetation shows a marked contrast to the relatively straightforward echolocation in open air. On the one hand there is the problem that every single surface reflects multiple disturbing echoes which become superimposed due to the different spatial distances ('noise'). On the

other hand, the reflecting surfaces are far larger than those of a prey item; thus the 'noise' is much more intense. Rhinolophids and hipposiderids have specialized their echo-locating system to high-frequency and constant frequency sounds (according to species—up to 156 kHz). With these they manage to discriminate the smallest differences in frequency and in this way filter out echoes of insects in motion from the very complex and loud disturbing 'noise' (Habersetzer *et al.* 1984). Adaptations in the helical structure of the inner ear (cochlea) in the form of an increase in volume can be found particularly in those areas of the cochlear duct where these echo frequences are processed (Figs 279–281). Apart from this there is an elongation of the cochlear duct (and therefore also of the basilar membrane) in the form of an 'acoustic fovea' (Neuweiler *et al.* 1980; Vater *et al.* 1985).

While among the extant bats, the high-flying species (emballonurids, molossids, rhinopomatids, and some vespertilionids) show a wide morphological variability of the wings and rather uniform echolocation sounds, the bats that fly closer to the ground (megadermatids, rhinolophids, and hipposiderids) show, in contrast, greater similarity in their wing morphology but completely different echolocation strategies (Habersetzer 1986). In the latter only the adaptation of the echolocating system to the difficult physical conditions among the vegetation makes successful active echolocation possible. In *Palaeochiropteryx tupaiodon*, which is, with regard to wing morphology, extremely specialized, we should therefore expect a similar acoustic morphological adaptation of the inner ear. But this is not the case (Figs 284–286). *P. tupaiodon* shows a construction of the inner ear which is unremarkable and which is shared by extant species of the middle and higher air space which are acoustically unspecialized. For the conditions of echolocation close to the ground and in the foliage for *P. tupaiodon* we can imagine only a passive acoustic location of prey insects. Only in rare cases in which the bat is closer to the vegetation than the insect would active echolocation (in front of an open

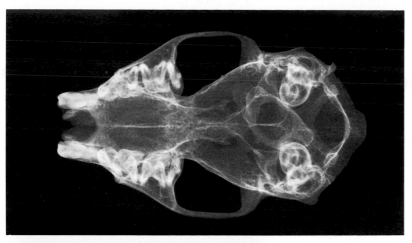

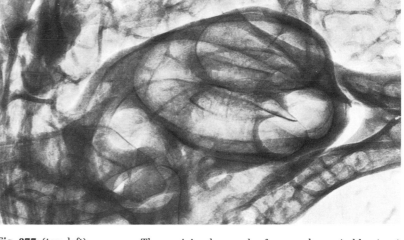

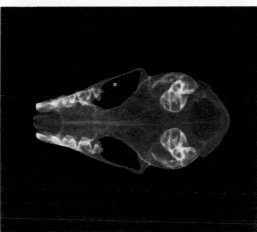

Figs 277–281: The hearing apparatus of extant bats shows adaptations for species-specific sonar. X-ray photographs. (Skulls: magnification, 3×; inner ears: magnification, 20×).

Fig. 277: (top, left). *Taphozous nudiventris kachhensis* is a fast- and high-flying sheath-tailed bat which uses at most 20–24 kHz echolocation sounds. Comparison with the teeth (which absorb X-rays strongly) shows that both helical inner ears consist of massive bones.

Fig. 278: (top right). For this photograph the back part (occiput) of the skull was lowered by 37°. This provides a view into the first (basal) half turn of the cochlea of *T. nudiventris kachhensis*. Between the two bone spurs which protrude into the canal of the cochlea (primary and secondary spiral laminae), the thin structure can be identified as the soft-part remains of the basilar membrane. The sensory cells which transduce sounds into stimuli for the auditory nerves are located on the basilar membrane. The spiral X-ray shadow of the separating bone reveals that the cochlear canal of this bat narrows continually from the base to its tip.

Fig. 279: (left). *Hipposideros fulvus*, an Old World leaf-nosed bat, hunts close to the ground and in the foliage with very high pitched echolocation sounds (157 kHz). It flies slowly, with a hovering flight. Its inner ears differ significantly from those of other species. The volume of the basal turn is very much enlarged at the expense of the other paths of the canal. This is visible in the survey radiographs. The overall size of the cochlea is not enlarged relative to *Taphozous nudiventris kachhensis*.

Fig. 280: (centre). In *Hipposideros fulvus* the cross-sectional area of the upper part of the cochlear canal (canalis spiralis) is twice as large as at the basal turn in other species. The distance between the two bony lamellae is only 0.05 mm, and thus the basilar membrane of this species is also very narrow. From this angle the cross-sectional area of the third half-turn—a full spiral turn higher—is clearly apparent, and distinctly shows the small volume remaining at the top of the spiral.

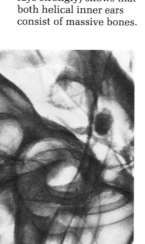

Fig. 281: (right). This X-ray shows the same inner ear together with the semicircular canals of the vestibular organ. The cochlea was tilted by 30°, bringing the second and fourth half-turn on the opposite side clearly into view. The combined evidence of this X-ray and that in Fig. 280 demonstrates that the morphological adaptations of the inner ear of *Hipposideros fulvus* are most pronounced in the basal section, i.e. the part where high frequencies are transduced.

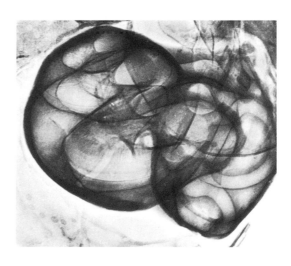

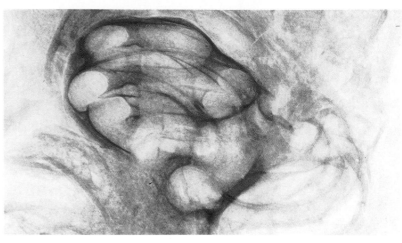

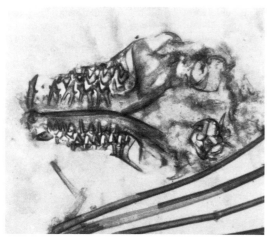

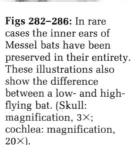

Figs 282–286: In rare cases the inner ears of Messel bats have been preserved in their entirety. These illustrations also show the difference between a low- and high-flying bat. (Skull: magnification, 3×; cochlea: magnification, 20×).

Fig. 282: (top, left). General view. 2.5 spiral turns can be distinguished in the cochlea of *Hassianycteris messelensis*. Many extant bats have a similar shaped cochlea, although the cochlea of *H. messelensis* is smaller. However, this does not exclude echolocation; some extant flying foxes have even smaller cochleae and echolocate. The fine structure of the inner ear therefore appears to be more important.

Fig. 283: (top, right). The cochlea of *H. messelensis* exhibits similar proportions to that of *Taphozous nudiventris kachhensis*, with a gradual decrease in canal volume from base to tip. The bony spines from which the basilar membrane is suspended are prominent. With a well-defined outer bone lamella this apparatus also closely resembles that of Recent echolocating bats. Thus even *H. messelensis* probably located its prey with rather low-frequency ultrasonic sounds.

Fig. 284: (left). The inner ears of *Palaeochiropteryx tupaiodon*, a bat that flies slowly and close to the ground, almost always show major structural defects and fractures in survey skull radiographs. The cochleae have shifted from their natural position during fossilization, which makes their investigation more difficult, but a rotation of the fossil in the beam of the X-ray apparatus can allow a position to be established which is comparable with that of other X-rays.

Fig. 285: (centre). For this picture the fossil was rotated by 22.5° in both spatial planes. The (top) cross-sectional area of the first half turn constitutes 9.7 per cent of the entire cross-sectional area of the cochlea. Apart from this we can also distinguish here the strongly formed primary and secondary spiral laminae between which extended, during the bat's life, a 0.11 mm wide basilar membrane. The width of the basilar membrane was, according to this, 5.5 per cent of the width of the cochlea. This corresponds with the proportions of bats hunting in the open air, unobstructed by vegetation.

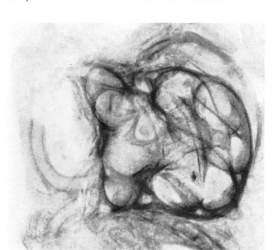

Fig. 286: (right). A different sample exhibits a continuous moderate loss of volume of the cochlear canal within 2.5 spiral turns. This is revealed by the X-ray shadow of the bony separation wall extending from the base to the apex. There are no hints of the type of specialization found in *Hipposideros fulvus* (cf. Fig. 280) and related species. *Palaeochiropteryx tupaiodon* was probably not capable of comparable echolocation of prey insects.

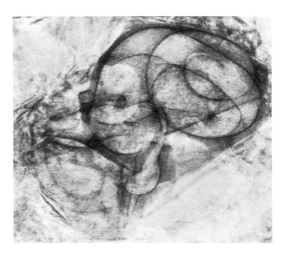

background) be possible. The hunt across the free water surface of Lake Messel also offers such an echolocation situation, but it is not typical for a bat with slow hover flight. *P. tupaiodon* represents in the strong contrast—extremely specialized in terms of wing morphology, but acoustically unspecialized—a combination of characters that no longer exists. No comparable Recent species exists and we can therefore suppose that, in their early evolution, bats first occupied ecological niches without prior adaptation of the echolocation system.

Wing morphology and food analyses show, independently of one another, that *P. tupaiodon* hunted for insects in the lower air space. Following the initially introduced hypothesis that toxic gases over the lake caused the bats to fall into the water. *P. tupaiodon*, as a low-flying hunter, would be far more endangered than all the other species. And, indeed, *P. tupaiodon* acounts for 74 per cent of all bat finds while, as yet, only one specimen (0.3 per cent) of the largest species, *Hassianycteris magna*, has been secured. Because these relationships certainly do not reflect the real population densities of the tropical/subtropical fauna of Messel, the toxic gas hypothesis appears to us to be the most likely explanation for this unique fossil bat site.

15

Our closest relatives: the primates

One of the peculiarities of Messel is that members of that order of mammals to which we ourselves belong, the primates, are particularly rare. This is unusual because primates at other sites of that age in Europe are by no means rare. This is true both in terms of the number of their species as well as of the number of individual finds. This fact is all the more striking because in the Eocene the environs of Lake Messel with a luxuriant rain forest and a tropical–subtropical climate would have offered a suitable habitat for a rich primate fauna.

Our closest relatives: the primates

JENS LORENZ FRANZEN AND
WIGHART VON KOENIGSWALD

Since the discovery of the fossil site there have been, in more than 100 years, only four finds of primates. All of them come from the most recent period of intensive scientific excavations.

The team of the Hessisches Landesmuseum, Darmstadt, was successful in recovering the first find in 1975. It was recognized as such, however, only during preparation in 1978 and a report was published in the following year (Koenigswald 1979). It is the lower part of a skeleton and consists of the pelvis and hind limbs (Fig. 287). To the skeleton belongs also a penis bone (baculum) of an unusual size for primates, which allows us to conclude that this was a mature, male individual. The entire remaining skeleton, including the presumably long vertebral column of the tail, was obviously already missing when the specimen was embedded into the sediments (Lippmann and Wiemer 1979; Koenigswald 1979).

In its characteristics the skeleton resembles the Lower Eocene North American prosimian *Notharctus* more than the Upper Eocene European *Adapis* and related forms. The big toe was particularly strong and could be opposed to all the other toes, as an extremity for grasping (Fig. 288). All toes bore already, as characteristic for most extant primates, flat nails instead of claws. The nail of the second toe was, however, narrower and was more strongly bent; it was obviously used, as by extant prosimians, as a grooming digit during body care. Because skull and teeth are missing, a more exact determination is, unfortunately, not possible at present; one can only say that it belonged to the Adapiformes.

The second primate find in Messel was a result of the Senckenberg excavations of 1982 (Franzen 1983, 1987). Here exactly that part is

preserved which is missing in the Darmstadt specimen, namely, the entire vertebral column of the back, the largest part of the chest, the right fore-arm together with the scapula, as well as, most importantly, the whole skull including the entire dentition (Figs 290 and 291). In addition to the hind-limbs and the

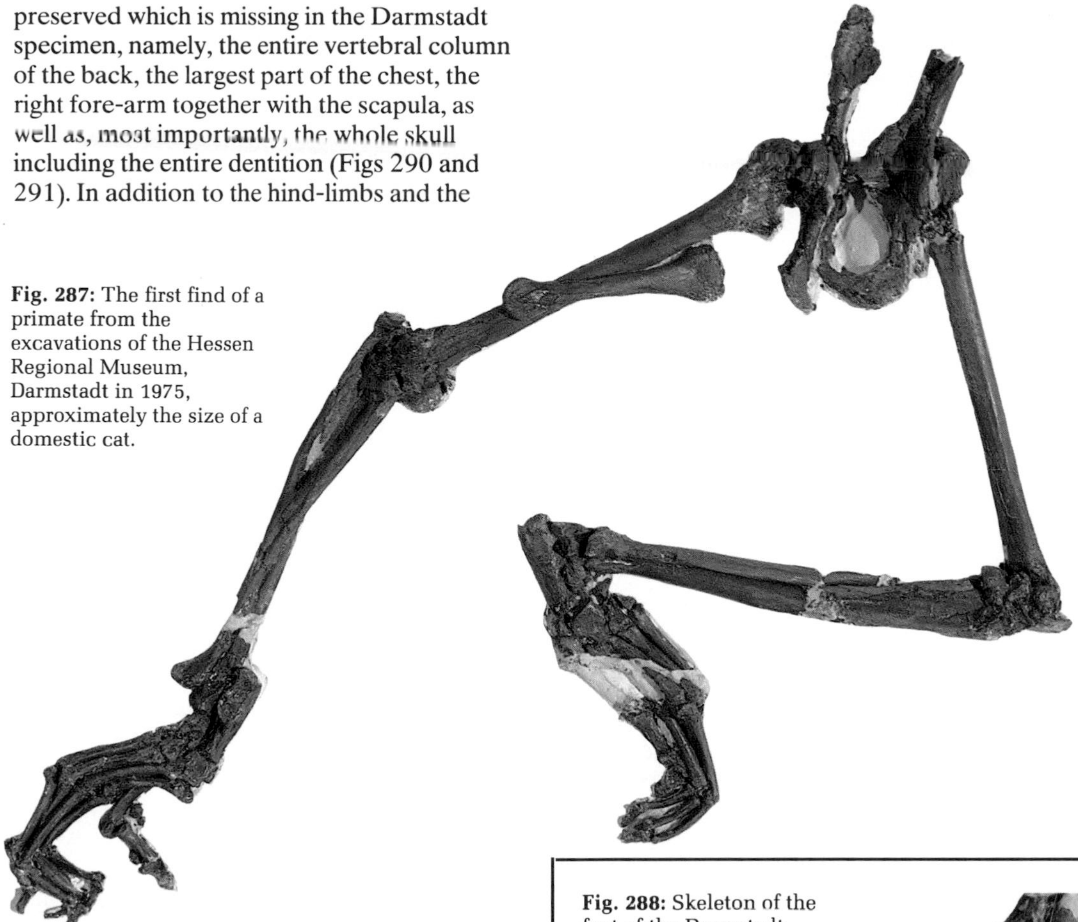

Fig. 287: The first find of a primate from the excavations of the Hessen Regional Museum, Darmstadt in 1975, approximately the size of a domestic cat.

pelvis, the vertebral column of the tail is also missing. In the area of the head and the hand, long hairs are preserved as black shadows 'bacteriographically' (see Chapter 25, 'From excavation to exhibition piece'). The so-called gut contents are, unfortunately, not preserved.

We can definitely exclude the obvious possibility that the Darmstadt and Senckenberg specimens could be parts of one individual. According to proportional measurements on extant lemurs the measurements on the Senckenberg primate find are not only 30–50 per cent smaller than would have been expected from the Darmstadt partial skeleton, but also the Senckenberg bones show, above all, still open growth structures. The dentition also still includes some of the milk teeth and thus shows that, in

Fig. 288: Skeleton of the foot of the Darmstadt prosimian. The flat, obviously nail-bearing digits of the toes and the opposable big toe are typical of primates.

contrast to the Darmstadt specimen and the subsequent Karlsruhe specimen, the Senckenberg primate is a juvenile.

Evidence that the Senckenberg find is also a primate comes from the terminal phalanges flattened to carry nails, from the opposable thumb, and also from the orbit which is lined at the back by a bony bar. An exact determination is possible in this case, because, in addition to the juvenile dentition, part of the permanent dentition, particularly the molars, is preserved as well. This is a representative of the genus *Europolemur* and is, indeed, a new species. According to the degree of development of its dentition, it stands in terms of time and of evolutionary development between *Protoadapis recticuspidens* from the

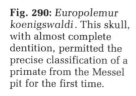

Fig. 290: *Europolemur koenigswaldi*. This skull, with almost complete dentition, permitted the precise classification of a primate from the Messel pit for the first time.

Fig. 291: *Europolemur koenigswaldi*. Senckenberg excavation, 1982: torso, skull, and right fore-limb. Typical of primates are the nail-bearing fingers, opposable thumb, and bony bar behind the orbit. The animal was half the size of a domestic cat.

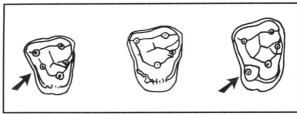

Fig. 289: Phylogenetic evolution of the occlusal pattern of the second upper molar from the Lower Eocene *Protoadapis recticuspidens* (left) via *Europolemur koenigswaldi* from the Lower Mid-Eocene of Messel to *Europolemur klatti* from the Upper Mid-Eocene of the Geiseltal (right). Note the evolution of the distal inner cusp (hypcone) after the cusp in between (metaconulus) became reduced.

Lower Eocene of the Paris Basin and *Europolemur klatti* from the Upper Mid-Eocene of the Geiseltal (Fig. 289). In honour of the famous researcher of fossil primates and discoverer of Javanese early man, Gustav Heinrich Ralph von Koenigswald (1902–1982), the new species from Messel was named *Europolemur koenigswaldi* (Franzen 1987).

What kind of primate was it? What did it look like and how did it move? Even though the entire distal half is missing, the remaining skeleton gives a series of clues that allow the conclusion that *Europolemur* was a fossil prosimian which probably resembled most closely the sifaka (*Propithecus*) and the extant lemurs of Madagascar (Fig. 292). It was, however, much smaller and approximately only half the size of a domestic cat. Like the sifaka, one can imagine *Europolemur* to have been active during the day, climbing through trees with its grasping hands (and also

presumably grasping feet). The structure of the shoulder region suggests that *Europolemur*, like some extant prosimians, could cling, torso erect, to the trunk, with occasional leaps from tree to tree.

In spite of all the similarities, however, it would be wrong to see in *Europolemur* a close relative of today's isolated lemurs of Madagascar (Fig. 293). The entire evolution of the primates, including that of our own ancestors, in the early Tertiary had still not gone beyond the stage that the prosimians of Madagascar have preserved to the present. It is, however, not possible to recognize definite ancestors of the higher primates among the fossil prosimians. These ancestors may possibly be expected in Africa where the first higher primates are preserved from the Lower Oligocene (*c.* 30–35 million-year-old strata in the Oasis Fayum in northern Egypt). In their body structure the forms that led to these first higher primates would not have differed fundamentally from the prosimians of Messel. If, with this in mind, we visualize the further development that led, in the following 49 million years, to the evolution of man, then we obtain an impression of the amount of time which has passed since the Eocene animal- and plant-world of Messel.

The excavation team of the Regional Natural History Collection, Karlsruhe, succeeded in making the third Messel primate find in 1984 (Koenigswald 1985). This is again a primary partial skeleton of an adult male. It resembles the Darmstadt find in many aspects and includes pelvis, penis bone, and a fragmentary hind-limb loosely connected to the skeleton. The Karlsruhe specimen is approximately 10 per cent smaller, however. An identification of genus and species is not possible at present.

A fourth primate specimen could, at last, be recognized as such recently, only after its preparation (Franzen 1988). This find also comes from the Senckenberg excavations of 1982. This is also a primary fragment. It consists of a right fore-arm with hand (Fig. 294). An opposable thumb and fingers that obviously bore flat nails provide the characteristics that are typical for primates. The individual bones are considerably larger than in *Europolemur koenigswaldi*. Regarding their size they would fit well with the Darmstadt and Karlsruhe primate finds at Messel. An exact determination, however, cannot be accomplished in this way.

It is regrettable that a site like Messel has to date added only a few and, moreover, only incompletely preserved primate remains. What a contribution would be made by complete

Fig. 292: The primate finds of Messel are still too incomplete to permit reliable reconstructions. The extant diadem sifaka (*Propithecus diadema*) represents a comparable life-form.

197

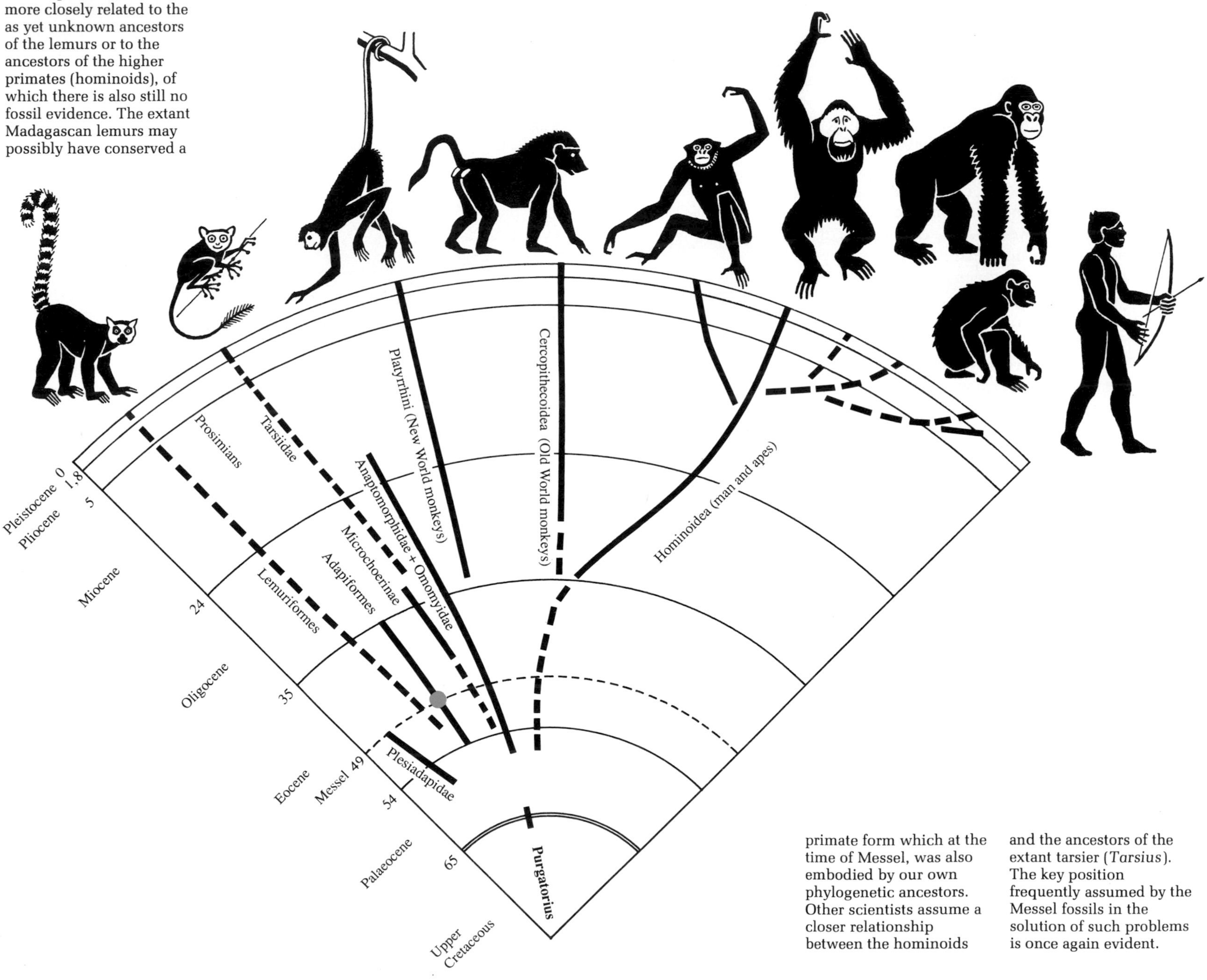

Fig. 293: Very much simplified primate phylogenetic tree. The green dot marks the position of the Messel finds. At present, it is still debatable whether the Messel prosimians were more closely related to the as yet unknown ancestors of the lemurs or to the ancestors of the higher primates (hominoids), of which there is also still no fossil evidence. The extant Madagascan lemurs may possibly have conserved a primate form which at the time of Messel, was also embodied by our own phylogenetic ancestors. Other scientists assume a closer relationship between the hominoids and the ancestors of the extant tarsier (*Tarsius*). The key position frequently assumed by the Messel fossils in the solution of such problems is once again evident.

Prosimians

Tarsiidae

Platyrrhini (New World monkeys)

Cercopithecoidea (Old World monkeys)

Anaptomorphidae + Omomyidae

Microchoerinae

Adapiformes

Lemuriformes

Hominoidea (man and apes)

Plesiadapidae

Purgatorius

Pleistocene 0 1,8

Pliocene 5

Miocene

24

Oligocene 35

Eocene

Messel 49

54

Palaeocene 65

Upper Cretaceous

198

skeletons, possibly with gut contents and soft-body contours to the elucidation of our own descent! But we must not lose hope. One day such finds will be recovered if we don't bury our future as well as our past under refuse!

Fig. 294: The fourth find of an adapiform primate from Messel: an isolated hand and right fore-limb from the 1982 Senckenberg excavations. It is of a comparable size to the primate finds of the Darmstadt and Karlsruhe museums. An exact determination, however, is not possible as yet.

16

Pangolins: almost unchanged for 50 million years

The pangolins (order Pholidota, comprising the single family Manidae) occupy a special place among extant mammals. They are distinguished by a singular mixture of very primitive characteristics with notable adaptations to a highly specialized feeding strategy, the eating of ants and termites. The primitive characters suggest an early origin for the pangolins, in the late Cretaceous (Starck 1941). This implies that their decisive biological adaptations were already developed in the earliest periods of the Tertiary. Evidence for this assumption came for the first time with finds in Messel of *Eomanis waldi* Storch 1978. These are, at the same time, the geologically oldest and the most complete known finds of pangolins (Storch 1978*a*, *b*).

Pangolins: almost unchanged for 50 million years

GERHARD STORCH AND
GOTTHARD RICHTER

Previous fossil evidence of pangolins was rare and only fragments had been preserved. The first specimen was described by Quenstedt (1885) from the Middle Miocene of Solnhofen, but he incorrectly attributed the isolated bones to an otter. In 1904 Schlosser compared these finds with pangolins and aardvarks and placed them within the edentates which, according to the understanding of that time, included the pangolins. In 1894, Filhol had already described two fossil genera of manids from the famous French fossil sites of the Quercy. After a somewhat confusing 'detour' in 1906 when F. Ameghino placed the Solnhofen finds, which presumably derived from the same individual, among the armadillos and pangolins, the relationship of these European specimens to the manids was re-established (Dehm, Helbing, Viret, and W. von Koenigswald). Only a few individual bones or fragments thereof were, however, added to the original finds. All these finds can be allocated to the genus *Necromanis* Filhol 1894, whose geological distribution extends from the late Eocene to the late Miocene. Some further finds, some of very recent date and not yet published in their entirety, are of Tertiary pangolins from three further continents: Asia (Inner Mongolia, China); North America (Wyoming, USA; Emry 1970); and Africa (Fayum Province, Egypt) where their earliest occurrence was in the early Oligocene. The reason why fossil pangolins are everywhere extremely rare might be, most of all, their complete lack of teeth. The usual proof in palaeontology, by means of the dentition, does not, therefore, exist. Within a given local fauna the pangolins were, in addition, probably always sparsely

Fig. 295: *Eomanis waldi*, a completely preserved skeleton of the Messel pangolin. Note the massive forelimb and the very strong clawed phalanx of the middle finger. Head and body length, 31 cm; length of tail, 19 cm.

represented, just as is true in their present areas of distribution in South-east Asia and Africa south of the Sahara.

Eomanis waldi (the Messel pangolin) was a strongly built and plump 50-cm-long animal (Figs 295–299). A handful of finds have become known to date. The Messel pangolin was already adapted to a diet of termites and ants and showed, in comparison with the extant species, an already surprisingly high degree of specialization. This emphasizes the great phylogenetic age of the pangolins combined with their adoption of an ecological niche which was already completed very early in their evolution. On the other hand this increases the difficulties of judging relationships with other ancient groups of mammals that are also specialized for a termite and ant diet, for example, the ant-eaters (and thus the order Edentata). The question then arises as to what extent adaptations toward the same goal have evolved independently (convergently) rather than being inherited from common ancestors. The complete finds from Messel allow us to draw palaeobiological conclusions, in particular with regard to food intake, foraging, and also protection and defence. As regards the food items, however, conclusions can be based only on skeletal characters of the masticatory apparatus. The analyses of the gut contents are, in pangolins, particularly controversial and still present a puzzle for us.

Small insects must have served as food for *Eomanis*. The lower jaws are, as in extant animals, extended, thin, and toothless bony

Fig. 296: Type skeleton of *Eomanis waldi* in dorsal view. Before preparation the trunk was surrounded by masses of a coal-like, crumbly substance which suggested horny armour scales. Total length, 47 cm.

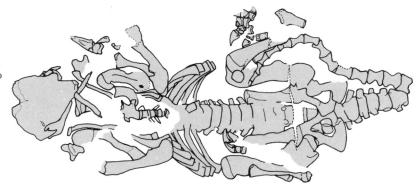

Fig. 297: The drawing records all the skeletal elements which can be isolated and identified from the fossil. To the left the ventrally bent skull; to the right the vertebral column of the tail, folded anteriorly.

Fig. 298: Translation of the same skeleton from a dorsal to a lateral view. Empty spaces on the hands, feet, and nape show where bones were not preserved in the fossil.

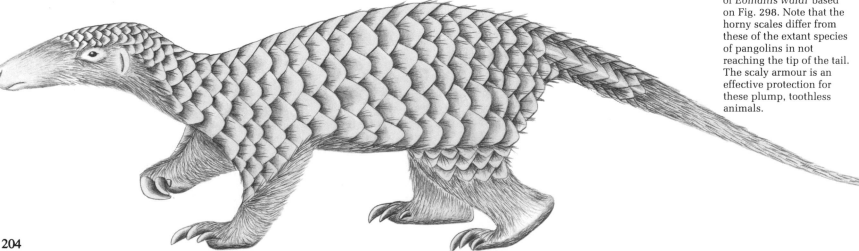

Fig. 299: Reconstruction of *Eomanis waldi* based on Fig. 298. Note that the horny scales differ from these of the extant species of pangolins in not reaching the tip of the tail. The scaly armour is an effective protection for these plump, toothless animals.

bars (Fig. 298); they even conform in such distinct details as the sharp upper margin and the point of the lower jaw symphysis which is separated by a bony fold. The upper jaws are also toothless; the snout is extended and wedge-shaped. Mastication was impossible. The plump condyles of the lower jaw, which were only loosely taken up by the skull, even exclude the possibility that the mouth could be opened to some extent. The food was, as in extant pangolins, taken up with the long, sticky tongue and mechanically broken down by a grinding organ in the stomach. Although the muscle processes on the lower jaws are a little more strongly developed than in extant species, they also allow us to conclude that the chewing musculature generated only a weak force component for pressure. In contrast, the horizontal component, which acted anteriorly, played a much larger role. This is characteristic for mammals whose jaws have lost the means for breaking up food and whose tongue has assumed the function of food intake (Storch 1968). Social insects are an obvious food source. *Eomanis* is not a small animal and, therefore, large numbers of insects had to be available that could be swallowed whole, for example, ants and termites.

The colonies of the insects are, however, not immediately accessible. Their dwellings must at least be scratched open and, in some termite mounds that are as hard as concrete, breaking up is required. The Messel pangolin is distinguished by a multitude of characters as a good scratching digger and all these adaptations must be interpreted in connection with foraging. The bones of the limbs are short and robust. They offer large muscle attachments, capability to sustain great mechanical stress, advantageous leverages, and—with a broad bone contact at the joints—protection from sprains. The fore-limb bone structure in particular allows one to assume that a strongly developed musculature was present (Fig. 295). The elbow process, as an attachment of the extensor muscles, is long and broad; the powerful crest of muscle attachment (crista deltopectoralis) extends very far down on the humerus and therefore provides good leverage for the pectoral

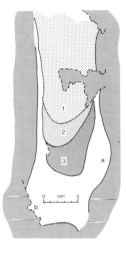

Fig. 300: Three scales firmly addressed on the femur (a, muscle process; b, neck of the condyle).

Fig. 301: Horny scales, overlapping like roof tiles, of *Eomanis waldi* are here substantially preserved.

muscle (far away from the shoulder joint around which it rotates the humerus); the outer and inner points of attachment of ligaments on the humerus are very broad (serving the strong lower-arm pivots, flexors of fingers, and extensors of the hand). The middle finger, and particularly the claw-bearing phalanx, dominate the hand so that the hand could serve as a 'pickaxe' on the harder substrate. The pointed, curved terminal phalanges, furrowed underneath, certainly bore very strong, horny claws; unlike in extant animals, though, their tips are not yet split for a better anchoring of the horny claws.

The mode of ingestion and foraging lead to slowness, lack of teeth, and, therefore, defencelessness. In contrast to the giant ant-eater with its pickaxe-like digging style, the development of the fore-limbs has not reached a stage where they would suffice as effective weapons for defence. Protective devices are therefore necessary.

The body cover of the pangolin, with large horny scales that overlap one another like roof tiles ('pine cone animals' in the German vernacular), is unique among extant mammals. The individual scales originate from the accumulation of horny substance in the epidermis, forming a flattened, backward-pointing cutis papilla. The sharp-edged, smooth and erectile, horny scales afford protection both from predators and from masses of attacking insects which serve as food. In addition, extant animals have the ability to roll up, forming a ball, and to wrap the fully-scaled tail around the body.

For *Eomanis* such scaly armour was first disclosed indirectly from the skeletal characteristics of the generally dorsally or ventrally positioned fossils in the oil shale (Figs 295 and 296)—whose curved body shell left an imprint in the soft deposits on the lake floor—as well as from the encapsulation of the unprepared finds in a coal-like crumbly substance in the area of the body (Storch 1978*a*). But later on it was possible to demonstrate horny scales directly, as preserved scales were found (Figs 300–303). In arrangement, appearance, and structure these correspond with the scales of extant species of *Manis*, but they are relatively small (Koenigswald *et al.* 1981).

Pangolins were, therefore, already present with all their major adaptations 50 million years ago, and differences in the degree of specialization, when compared with Recent species, are small. Such differences concern, for example, the extent of scale cover. The tail of extant species is broad, flat, and strongly scaled to its tip. Accordingly, the vertebral column of the tail bears, along its entire length, strong transverse processes as a foundation for the broad form, the large scales, and strong musculature. For the rolled-up animal the tail becomes an armoured shield, a third leg during the scratch-digging with both forelegs, and an aid in pushing as a grasping organ. These protecting and supporting functions largely do not apply to *Eomanis*. Short transverse processes are present only on the proximal nine tail vertebrae and the larger part of the distal area of the tail must have been round and free of scales (Fig. 299). *Eomanis* was certainly also less well able to roll itself up. In comparison with extant animals, the short and clumsy lumbar vertebral column and the long pelvis present obstacles to rolling up in a ball. The pushing back and expulsion of soil with the hind-feet when digging dens must also have been more strenuous for *Eomanis* than for extant species. The bone process, to which the extensors of the hip joint are attached, is still positioned in the upper half of the femur. In extant animals the attachment has moved to the vicinity of the knee joint, thus increasing the effectiveness of the femur extensors.

Taking into account all that we have learned, the Messel pangolin lived as a clumsy plantigrade on the ground, where, by means of digging and scratching, it opened up the mounds of social insects and also dug dens. According to the structure of the skull, its eyes must have been quite small. The sense of smell may have been dominant in the location of food.

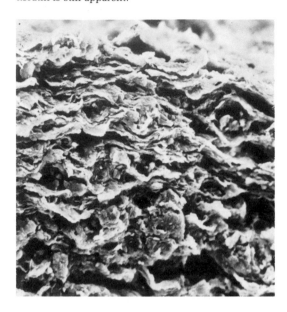

Fig. 302: Transverse section of a horny scale from *Eomanis waldi*. The original layering of the shrunken and partially destroyed lamellae of keratin is still apparent.

Fig. 303: Transverse section of a horny scale from a recent pangolin (*Manis javanica*). The arrangement of the keratin lamellae is approximately as in Fig. 302.

The reconstruction of biological adaptations to myrmecophagy—the eating of ants and termites—appears compelling and correlations of *Eomanis* with Recent species of *Manis* do often exist down to the finest detail of the skeleton.

Because of the unequivocal anatomical findings, the results of investigations of the gut contents should have been predictable. It was clear what the animal should have been eating (social insects), and it was even clearer what it could not have been eating, namely, plants or parts of plants of any kind. Apart from soft fruits, plant food can be easily obtained, but is processed with much more difficulty. It has to be torn off, gnawed off, or grated; plant cells have to be 'broken open' mechanically (by sustained mastication) or chemically (by cellulases). The toothless *Eomanis* with the narrowed opening of its mouth was not capable of any of this.

In all five specimens of *Eomanis* that have been investigated so far, 'gut contents' were discovered, that is, a dark-grey–dark-brown coal-like mass in the area of the last ribs and distal to them. A larger amount of this mass is present in four animals and, much reduced, in a fifth animal. The substance is shot through with numerous particles of quite variable shape and size, with a deep black coloration. Due to the numerous coarse grains of sand, which are missing in the surrounding sediment, it is particularly likely that these are gut contents. (Such sand grains or small stones help the distintegration of insect food in the stomach and are always present in recent pangolins and ant-eaters.)

But here ends the correspondence between extant and fossil pangolins. Only in one specimen were any insect remains (unidentifiable fragments of chitin cuticles) found at all, and here only a limited amount. Apart from these—and in all the other animals—the gut contents consist of plant material. Leaf cuticles, pieces of leaves with cell preservation, fragments of stalks, splinters of wood, etc. are available in great number, that is, just the sort of particles on which *Eomanis*, according to its structure, could not have fed.

This is the first time—and luckily the only time—that findings of Messel research, related to functional morphology and to palaeoecology, clearly contradict one another and all attempts to interpret this phenomenon have, up to now, been unsuccessful. The possibility of disintegration of chitin in the stomach through the action of formic acid (provided by ants which had just been ingested) does not apply according to our investigations. The floating of plant material into the carcass at the bottom of the lake is possible, but requires that all specimens were embedded with empty stomachs, which in turn would be unusual (Richter 1987). In view of such controversial results the palaeoecologist must either give up, or suggest a hypothesis which combines the seemingly incongruous evidence according to the motto 'the more daring, the better'. Figure 304 represents the artistic elaboration of such a hypothesis.

Fig. 304: Was this how ant-eating evolved? *Eomanis waldi*, in practice already myrmecophagous, but 'psychologically' still vegetarian, managed to survive in the Messel jungle by lying in wait in the paths of leaf-cutter ants in order to steal the leaf segments that were already cut to size. The worm-like sticky tongue that had already evolved is here of greatest importance. Further evolution towards the recent exclusively ant-eating pangolin is almost a matter of course. At first, only occasionally, ants which clung to their leaf segments, were seized by the tongue and ingested with the plant material. They proved, however, to be such a tasty and nutritious food that they—originally only an accompaniment to the catch—quickly became the principal and finally the only source of food.

207

17

The ant-eater *Eurotamandua*:

a South American in Europe

In 1974 the amateur palaeontologist Dr G. Jores excavated the complete and three-dimensionally preserved skeleton of the Messel ant-eater *Eurotamandua joresi* Storch 1981. To date, no further specimens have been found, as is true for a series of other mammal species described from Messel, but it is clear that further surprises can still be expected from this irreplaceable fossil site.

The ant-eater *Eurotamandua*: a South American in Europe

GERHARD STORCH AND
GOTTHARD RICHTER

The surprising similarity of the Messel fossil to skeletons of extant ant-eaters and, in particular, to that of the collared ant-eater (*Tamandua tetradactyla*) is visible at first glance (Fig. 305). In order to accept the fossil as an ant-eater, however, it is first necessary to overcome an intervening obstacle. Ant-eaters (suborder Vermilingua), armadillos (Cingulata), and the two-toed and three-toed sloths (Pilosa) today represent the order Edentata, which once was much more varied and which characterizes, as does no other group, the present and past faunas of South America. It was regarded as a safe assumption that the Edentata had originated there, had been able to evolve there for 60 million years during the isolation of South America in the Tertiary, and were only able, in the late Tertiary, to leave their home continent and head towards North America. Evidence of Tertiary edentates outside South America was proved to be spurious. In Europe the armour plates of the anguid *Placosaurus* and of the leather-turtle *Psephophorus* had led to the misidentification of armadillo dermal plates and to the description of new species of armadillos. F. Ameghino (1906) had, moreover, arbitrarily attributed a skull fragment and several bones of pangolins to the armadillos. The only exception is *Ernanodon antelios* Ding 1979 from the late Palaeocene of south China which was described scientifically shortly before *Eurotamandua*. This strange fossil mammal possesses certain similarities with armadillos but also with sloths, and is probably an edentate. Thus, even if an ant-eater in Messel appears, at first glance, absolutely 'exotic' zoogeographically (but see Chapter 27, 'The Messel fauna and flora: a

Fig. 305: The ant-eater *Eurotamandua joresi* belongs among the mammalian rarities at Messel. Head and body length, 50.5 cm; length of tail, 35.5 cm.

biogeographical puzzle'), its characteristics are unequivocal and this alone is decisive. Recently, Storch and Haubold (1989) discovered a *Eurotamandua*, also from the early Middle Eocene Geiseltal site.

Eurotamandua is not only the sole find of an ant-eater outside South America; it is also by far the most complete and, far and away, the geologically oldest one. The evidence to date in South America stretches only back into the early Miocene (the Santa Cruz Formation of Argentina). Geographically the late Miocene La Venta fauna in Columbia covers the most northern Tertiary occurrence. The finds are

very rare and consist only of isolated skeletal elements, but, because these originate mostly from different regions of the body and so can scarcely be compared with one another, it is difficult to judge even the most recent phylogeny of the ant-eaters.

This is surprising, because a multitude of primitive body characters that have been preserved until now suggest that the edentates should be traced back to a first split from higher, placental mammals and that their origin should be looked for a long way back in the Cretaceous. Such characters prompted M. C. McKenna, in his classification of 1975, to place

211

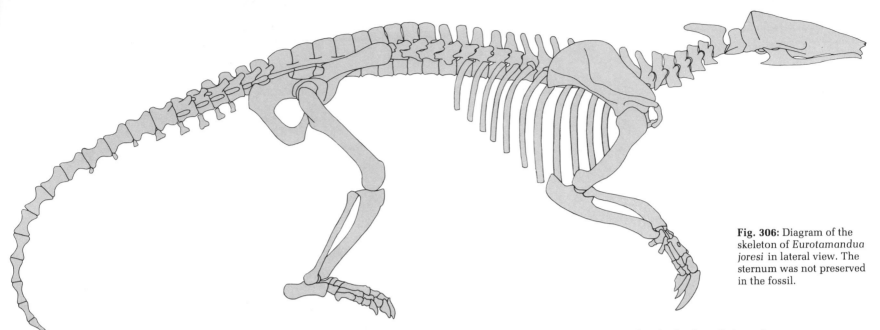

Fig. 306: Diagram of the skeleton of *Eurotamandua joresi* in lateral view. The sternum was not preserved in the fossil.

the edentates opposite all other placental mammals, in sister-groups of equal rank. Also, the splitting of the edentates into their different families, and therefore the origin of the ant-eaters, should, on the basis of biochemical evidence, have occurred during the Cretaceous (Sarich 1985; de Jong *et al.* 1985). Our poor knowledge of the early phylogeny of ant-eaters is due to gaps in the fossil record. As in

pangolins, identification is made more difficult due to the animals' lack of teeth. On the other hand, a tropical environment in which the ant-eaters might have evolved at first, normally offers, as can be imagined, bad preconditions for fossilization.

Eurotamandua joresi was a very strong and massively built animal of approximately 90 cm in length. Its excellent preservation leaves no

room for doubt that all those features are present that distinguish the order Edentata, as well as all those that specifically characterize the suborder Vermilingua. The extreme adaptations to the myrmecophagous mode of feeding—the eating of ants and termites—are already perfectly developed (Storch 1981). *Eurotamandua* resembles the extant collared ant-eater (*Tamandua*) strongly. The collared

Fig. 307: Reconstruction of *Eurotamandua joresi*, based on the accurate skeletal drawing of Fig. 306.

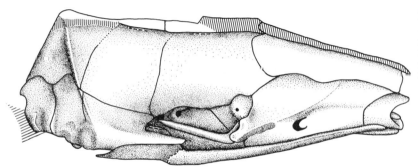

Fig. 308: The skull of *Eurotamandua joresi* in plan view. The tubular toothless skull is characteristic of the ant-eater family

myrmecophagids (Storch and Habersetzer 1991).

The hand is most specialized (Fig. 308). The middle finger dominates in terms of bulk and length. The clawed phalanges of the second and third fingers are very long and strong; those on the outside of the hand, by contrast, are small. This suggests that *Eurotamandua*, just like the extant *Tamandua*, shifted the

ant-eater lives in trees as well as on the ground and is, therefore, less specialized in habit than both its living relatives, the purely ground-dwelling giant ant-eater (*Myrmecophaga*) and the exclusively arboreal pygmy ant-eater (*Cyclopes*).

The xenarthrous articulations are additional joints of the last thoracic and lumbar vertebrae, which contribute to the strengthening of the posterior area of the trunk. They are a significant character of the edentates. These vertebrae possess a complicated pattern of numerous bone processes. The xenarthrous articulations are situated below the normal vertebral articulation (Fig. 310). A lateral process of a vertebra, the so-called

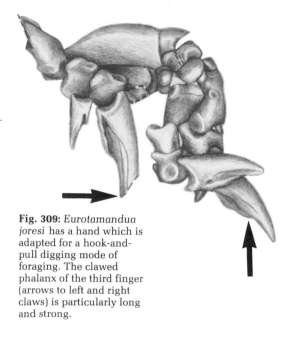

Fig. 309: *Eurotamandua joresi* has a hand which is adapted for a hook-and-pull digging mode of foraging. The clawed phalanx of the third finger (arrows to left and right claws) is particularly long and strong.

anapophysis, extends backward and is interlocked below and on top with one additional articulation area each, of the succeeding vertebra. The pelvic region is further consolidated. The sacrum consists of five fused vertebrae: of these, the three proximal ones are connected, via their transverse processes, with the ilium of the pelvis. Exhibiting another edentate character, the most distal vertebra of the sacrum still has a second ossified connection to the pelvis, namely, with the ischium. By this means a characteristic window, framed all round by bones, is formed between the vertebral column and the hip joint. An important edentate character of *Eurotamandua* is also found in the structure of the shoulder blade. Its outer face does not show—as is usual—only one strong longitudinal crest, but bears two of these, leading to an enlargement of the attachment base for the musculature.

Other conspicuous and just as strange features are characteristic only of the family Myrmecophagidae, including the extant *Myrmecophaga* and *Tamandua*. The skull of *Eurotamandua* is elongated and tube-shaped and the lower jaws represent very long, thin bony bars. The teeth are entirely absent (Fig. 309). But also in anatomical details there is a correspondence with living ant-eaters. The premaxiliary, for example, is very small and the parietal bone extends far down the wall of the undivided, shallow orbito-temporal depression. The lower jaw symphysis is slightly blade-shaped and bent down; this is quite different from that of the, also toothless, pangolins. *Eurotamandua* already possesses the highly diagnostic ear features of extant

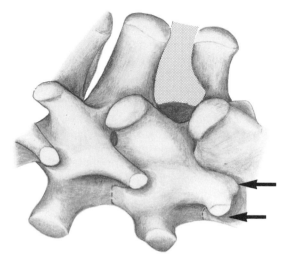

Fig. 310: Vertebrae from the trunk of *Eurotamandua joresi*, with multiple bone processes. The additional xenarthrous articulations are marked by arrows. Drawn in an oblique dorsal view; the bodies of the vertebrae are not visible.

weight of its body, when walking, to the outer rim of the hand, while turning the second and third fingers with their armour of long, horny claws, back towards the palm. The middle finger exhibits extreme specializations that essentially lead to a curtailment of mobility in the joints and that are certainly connected to the hook-and-pull digging mode of foraging. Powerful and fast pouncing with a strongly flexed hand involves, on impact with hard insect mounds, an element of danger from sprains. The massive middle hand bone bears an insertion scar of extensor muscles that is also characteristic for *Tamandua* and

Myrmecophaga and, at the end, a large, medial crest to steady the control of the joint. The short and compact basal phalanx of the third finger has no pivot joint toward the middle phalanx but is almost completely interlocked with the latter by means of differentially orientated articulation facets.

Eurotamandua and *Eomanis* from Messel are the two oldest and most complete finds of ant-eaters and pangolins known. Morphological adaptations towards the same goal of a highly specialized feeding strategy are very extensive particularly in both these groups; this, of course, increases the difficulty of assessing possible relationships between them. In the features that were introduced above as being characteristic of the group, however, *Eurotamandua* correlates already with the Recent ant-eaters (Figs 306 and 307), just as *Eomanis*, on the other hand, correlates with the Recent pangolins, so that the Messel fossils give just as little hint of a closer relationship, as the most recent findings of zoology and biochemistry. These two remarkable mammals have evolved independently of one another.

Just like the present-day *Tamandua*, *Eurotamandua*, aided by powerful arm and shoulder muscles, must have torn open nests of ants and termites and, before their soldiers could unite in defence, quickly ingested the teeming insects with its long, sticky tongue. The whole power of the blows was, as with a pickaxe, practically concentrated on one point, namely, the tips of the powerful, strongly flexed middle fingers. Such a tool is not only very important in gaining access to food, but also represents a powerful weapon with which deep wounds can be inflicted on a comparatively large aggressor. The prey was certainly tracked down by using the sense of smell. It had to be available in sufficient volume throughout the year—extant collared ant-eaters swallow approximately 9000 ants per day (Montgomery 1985).

This single specimen of *Eurotamandua joresi* fortunately contained a relatively large amount of gut contents. The flaky, dark-brown mass consists to a small degree of sand grains (as used in the stomachs of Recent ant-eaters for the mechanical disintegration of insects) as well as of plant material, among this a few 'fresh' leaf segments (with cell preservation). Mixed in with these plant and mineral remains are densely packed layers of chitin cuticle with predominantly similar scaly patterns (Figs 311 and 312). Closer hints as to the insects eaten are not obtained from such cuticles, because varying scale patterns can occur on different parts of the body of the same insect (Figs 313–315). Numerous very thin (and therefore 'crumpled' in appearance) cuticles (Fig. 316) could, according to comparative investigations, originate from termites (Figs 313–315) which have a soft-skinned abdomen, but this cannot be proved without finds of head capsules and mouth-parts.

Besides a few leaf segments and unidentifiable splinters of wood, the gut contains numerous small lumps of a substance which partly has a woody structure, and partly is without structure (Fig. 317). In this it resembles strikingly the wall material of a nest of extant tree termites (Fig. 318). Tree termites use splinters of wood, cemented with a sort of saliva, to build their nests, and parts of such nest walls can be ingested—together with the nest-dwellers—by an ant-eater.

If the finds are interpreted carefully one can say confidently that the investigated *E. joresi* specimen had eaten large amounts of insects, that these insects represented, presumably, few species, that the many thin cuticles could come from termites (but also from insect larvae), and that, finally, one type of particle found very frequently in the gut contents corresponds, in

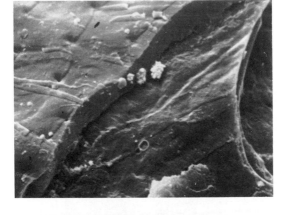

Fig. 311: Gut contents of *Eurotamandua joresi*. Multiple layers of insect cuticle of unknown origin. SEM: magnification, 1700×.

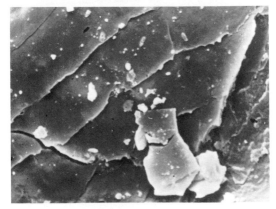

Fig. 312: Gut contents of *Eurotamandua joresi*. Fragment of a scaly insect cuticle. SEM: magnification, 1700×.

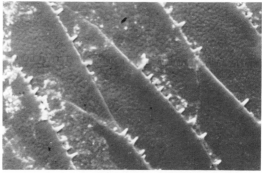

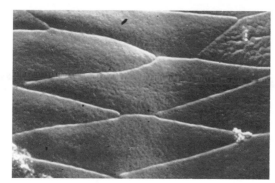

its structure, with the wall material of recent tree-termite nests. The feeding picture that arises from this, although it may be speculative in some parts, corresponds with the investigated group of animals exactly and should not be refuted by other (hoped for) finds, but will probably be differentiated and broadened. Just as this specimen had plundered, presumably, a tree-termite nest, another specimen could have emptied a mound of earth termites or of ants, which could probably be proved in the fossil gut contents.

Fig. 313: (left, above); **Fig. 314:** (left); **Fig. 315:** (right, above). Cuticular sculpturing on the abdominal segment of an unidentified species of extant termite. The variation dependent on angle of view (dorsal, lateral, or ventral) could be easily misinterpreted if the individual parts were investigated separately. SEM: magnification, 1600×.

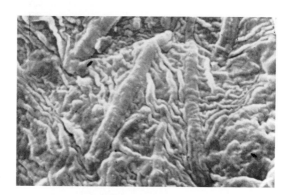

Fig. 316: Gut contents of *Eurotamandua joresi*. Thin, strongly crumpled insect cuticle with single hairs. It probably belonged to a termite (abdomen) or a thin-skinned insect larva, and is very common in gut contents. SEM: magnification, 4200×.

Fig. 317: Gut contents of *Eurotamandua joresi*. Fragment of a substance which is relatively common in the material investigated and in which pieces of woody plant tissue are cemented together by an unstructured mass. SEM: magnification, 165×.

Fig. 318: Fragment of the wall of a recent tree-termite carton nest. Here too we find the same mixture of splinters of wood and unstructured cementing substance. SEM: magnification, 65×.

18

Rodents: at the start of a great career

The characteristic feature of rodents is the pair of incisors in each of the upper and lower jaws, whose continuous growth promotes a strong grinding action, without their outer appearance being altered. This feature had already evolved by the early Tertiary, at the boundary between the Palaeocene and the Eocene, when the first rodents appeared. However, at that time, neither the tooth enamel nor the mastication muscles had the differentiated structure that is found in modern rodents. These earliest rodents belong to the family Paramyidae which survived in some phylogenetic lineages until the end of the Eocene. Two of these lineages are represented in the Messel oil shale, by the genera *Ailuravus* and *Microparamys* (Wood 1976). The third genus of rodents from Messel, *Masillamys*, belongs to the family Pseudosciuridae (Hartenberger 1968), an already further differentiated group which assumed its greatest importance in the Oligocene.

Rodents: at the start of a great career

WIGHART VON KOENIGSWALD,
GERHARD STORCH, AND
GOTTHARD RICHTER

Ailuravus macrurus is the largest of the Messel rodents. The animal is documented by numerous skeletons, in part with excellent preservation of silhouettes of the fur and gut contents (Figs 319 and 321). It was first described by Weitzel (1949) and later investigated afresh in relation to its dentition by Tobien (1954) and Wood (1976).

Ailuravus is far larger than a squirrel and, with its *c.* 9-cm-long skull, it approaches the size of a marmot. However, the 40 cm head and body length of *Ailuravus*, like its limbs, is distinctly shorter than that of the marmot. Nevertheless, in its tail length of approximately 60 cm, *Ailuravus* scores over the ground-living marmot.

The massive head of *Ailuravus* is marked by bulky muscle attachments on the roof of the skull and below the cheek-bones. The arrangement of the mastication musculature corresponds with the primitive ('protrogomorph') type (Fig. 320). The arms are shorter than the legs by approximately one-third and permit free movement in all directions. In the hand, the thumb is strongly reduced and the remaining fingers bear sharp claws which resemble those of the squirrel. The legs are strong and the digits of the foot can be spread so that they allow a secure grip. On the toes sharp claws, characteristic for tree-climbers, can also be observed. The tail is of great importance for its type of locomotion. It is strongly tufted, particularly in its distal part (Reisinger 1982); this does not significantly change its weight as a balancing organ, but contributes important steering qualities when jumping.

According to its preserved bone morphology, *Ailuravus* should have therefore

Fig. 319: *Ailuravus macrurus*, large rodent prepared completely free from the matrix. Note the large incisors in the upper and lower jaw. Total length, 54 cm.

been arboreal, moving perhaps, like the extant oriental giant squirrel (*Ratufa*) from the Indo-Malaysian area, with large leaps through the tree canopy (Weitzel 1949).

This biological reconstruction appears to be supported by the evidence from gut contents. *A. macrurus* was a leaf-eater, perhaps even more so than the leaf-eating, early horse whose gut contents sometimes include large amounts of seeds as well as leaves. The leaf remains in *Ailuravus* gut contents are exclusively cuticles

with a diameter of 3 mm. They are relatively large and excellently preserved. However, while leaf margins are frequent, petiole fragments, fruits, or seeds are completely absent. To this extent *A. macrurus* was a food specialist, although this does not necessarily hold for the types of plants eaten. In each of the two specimens whose gut contents were investigated, leaves of different plant species occur mixed together. Although leaves of one given plant predominate in the gut contents of each animal, in each instance they are of a different species. In one case, they are leaves with a very thin cuticle in which only the leaf margin is thickened, and have large epidermal cells (leaves from a member of the Lauraceae; Wilde 1989; Fig. 322). In the other case, the leaves have a very thick cuticle and relatively small epidermal cells (leaves of *Polyspora saxonia* (Thecaceae); Wilde 1989; Figs 323 and 324).

A. macrurus was, therefore, exclusively a leaf-eater, but did not require the leaves of one specific plant species or genus as the koala does today, for example. Nevertheless, one plant species distinctly dominates in each of the gut contents. How can this be interpreted in a biotope in which the flora was so rich in species and diverse as at Messel (see Chapter 4, 'The vegetation—fossil plants as witnesses of a warm climate')? Food analyses indicate that *A. macrurus* was, indeed, arboreal, that is, it lived in trees just as Weitzel (1949) postulated on the basis of its anatomical characteristics. Only an arboreal leaf-eater will have in its stomach, preferably or even exclusively, the leaves of the tree species which it has eaten last. According to Dr Wilde (personal communication), the food plants are certainly leaves from trees or tree-like bushes.

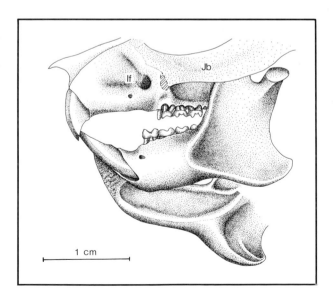

Fig. 320: *Microparamys parvus*, upper and lower jaw. The primitive (= protrogomorphous) mastication apparatus is characterized by a small infraorbital foramen (If) and an insertion of the chewing muscle (masseter muscle) which is restricted to the lower side of the cheek bone (Jb).

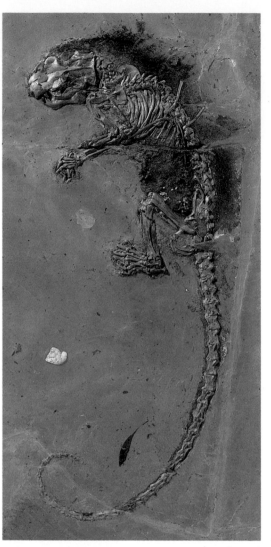

Fig. 321: *Ailuravus macrurus* with dense fur on the back. Other specimens show a bushy tail. Total length, approximately 1 m.

Ailuravus has a parallel form in the North American genus *Leptotomys* (Wood 1976). Both forms hark back to common ancestors within the paramyids which occurred in both North America and Europe due to the intensive faunistic exchange via the Arctic (see Chapter 27, 'The Messel fauna and flora—a biogeographical puzzle'). The fact that different genera evolved in the middle Eocene in North America and in Europe shows that the land bridge did not permit further exchange. The rapid pace of rodent evolution enables such estimates of time, whereas crocodiles and some birds, for example, *Diatryma*, can still be assigned to the same genus long after the geographical separation of faunas.

The rapid evolution of the rodents makes them excellent zone fossils. On this basis Haupt (1911) recognized that the Messel Formation, which had until then been assigned to the Oligocene, had to belong to the Eocene because he recognized *Ailuravus* in addition to the small early horse *Propalaeotherium*.

Although *Microparamys parvus* (Fig. 320) is the smallest of the rodents known so far from Messel, even this species is still distinctly larger than the modern European wood and house mice. The length of the skull is approximately 35 mm, the head and body length 13–14 cm, and there is a 12.5-cm-long tail. Larger than *M. parvus*, but distinctly smaller than *Ailuravus*, are the two species of *Masillamys*: *beegeri* and *krugi*. Their skulls are approximately 47 mm long, their head and trunk lengths are *c.* 17.5–20 cm, and the tails are only slightly longer.

At first glance, it is apparent that *Microparamys* and *Masillamys* could not have rushed through the canopy with the powerful, long leaps that we assume for *Ailuravus*. Their sparsely haired tails could neither serve as rudders nor as parachutes when leaping. The body proportions even more clearly exclude such a mode of locomotion. The legs of *Microparamys* and *Masillamys* (Fig. 326) are extraordinarily short: if their leg lengths are compared to their trunk lengths (thoracic and lumbar vertebral column), then the relative values for the hind legs correspond with those of extant digging species of voles which live predominantly underground. The arms and

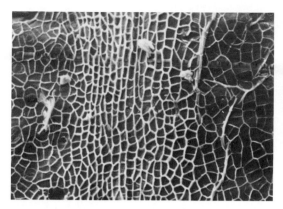

Fig. 322: Gut contents of *Ailuravus macrurus*. Cuticle of a member of the Lauraceae. Left: Underside of leaf with stomata; middle: leaf margin; right: upper surface of leaf without stomata. SEM: magnification, 85×.

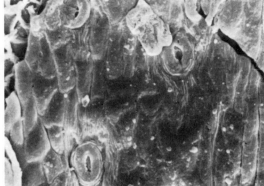

Fig. 323: Gut contents of *Ailuravus macrurus* (animal shown in Fig. 324). Leaf cuticle of *Polyspora saxonia*. 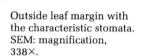 Outside leaf margin with the characteristic stomata. SEM: magnification, 338×.

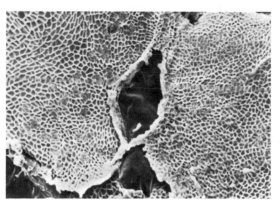

Fig. 324: Gut contents of *Ailuravus macrurus*. Cuticle from the dominant type of leaf found in this stomach (*Polyspora saxonia*). The leaf is strongly cutinized. SEM: magnification, 70×.

hands are even shorter than in these specialized underground-dwellers. The European red squirrel, by contrast, has relatively long and powerful arms and legs (for measurements see Schmidt-Kittler and Storch 1985).

Over the course of time, rodents have assumed very different life-styles, which include climbing and jumping through the branches, running and hopping on the ground, and digging and swimming. These life-styles are particularly expressed in the proportions of their limbs. Despite their similarity to the voles

in their short extremities, one cannot assign both species to a subterranean habitat as a matter of course.

The relatively long tails alone would be unusual for inhabitants of narrow passageways. Moreover, adaptations to digging with the front feet (long claws, prominent muscle scars on the skeleton of the arm) or with the aid of the jaws (incisors inclined strongly anteriorly) are absent. What alternative remains? In view of Lake Messel, swimming could be considered. Among the rodents, not only the specialists but also the 'all-rounders', with generalized

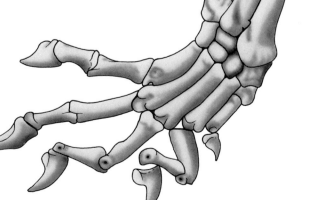

Fig. 325: Isolated arm of an as yet unidentified rodent. Note the considerably reduced thumb (at the lower margin of the hand), characteristic of the early rodents. Length of humerus, 7 cm.

Fig. 326: *Masillamys beegeri*, a young animal with an incomplete tail and excellently preserved soft-body outline. Total length, 17 cm.

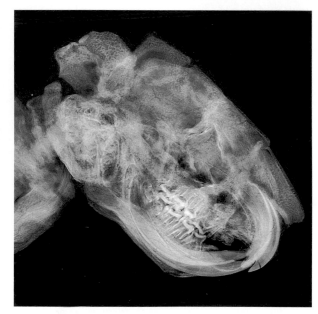

Fig. 327: X-ray photograph of *Masillamys beegeri* (the animal shown in Fig. 326). The rootless incisors, anchored deeply in the jaws, are striking.

proportions such as the brown rat, are excellent swimmers. The striking body proportions of *Microparamys* and *Masillamys* which deviate from these suggest, however, that a very definite mode of locomotion was paramount. For instance, the Australian–New Guinea water rat (*Hydromys chrysogaster*) is specialized for a life in water. In this swimming body form, the hind-legs carry out the main work of propulsion; the foot and shank are correspondingly lengthened relative to the upper thigh (Schmidt-Kittler and Storch 1985). The Messel genera all have shorter hind-legs and, most importantly, roughly equal lengths of upper thigh, shank, and foot that exclude them from being specialized water-dwellers. There are, finally, also no skeletal characteristics suggesting a running or jumping locomotion on the ground. The short extremities would argue against this.

We must not expect that all adaptations of the mammals of Messel need to have counterparts in extant species—even modes of locomotion have obviously become extinct (see Chapter 13, 'Primitive insectivores, extraordinary hedgehogs, and long-fingers').

On the other hand, extant types of adaptation compared here do not necessarily fill all possible ecological niches. Arborealism—living in trees—can be realized perfectly by such different mammals as ambling sloths and leaping primates. The squirrel type is not the only solution among the rodents.

Among the living mice and rats there are a number of climbing forms which inhabit tropical rain forests, which correlate well in certain characteristics with our species at Messel. In these, the hands and feet are also short but rather wide. Their tail length does not deviate much from their head and body length and the claw-bearing digits are relatively narrow and high. For instance, *Microparamys* and *Masillamys*, may have corresponded in their life habits most closely with climbing rats (*Tylomys*), Malay tree rats (*Pithecheir*), or Philippine forest rats (*Carpomys*). These extant rodents are all confident climbers that, however, do not move through the branches by leaping.

19

Carnivores: agile climbers and prey catchers

The absence of carnivores from the ecological balance of a modern fauna is unimaginable. Palaeontologists were surprised, therefore, that, for a long time, they found no evidence of carnivores in the Messel oil shales. Two orders of mammals in particular were absent, which would normally have been expected in the Middle Eocene: namely the representatives of the so-called primitive carnivores (Creodonta), and those of the ancestral group of the modern carnivores (Carnivora). Finally, in 1980 a complete skeleton became known, which had been discovered in 1974 by the private collector Otto Feist of Mühltal, near Darmstadt.

Carnivores: agile climbers and prey catchers

RAINER SPRINGHORN

As a basis for better differentiation in respect to other orders of mammals, but also in order to establish a nomenclature, a short description of the generalized dentition of the carnivore (Fig. 331) will be presented here. Lower Tertiary Creodonta and Carnivora generally possess the complete dentition of mammals, which consists, on a given side of a jaw, of three incisors, one fang or canine, four premolars, and three molars. Incisors and canines together

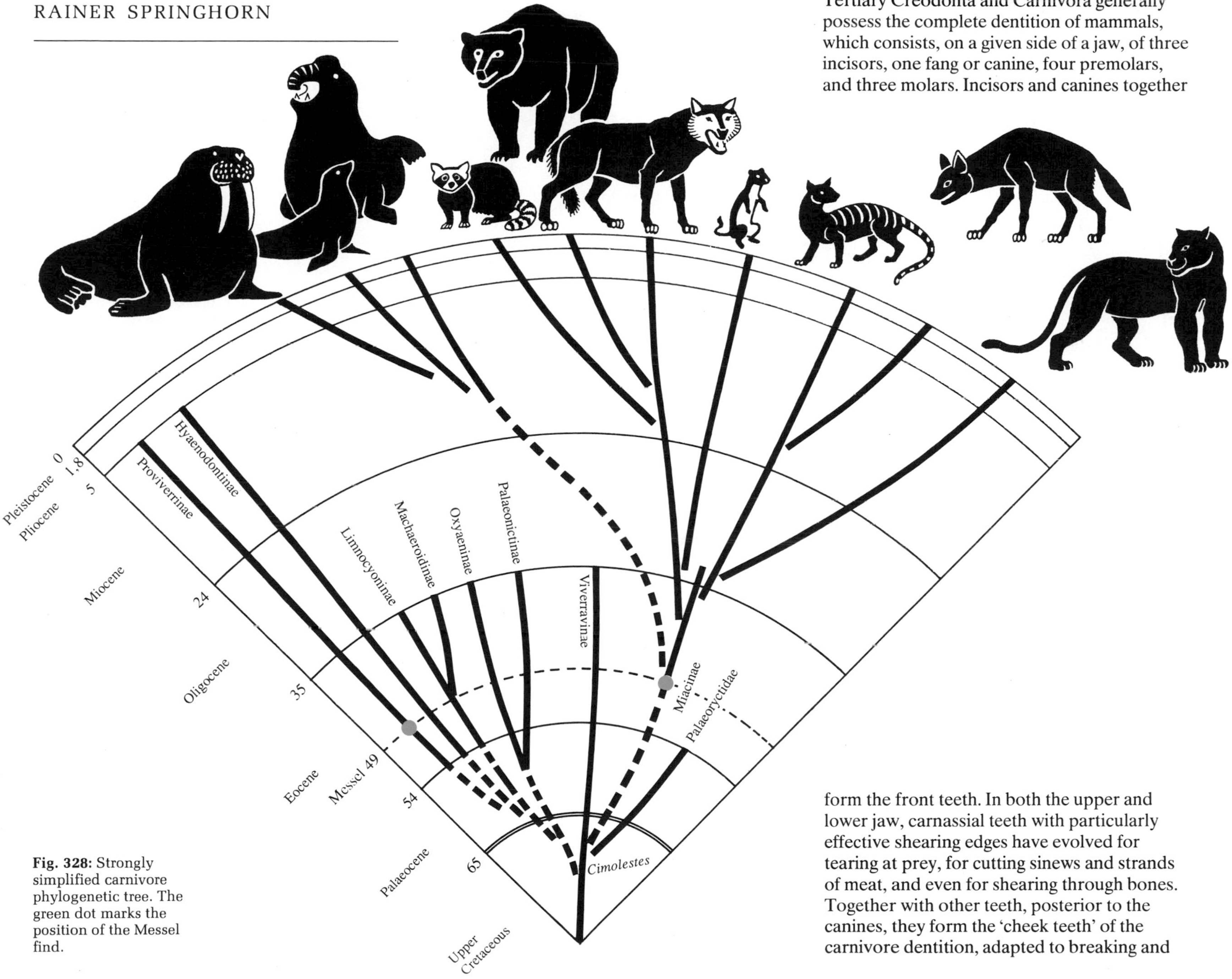

Fig. 328: Strongly simplified carnivore phylogenetic tree. The green dot marks the position of the Messel find.

Labels in figure: Pleistocene 0 1.8, Pliocene 5, Miocene, Oligocene 24, Eocene 35, Messel 49, 54, Palaeocene 65, Upper Cretaceous; Hyaenodontinae, Proviverrinae, Limnocyoninae, Machaeroidinae, Oxyaeninae, Palaeonictinae, Viverravinae, Miacinae, Palaeoryctidae, Cimolestes

form the front teeth. In both the upper and lower jaw, carnassial teeth with particularly effective shearing edges have evolved for tearing at prey, for cutting sinews and strands of meat, and even for shearing through bones. Together with other teeth, posterior to the canines, they form the 'cheek teeth' of the carnivore dentition, adapted to breaking and

225

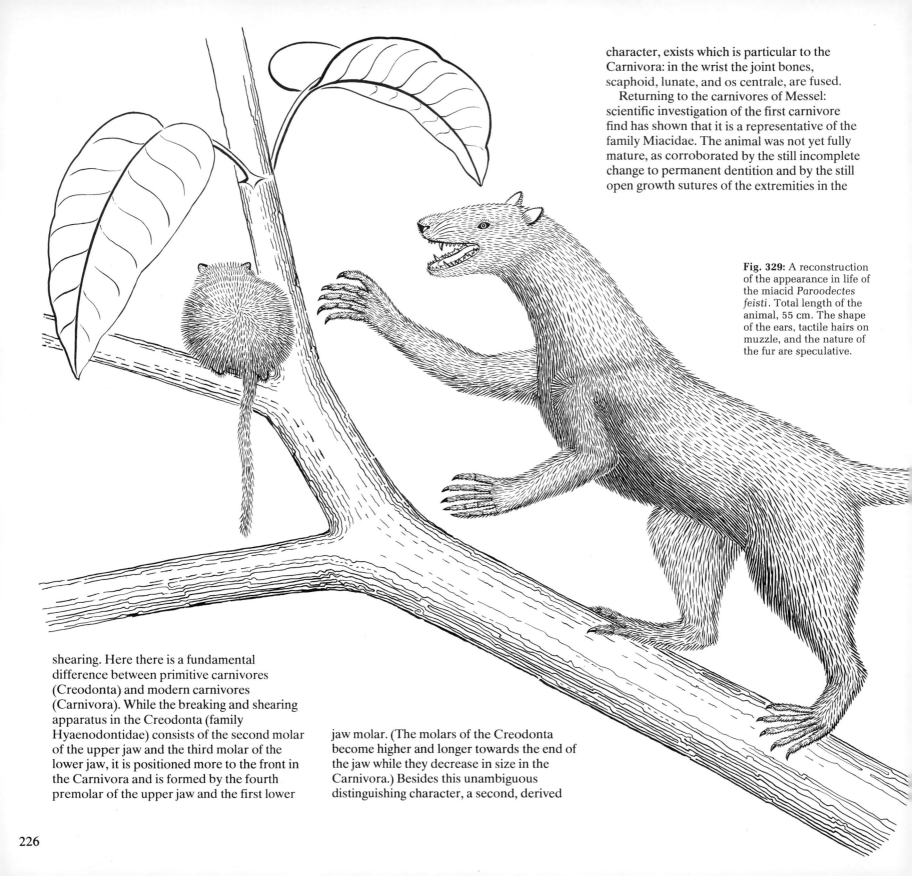

character, exists which is particular to the Carnivora: in the wrist the joint bones, scaphoid, lunate, and os centrale, are fused.

Returning to the carnivores of Messel: scientific investigation of the first carnivore find has shown that it is a representative of the family Miacidae. The animal was not yet fully mature, as corroborated by the still incomplete change to permanent dentition and by the still open growth sutures of the extremities in the

Fig. 329: A reconstruction of the appearance in life of the miacid *Paroodectes feisti*. Total length of the animal, 55 cm. The shape of the ears, tactile hairs on muzzle, and the nature of the fur are speculative.

shearing. Here there is a fundamental difference between primitive carnivores (Creodonta) and modern carnivores (Carnivora). While the breaking and shearing apparatus in the Creodonta (family Hyaenodontidae) consists of the second molar of the upper jaw and the third molar of the lower jaw, it is positioned more to the front in the Carnivora and is formed by the fourth premolar of the upper jaw and the first lower jaw molar. (The molars of the Creodonta become higher and longer towards the end of the jaw while they decrease in size in the Carnivora.) Besides this unambiguous distinguishing character, a second, derived

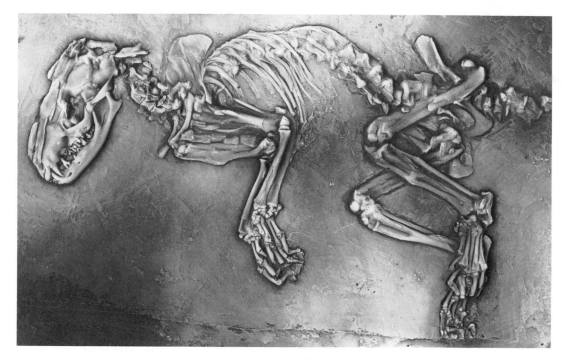

long bones. In spite of this, with a length of scarcely 55 cm from the snout to the tip of the tail, the animal is the largest carnivore from Messel to date. On the basis of the close similarity of its dentition with that of the North American genus *Oodectes*, and in honour of the discoverer, the species was called *Paroodectes feisti* (Springhorn 1980).

Some characters of the skeleton are evidence for the primitive nature of this carnivore. The dentition possesses a complete tooth formula and shows, during the change of teeth, parallel functions of the 'breaking and shearing' milk and replacement dentition. The clavicles, generally completely vestigial in extant carnivores, are strongly developed. They ensure the stabilization of the shoulder girdle and assist climbing. The fact that the three wrist bones are not entirely fused may relate to the youthfulness of the animal. Finally, a small third trochanter is developed on the femur.

Fig. 330: Complete skeleton of the miacid *Paroodectes feisti*. The length of the illustrated portion of the skeleton is 25 cm. Only the first five vertebrae of the completely preserved tail are illustrated. Skeleton sprayed with ammonium sulphide to achieve greater contrast.

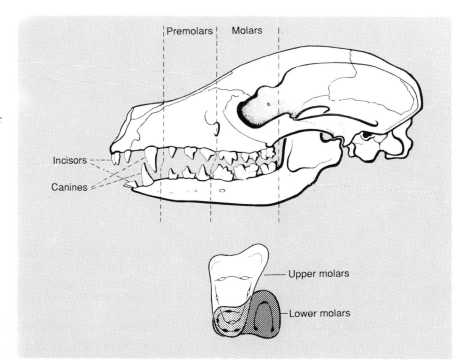

Fig. 331: Skull of a Recent southern coati (*Nasua nasua*) showing the different sections of carnivore dentition (above). Upper (white) and lower jaw molars (shaded) in surface view. The edges and cusps show the areas of closing and shearing when the jaw is closed (below).

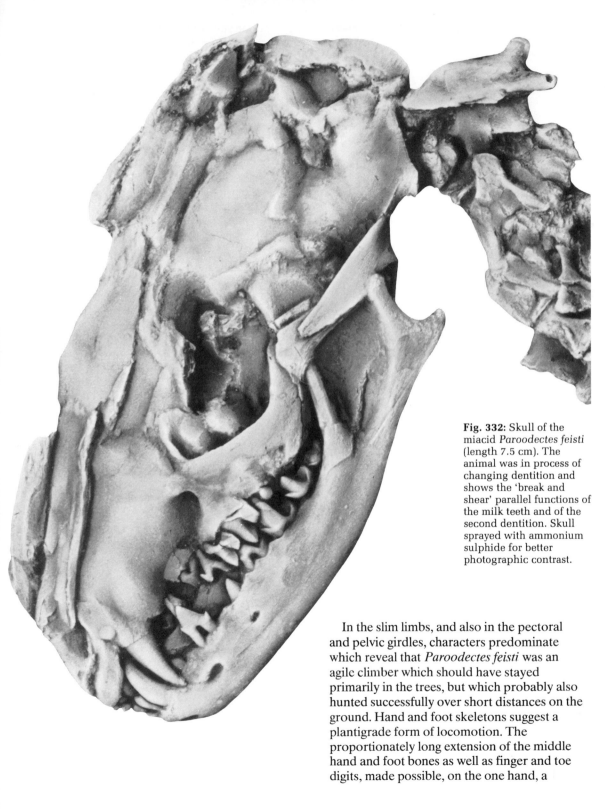

spreading of paws and gripping of branches and, on the other hand, increased the catapult-like action when jumping. The strong, short claws were retractable and ensured reduced frictional effects upon running.

The very long tail of the animal, measuring far more than half of its entire length, was not used for grasping and holding tight to a branch as in some modern New World monkeys but must have served for keeping its balance and as a rudder during leaps from branch to branch.

The preferred prey of *Paroodectes feisti* were, most likely, tree- and ground-dwelling insectivores, rodents, and primates. It possibly also lay in ambush, sitting in trees, for hooved animals, pouncing, preferentially, on their young.

In 1970 the private collector Ullrich Kessler from Darmstadt recovered an individual of a second species of carnivore from the Messel pit. Data on this were, however, not published until 1982 under the name *Miacis? kessleri* (Springhorn 1982). In the meantime, two further animals of this species have been described (Springhorn 1985), of which one—found in 1982 during the Senckenberg excavation—shows vestiges of the fur and of the hair on the tail. All three skeletons are evidence for the fact that *Miacis? kessleri* was a very small species of carnivore of approximately 21 cm total length.

All three specimens are young animals which have either not yet started or not yet concluded their change of teeth. In the X-ray the different stages of this process can be documented. The upper fourth premolar or the milk dentition has an extraordinary shape for a member of the Carnivora. With two outer cusps and an additional distal cutting edge, it resembles the replacement dentition molars of the Creodonta. In so far as this structure, obtained from X-ray analysis, can be confirmed, this tiny character could be taken as argument that Carnivora and Creodonta had common ancestors at a geologically earlier time (probably in the late Cretaceous).

The skull of the miacid is relatively long. This is emphasized by a scarcely raised forehead, the merest hint of a muscle attachment on the cranium, and a very slim lower jaw.

Fig. 332: Skull of the miacid *Paroodectes feisti* (length 7.5 cm). The animal was in process of changing dentition and shows the 'break and shear' parallel functions of the milk teeth and of the second dentition. Skull sprayed with ammonium sulphide for better photographic contrast.

In the slim limbs, and also in the pectoral and pelvic girdles, characters predominate which reveal that *Paroodectes feisti* was an agile climber which should have stayed primarily in the trees, but which probably also hunted successfully over short distances on the ground. Hand and foot skeletons suggest a plantigrade form of locomotion. The proportionately long extension of the middle hand and foot bones as well as finger and toe digits, made possible, on the one hand, a

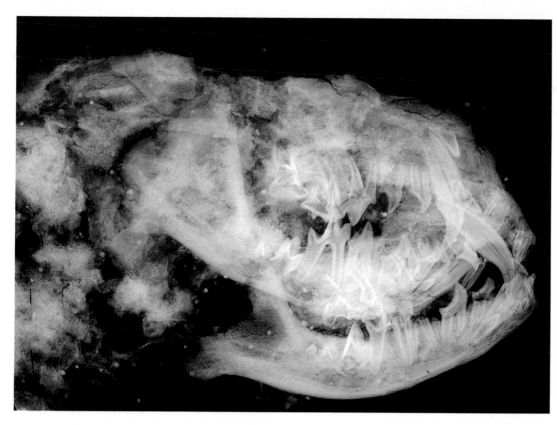

This species of Carnivora also shows well-developed clavicles. Structures for muscle attachments, ridges for the attachment of sinews on the humerus, the development of the elbow, characters of ankle elements as well as of the paws adapted to plantigrade motion, all suggest that *Miacis? kessleri* was not a very good runner, but that it was very well adapted to life in the branches of trees. The short lower limbs of anterior and posterior extremities contribute to this view because, with their help, the small predator was able to press its body close to its support in order to maintain its balance. The bones of the fore and, respectively, the hind middle foot (metapodia), as well as the finger and toe digits, are elongated, even if not as strongly as in *Paroodectes feisti*. Armed with strong claws, they preadapt *Miacis? kessleri* for climbing.

The bushy tail measures somewhat more than one-third of the entire length of the animal and would have been used as a rudder in leaping.

To date, only meagre vestiges of fur of a single immature animal are recorded and a

Fig. 333: X-ray photograph showing the skull structure of *Miacis? kessleri*. The young age of the animal is indicated by its complete set of milk teeth. In addition, the first elements of the second dentition are visible as embryonic teeth. The left lower jaw was compressed under the right lower jaw. With the exception of the canine, the dentition of the left upper jaw—behind the dentition of the right lower jaw—is only very indistinctly visible.

Fig. 334: (left); **Fig. 335:** (right). Incomplete skeleton of *Miacis? kessleri*. The left forelimb was presumably detached after the corpse was embedded in the sediment and now lies above the base of the tail. The specimen is remarkable because it shows vestiges of fur in the areas of nape, chest, and tail. Length, *c.* 21 cm.

reconstruction of the pelt of a mature animal is, therefore, not admissible. It is possible that in the young animal only vestiges of the underfur remained, while the longer silhouette fur had not yet grown. The reconstruction of *Miacis? kessleri* attempted here (Fig. 336) is contrasted with another of Savage and Long (1986, p. 75). These authors assume a longer hair cover that corresponds with that of the Recent pine marten and has a pattern of dark stripes.

Regarding its nutrition, it can be assumed that the little miacid, as a nest-robber, took eggs and flightless chicks. Inexperienced, albeit just able to fly, nestlings could be regarded as prey just as much as small, arboreal reptiles and insects.

The primitive carnivores (Creodonta) in the Lower Tertiary experienced a moderate evolution of different types of adaptive (radiation). Within the two families: Hyaenodontidae and Oxyaenidae, a differentiation into six subfamilies was achieved. At Messel only one of these is represented with only one species. *Proviverra edingeri* (Springhorn 1982) is a representative of the subfamily Proviverrinae within the Hyaenodontidae. Two individuals are known. One was recovered in 1974 by a private collector from Leverkusen, Ernst Edinger; the second comes from the 1983 Senckenberg excavation.

Both animals had not yet completed their second dentition; the growth sutures on the long bones are, in part, still open. Both skeletal finds are incomplete, so that a reconstruction of the species is impossible. The specimen that was recovered first has a head and body length of 18 cm; because only the first five vertebrae are preserved from the tail skeleton, the total length cannot be given.

With the exception of the first upper premolar, there is a complete tooth formula. In the frontal dentition of the upper jaw are exceptionally large, canine-like, third incisors which are placed immediately in front of the actual canines. The upper molars have prominent outer cusps which are distally succeeded by a marked cutting edge and which are characteristic of the Creodonta. Besides the 'break and shear' effect described above, the fourth premolar of the upper jaw and all other molars have, in addition, a subordinate cracking effect. The skull of the animal is slim; in the area of the cranium and the back of the head only a low crest for muscle attachment is displayed.

The characteristics of the limbs do not point unequivocally to a specialized runner. The species also cannot be identified as a climber. The individual radii of the paws exhibit a slightly splayed arrangement; when walking they must have touched the ground only to

such an extent that the middle hand and the foot bones had almost no contact with the ground. In contrast to the case of the Carnivora whose third and fourth toes of the front and hind paws are, approximately, of equal length, the middle toe dominates in *Proviverra edingeri*. The front paws are unusually proportioned in so far as the first and second toe are distinctly longer than toes four and five. This can possibly be traced back to an adaptation for digging and scratching (Springhorn 1988). An otherwise typical character of the Creodonta seems not to be developed in the proviverrids of Messel: there is no unequivocal evidence of a splitting of its claw tips.

An overview of all anatomical characters of the skeleton shows that *Proviverra edingeri* did, as a matter of preference, hunt in the undergrowth of the paratropical jungle of Messel, and during this hunt also scratched in the ground for insects and invertebrates as well as for small amphibia and reptiles.

It is true, with regard to the nutrition of all Messel predators, that, in addition to meat requirements, a not inconsiderable amount of plant food (predominantly fruit) was probably available. One must draw this conclusion with reference to the nutritional needs of modern carnivores in similar habitat conditions.

In summarizing this review of the Messel

Fig. 336: Reconstruction of the miacid *Miacis? kessleri* when alive. Total length of the animal, *c.* 22 cm. The basally rounded lower jaw, the low attachment of the ears, and the low forehead that exhibits almost no upward curve are characteristic for this small carnivore. The relatively short lower legs are very unusual.

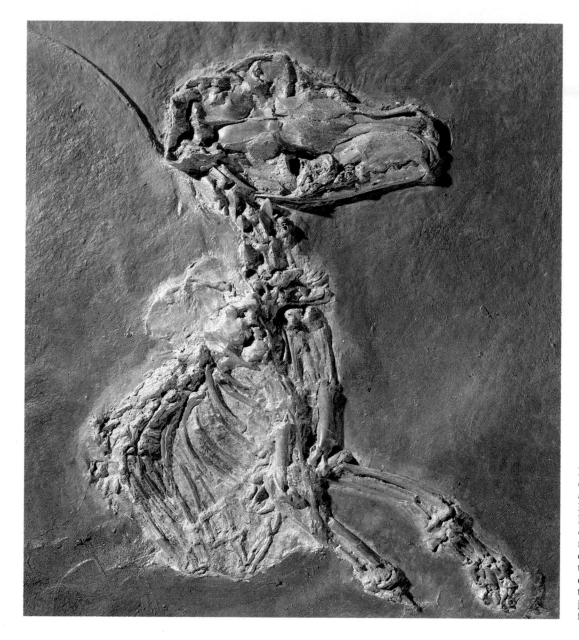

Fig. 337: Incomplete skeleton of the hyaenodontid *Proviverra edingeri*; fossil slab containing the right side of the body. The distal part of the body is largely missing even in the left (matching) fossil slab. Length of the partial skeleton, 17 cm. The skull was compressed into the plane of sedimentation such that the observer views the roof of the skull.

Fig. 338: Left frontal dentition with premolars and molars from the right lower law of the hyaenodontid *Proviverra edingeri*. A window was made in the back of the epoxy resin matrix. From the right side (Fig. 337). It can be clearly seen that the molars of the lower jaw increase in height from front to back.

carnivores two questions arise. Why have only young animals been found of the three known species, and why haven't more and, above all, larger forms been found?

The first question cannot be answered satisfactorily. There have been repeated deliberations on the cause of death and the thanatocoenosis of the vertebrate cadavers encountered in the sediments of Lake Messel.

Among the mammals, there is a significant preponderance of finds that, according to their relaxed skeletal position and the degree of disintegration, must be regarded as having drowned. This is true for all finds of carnivores. Whatever causes led to the drowning of the animals remain unknown. Perhaps the young animals, lacking experience and being weaker than the mature animals, fell victim to flood-

waters after heavy rains or, perhaps, while exploring the morass at the lake shore, they became stuck, drowned exhausted, and later drifted away. Scratch marks in the sediment from the death throes of the animals have not been reported, however, neither in adjoining skeletons nor in their vicinity.

The second question, relating to the paucity of species and to body size, is justified because

231

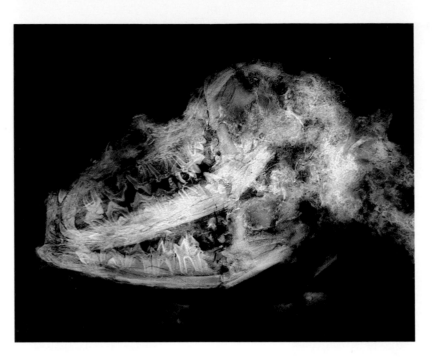

Fig. 339: X-ray of the skull of *Proviverra edingeri*. This is an almost fully-grown individual in which the change of dentition is almost complete. The right lower jaw was distorted to lie below the left. The dentition of the right upper jaw cannot be determined. Length of skull, *c.*5.3 cm.

from other, approximately contemporaneous fossil sites of Central Europe, mainly from the Geiseltal, Halle/Saale, Creodonta have been recorded which, with few exceptions, are distinctly larger than the Creodonta and Carnivora from Messel. But there the ecological conditions were different in that there was an open type of landscape which favoured a different species spectrum. Still, only three carnivores are few even for a paratropical jungle. It is, therefore, not presumptuous to predict that, in future, more species of carnivores will be recovered from Messel.

20

The arboreal *Kopidodon*, a relative of primitive hoofed animals

Teeth serve as an identity card for mammals. They give information about the individual age of the animal, usually allow us to determine genus and species, and, in addition, give unequivocal indications of relationships with other animal groups. Thus some descriptions of extinct families of animals appear to consist solely of a combination of the characters of the teeth of the upper and lower jaw. The finds of skeletons from Messel, however, open a window into the palaeobiology of these animals.

The arboreal *Kopidodon*, a relative of primitive hoofed animals

WIGHART VON KOENIGSWALD

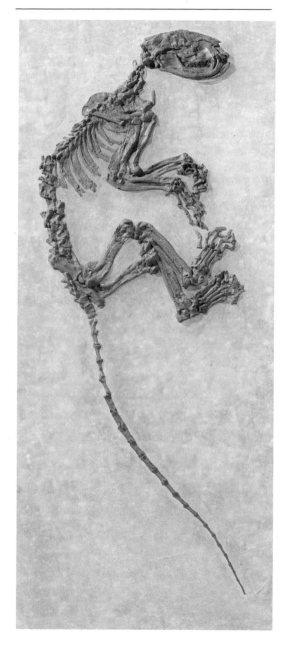

Fig. 340: *Kopidodon macrognathus*, an arboreal fruit-eater. The basal part of the tail has been incorrectly reconstructed. Total length, *c.* 115 cm.

Dental characters of the genus *Kopidodon* (Fig. 340) indicate that it is a member of the Paroxyclaenidae (Tobien 1969), a group that belongs among the relatives of early hoofed animals. However, this systematic classification does not reveal the mode of life of *Kopidodon macrognathus*; in what kind of habitat did it live? This question appears to be very simple: whether *Kopidodon* should be counted among the highly specialized swimmers, fast runners, slower plantigrades, burrowers, or climbers, either those that jump from branch to branch or those that glide with flight membranes, its mode of locomotion will be revealed by analysis of the skeleton. Bone-by-bone analysis of those parts of the skeleton involved with movement of the body can identify characteristics pointing to adaptations for one form of locomotion or another. One must keep in mind, however, that, to a certain degree, most mammals can walk and swim and often also climb to some extent. The goal (of an analysis) is to identify particular modes of locomotion that have been emphasized and are documented by diagnostic adaptations.

In the pectoral girdle of *Kopidodon* the clavicle is retained. As a rule this bone is reduced in specialized runners because the front legs swing only parallel to the body axis but do not need to be extended towards the side. The radius and ulna, the two bones of the lower arm, are completely separate, permitting a rotation of the hand. Resembling the condition in carnivores, the head of the radius is rounded; thus one can assume that *Kopidodon* had a great range of movement in its large hands. The phalanges have small lateral flanges to which strong flexor muscles for the individual fingers were attached. The burial position of the folded hands, which reflects complete relaxation of the muscles, suggests great flexibility (Fig. 343), far greater

than would be expected in the paw of a bear. The terminal phalanges, visible at the tip of each finger, were covered by even larger horny claws.

With large claws an animal can catch prey, dig, or hold fast to the bark of a tree while climbing. It is important to consider in what

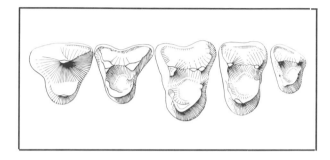

Fig. 341: *Kopidodon macrognathus*. The shapes of the molars are characteristic of primitive hoofed animals; they have no shearing edges.

manner the claws are enlarged. Large claws can indicate a predator, broad, flattened claws suggest a digger, and laterally flattened but high claws characterize a climber. *Kopidodon* has extremely narrow but very high claws that are typical of a tree-climber. For climbing, a wide range of extension of the arms is just as advantageous as an ability to rotate the hand.

When considering the hind legs, a similar picture unfolds: great mobility of the bones of the lower leg and, on all toes, comparable large claws with high cross-sections.

The strong tail fits well into this picture; it could have served to maintain balance as well as being used, occasionally, to support the animal. In spite of the considerable body size of *Kopidodon*, one can assume that this animal climbed in the canopy, because the skeleton contains evidence of diagnostic adaptations for climbing.

Fig. 342: Reconstruction of *Kopidodon macrognathus* foraging in the canopy.

There is no direct information about its diet, because, in contrast to other mammals from Messel, no stomach contents have been discovered as yet. The large canines might suggest a predator, but, because shearing crests are absent from the teeth, a specialization for meat-eating is unlikely (Fig. 341). Therefore, a diet based on leaves or fruits of trees is assumed. A peculiarity in the wear of the teeth indicates that it was a fruit-eater rather than a specialized leaf-eater (Koenigswald 1983).

Although this biological reconstruction of *Kopidodon* (Fig. 342) has a high degree of likelihood, one cannot exclude the possibility that this animal also ran on the ground and certainly was able to swim, as are almost all mammals. One has to be particularly careful with generalizations, especially when one wants to suggest that the life-style of one form was characteristic for all members of the family. In the present case, however, the arboreal habit can be assumed with some plausibility for the other members of the family Paroxyclaenidae, because they share the above-mentioned peculiarities of dental wear with *Kopidodon*. Only discoveries of skeletons of these other genera can verify this interpretation. The paroxyclaenids are known at present only from Europe.

The arctocyonids, known from skeletons discovered in Upper Palaeocene strata in the area of Rheims (Russell 1964), are phylogenetically closest to the paroxyclaenids. *Arctocyon* has very differently constructed teeth. They are broad for crushing—characteristic of plant-eater. Nevertheless, *Arctocyon* also has large canines. Basic similarities in the bone structure of *Arctocyon* and *Kopidodon* mirror the family relationships. But *Arctocyon*, with its much stouter bones, was certainly an essentially terrestrial animal that probably walked by bringing the entire soles of its feet to the ground.

Because all the forms closely related to *Kopidodon* are extinct, it is much more difficult to make intelligible the phylogenetic interrelationships. The family Paroxyclaenidae, which includes *Kopidodon*, was named after the genus *Paroxyclaenus* that is documented by only a skull fragment and associated lower jaw from the Eocene of Belgium. Besides the similar shape of the teeth this specimen shows the same, peculiar pattern of dental wear. Otherwise only a few fragments of jaws can be referred to this family. The ecological niche can only be determined from the specimens of *Kopidodon* from Messel.

The nearest relatives of the paroxyclaenids are the arctocyonids of the late Palaeocene, which looked rather like small bears and had significantly stouter bones. From the area of Rheims in France numerous isolated bones of the nominate genus *Arctocyon* are available and from them a skeleton has been reconstructed (Russell 1964). Although the basic similarity in the bone structure is obvious, the teeth of *Arctocyon* have a different morphology; the cusps are flat and broad, adapted for crushing fruit. *Arctocyon*, nevertheless, has large canines.

The arctocyonids are an important ancestral group in the evolution of mammals, and can be traced back to the late Cretaceous. They gave rise, on one hand, to the arboreal paroxyclaenids including *Kopidodon* and, on the other, to the typical primitive hoofed animals, the Condylarthra, which, finally, evolved into perissodactyls (Thenius 1969). But this evolutionary divergence must lie far back in time, because true perissodactyls are already known at the beginning of the Eocene. The typical early hoofed animals had small hooves and the proportions of the long bones show that they already were adapted to walking on their toes. That they have a close relationship with the arctocyonids and paroxyclaenids, which have claws, becomes evident from the corresponding positions of foramina (openings in the skull) at the back of the orbito-temporal fossa (the lateral depression in the skull for the eyeball and musculature for mastication) through which the main nerves of the brain exit (Russell and McKenna 1961). The position of these nerves in relation to one another, which can be documented in skulls by their foramina, has great systematic value because it is fixed very early during ontogeny. On the compressed

skulls from the oil shale these foramina unfortunately cannot be checked.

Recognizing these family relationships, the arctocyonids and paroxyclaenids were classified with the primitive hoofed animals in the order Condylarthra. Only through discovery of the skeletons at Messel did it become known that this very old order of mammals had invaded very different ecological niches and had even produced arboreal forms like *Kopidodon*.

Russell and Godinot (1988) recently proposed to remove the Paroxyclaenidae, including *Kopidodon*, from the Condylarthra to group them with the Pantolestidae which are represented at Messel by several skeletons of *Buxolestes* (see Chapter 13, Fig. 236). However, significant differences in the postcranial skeletons of these species suggest that the systematic position of *Kopidodon* may still be unclear.

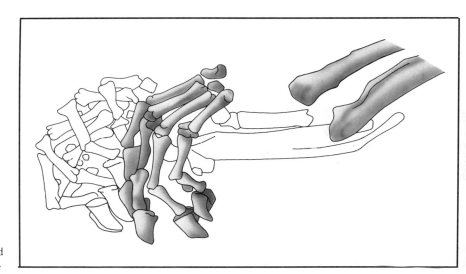

Fig. 343: Skeleton of the very agile hands of *Kopidodon macrognathus*, which had strong claws for climbing.

21

The Messel horse show, and other odd-toed ungulates

Messel became famous as a fossil site mainly through finds of primitive horses. Along with tapirs and rhinos, horses are at present the only representatives of the odd-toed ungulates, the mammal order Perissodactyla (Greek, *perissos*: odd-numbered and *daktylos*: toe). In contrast to the even-toed ungulates, they are characterized by the fact that in their fore- and hind-limbs only the middle toe is strengthened, and carries the main weight of the body. In modern horses this development has been brought to a point in the truest sense of the word. In these odd-toed ungulates, as also in zebras and donkeys, there is only one toe on each foot. The toe is connected to the remainder of the leg by strong, elastic tendons and ligaments, and in this way acts as an energy-retrieval apparatus during movement (Camp and Smith 1942).

The Messel horse show, and other odd-toed ungulates

JENS LORENZ FRANZEN

Fig. 344: A particularly fine specimen of the small primitive horse (*Propalaeotherium parvulum*) from the 1982 Senckenberg excavation.

Man feels particularly close to horses because they are domestic animals and even more because they are his companions in sport. As an aside, they are the only animals that are admitted to the Olympic Games. It is not surprising, therefore, that the primitive horses of Messel have always attracted particular attention. But they are, as we have learned in the meantime, not uncommon there.

Since the first publication by Haupt (1911), the remains of more than 70 individuals have been unearthed in the Messel pit. Among these are over 35 more or less complete skeletons which belong to two different species (Figs 344 and 345). *Propalaeotherium parvulum* was fully grown with a shoulder height of *c.* 30–35 cm, approximately the size of a fox terrier, while *Propalaeotherium hassiacum*, with a shoulder height of 55–60 cm, was about the size of an Alsatian.

From both species foals are known; from *Propalaeotherium parvulum* even several pregnant mares (Figs 347–349)! In addition, there are black silhouette-like 'body shadows' which trace the soft-body contours in exact detail down to the ends of the hairs (Franzen 1983) and, also, in a whole series of individuals, gut contents (Fig. 346). A

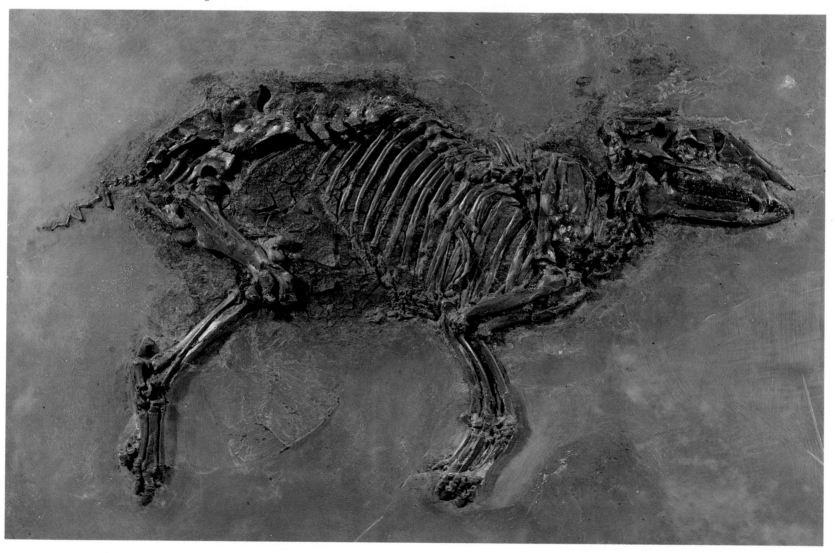

Fig. 345: Skeleton of the larger primitive horse (*Propalaeotherium hassiacum*) from the 1983 Senckenberg excavation. While *Propalaeotherium parvulum* attained approximately the size of a fox-terrier, *Propalaeotherium hassiacum* reached the size of an Alsatian.

Fig. 346: Food remains from the gut of a primitive Messel horse, viewed under scanning electron microscope (SEM). Underside of leaf showing cell walls and stomata. SEM: magnification, 210×.

comprehensive documentation such as this of such primitive horses is unique up to now. From the famous fauna of the Geiseltal and also from the entire North American continent the articulated skeleton of only one primitive horse respectively is known of comparable age and completeness, but without gut contents and without body contours. Messel is *the* fossil site of the world as far as primitive horses are concerned! This is, moreover, of great importance because the phylogenetic tree of horses is regarded as the prime example of evolution as recorded by fossils. There is

242

recognize them as relatives of extant horses. On the front legs they still had four hooves, on the hind legs three each—in all 14 hooves! Legs and neck were still quite short and the back was still strongly curved, reminiscent of the extant duiker antelopes. The *Propalaeotherium* (not the *Palaeotherium*!), just as we find it at Messel, evolved in Europe from *Hyracotherium* via *Propachynolophus*.

But, while in the Old World the horses became extinct at the end of the Eocene approximately 35 million years ago, their evolution in North America continued. The fundamental tendencies of this process consisted of the reduction of all lateral toes, the elongation of legs, back, and neck, and the development of high-crowned grinding teeth.

Fig. 347: Pregnant mare of the smaller species of primitive horse (*Propalaeotherium parvulum*). A complete, articulated skeleton of a fetus was found in the abdomen. Found by the Hessen Regional Museum, Darmstadt in 1986.

hardly a biology textbook which does not consider it in detail. The basic outlines of the phylogenetic tree, in the form of a chronologically and morphologically almost continuous series of complete skeletons, have been known since the middle of the last century from the American continent (Simpson 1977). Initially it was believed that evolution progressed in a straight line that led from primitive forms, with five small hooves on each extremity, to the Recent single-hoofed representatives, but we know today that the development has not proceeded so simply. At all times there were ramifications when evolution, mosaic-like, sometimes progressed more slowly, while at other times or in other parts of the body it proceeded more quickly. Sometimes one complex of characters evolved, then another. Again and again lines became extinct (Fig. 353).

While the major part of the evolution of the horse took place in North America, the development started in Europe, namely with the genus *Hyracotherium*. Its origin is unknown. At the beginning of the Eocene, *Hyracotherium* suddenly appeared, 54 million years ago, at first in Europe and somewhat later also in North America (Hooker 1980; Godinot 1982).

If one could meet these small animals today, which ranged in size from that of a pekinese to that of a fox terrier, one would hardly

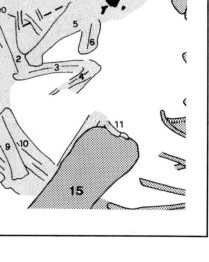

Fig. 348: (left); Fig. 349: (right). Detail from Fig. 347. Opposite: Explanatory drawing (from Koenigswald 1987) of the fetus: 1, region of skull with milk teeth (black); 2, right upper fore-limb; 3, right lower fore-limb; 4, mid-part of right fore-foot; 5, left lower fore-limb; 6, mid-part of fore-foot; 7, ribs; 8, pelvis; 9, right thigh bone; 10, right shin bone; 11, mid-part of right hind-foot. Mare: 12, lumbar vertebra; 13, pelvis; 14, right thigh bone (partly removed); 15, right shin bone.

243

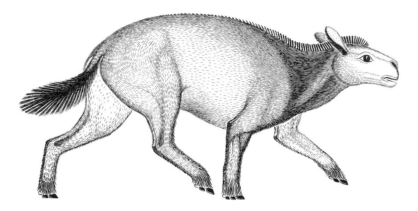

Fig. 350: The smaller primitive horse from Messel, *Propalaeotherium* *parvulum*, at different stages of locomotion.

Fig. 351: The primitive tapir *Hyrachyus minimus*, found by a private collector, is close to the phylogenetic ancestor of the rhinoceros.

After the direct land connection via the North Atlantic had broken down in the early Eocene, 'side-branches' of the ancestral horses returned in the early and late Miocene, respectively, 20 million and 12–11 million years ago, via the Bering Strait and the Asiatic continent, to Europe. But in Europe and in Asia they too became extinct after some millions of years. Only shortly before the Pleistocene, the great ice age, a last wave of now single-hoofed horses appeared, about 3 million years ago, in the Old World. This meant the salvation of horses as a whole because, after the end of the last ice age, these animals became extinct in North America approximately 8000 years ago. The cause of this is not known. The Spaniards, at the start of the seventeenth century, brought horses back to their homeland and here they spread once more across the entire continent in the course of only 80 years (Simpson 1951).

With this background what can the finds from Messel contribute to the understanding of horses' evolution? First of all, the Messel horses are still close to the phylogenetic origin of the entire family. They therefore offer us information about the habit as well as the life-style and locomotion of the earliest ancestors of horses in an amount of detail never before

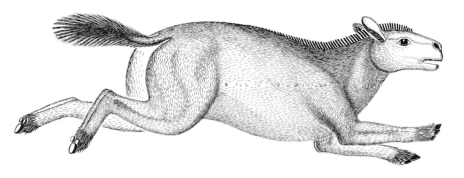

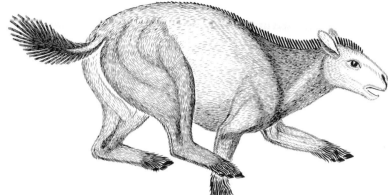

encountered (Fig. 350). In this context it is, secondly, of special importance that at Messel not only are complete skeletons with soft-body contours preserved but also that gut contents are available.

More than 100 years ago the Russian mammal palaeontologist Vladimir O. Kowalevski (1842–1883) pondered how the main trends of the evolution of horses might be explained, namely, the reduction of lateral toes and the development of particularly high-crowned, grinding cheek teeth. Palaeobotanist friends informed him that the environmental conditions had changed during the Cainozoic Era (Tertiary plus Quaternary), from the original predominance of jungles, towards more and more spreading steppes and savannahs. This, in 1876, gave Kowalevski the idea of considering the evolution of horses in a palaeobotanical context. Accordingly, the earliest ancestors of horses would have been omnivorous forest-dwellers, that changed in the course of their phylogenetic development into the typical grass-eating inhabitants of steppes and savannahs as embodied in the wild horses of today.

The theory of Kowalevski has not remained unchallenged. Fifty years later an American vertebrate palaeontologist, William Diller Matthew (1926), who worked at the American Museum of Natural History in New York, through a comparison of the dentition of primitive horses with that of similarly built extant mammals, postulated that the earliest ancestors of horses were probably not omnivorous but were rather browsers.

Only the successful find of a primitive horse by the excavation team of Senckenberg in 1975 decided this question. During preparation the area of the abdomen turned up as a grey, felt-like substance. J. L. Franzen, at first, supposed this to be the remains of the pelt. In test samples, which he subsequently investigated in collaboration with G. Richter, Richter discovered foliage leaves under the light microscope which seemed to belong to the gut contents (Franzen 1976, 1977). The preservation here was so excellent that even the cell walls and the structure of the stomata, could be recognized in detail. This can be seen particularly clearly under the scanning electron microscope (Fig. 346). There are no differences in quality between these and preparations of extant leaves! This is all the

Fig. 352: Reconstruction of the primitive tapir *Hyrachyus minimus*. Its body structure still very much resembled that of primitive horses, from which it differed fundamentally in its dentition.

245

more astonishing if one calls to mind that these leaves have been chewed by the little primitive horse and have been digested. Subsequently, they lay for 49 million years in the slowly hardening mud at the bottom of Lake Messel until they released their secret.

Later analyses, which were carried out on other preparations of the same as well as on many other specimens of primitive horses, have basically confirmed the result of the first analysis (Sturm 1978; Heil and Koenigswald 1979; Richter 1987). One specimen, however, does not fit this picture. This is a skeleton of *Propalaeotherium parvulum* found in Messel by the excavation team of the Hessisches Landesmuseum, Darmstadt, in 1982. A gourmet must have died here! In this animal the intestine was filled with grape-seed (Koenigswald and Schaarschmidt 1983).

In spite of this the decision had been made in favour of the theory of Matthew, at least as far as the evolution of the masticatory apparatus is concerned. The evolution of the locomotory system has, by contrast, developed rather

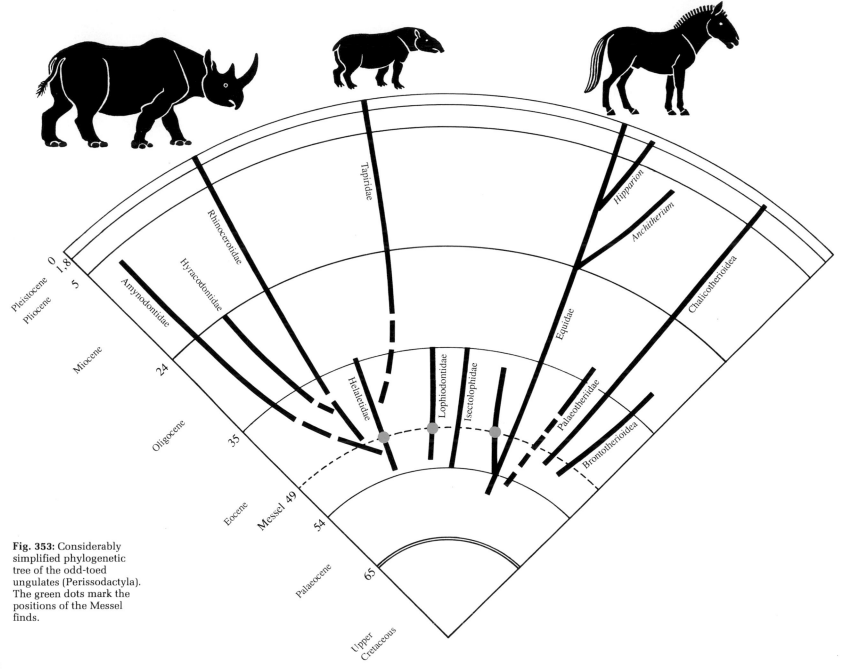

Fig. 353: Considerably simplified phylogenetic tree of the odd-toed ungulates (Perissodactyla). The green dots mark the positions of the Messel finds.

independently of a change of environment, in the sense of a gradual biomechanical optimization (Franzen 1984).

Two other groups of odd-toed ungulates have been recovered in Messel, apart from the primitive horses. The first is represented by a complete skeleton of *Hyrachyus minimus*. It was found in 1973 by private collectors and, as a putative find of a primitive horse, it immediately caused great excitement (Strübel 1974, 1975). It is, however, a representative of the family Helaletidae (Franzen 1981), which are primitive ancestors of tapirs and/or rhinoceroses (Thenius 1979; Prothero *et al.* 1986). The beautifully preserved specimen represents the first and, to date, only find of a skeleton of this genus from Europe (Figs 351 and 352). Complete skeletons of the genus *Hyrachyus* are, however, known from North America. Apart from the dentition, the skeleton of *Hyrachyus* still very much resembles that of the primitive horses. These animals were obviously still close to the common phylogenetic origin of all odd-toed ungulates (Fig. 353).

It was only possible recently to document a third group of odd-toed ungulates from a single tooth (Tobien 1988). The lower premolar which came to light during the excavations of the Regional Natural History Collection, Karlsruhe, is presently the only evidence at Messel of lophiodonts, primitive animals of considerable size resembling the tapir in their dentition. Individual species of this family in the Eocene had already attained almost the size of extant horses. This too is a peculiarity of Messel that the lophiodonts are so rare and have not been documented by a single skeleton as yet (but see foreword). Representatives of this family are elsewhere among the most frequently found mammals in the European Eocene. From the Geiseltal near Halle, for example, there have been several finds of skeletons (Fischer 1964; Krumbiegel *et al.* 1983). At Messel there is obviously a size limitation for finds of land mammals which did not live, like the crocodile *Asiatosuchus*, close to the border of the lake (Franzen *et al.* 1982).

22

Primitive even-toed ungulates: loners in the undergrowth

Even-toed ungulates form the second large order of hoofed animals, alongside the odd-toed ungulates. Nowadays they are among the most important mammals in economic terms. Pigs supply us with meat, cattle with meat and milk, sheep with wool. Even-toed ungulates play a special role in the Third World. Water buffalo work in the paddy-fields of South-east Asia as natural tractors; camels and dromedaries are still one of the major means of transport and also a source of milk and meat in the deserts of the Old World; whole populations, such as the Masai in East Africa or the Lapps in the subarctic regions of Scandinavia, rely for their existence entirely on even-toed ungulates, namely, cattle or, respectively, reindeer. Finally, even-toed ungulates in the form of wild-boar, deer, stag, and moose provide sport for hunters or, as antelopes, giraffes, and vast herds of gnu, delight the eye of the tourist on safari.

Primitive even-toed ungulates: loners in the undergrowth

JENS LORENZ FRANZEN AND
GOTTHARD RICHTER

When compared with the odd-toed ungulates, the even-toed ungulates are mainly characterized anatomically by supporting each of their extremities on two strengthened middle toes. The name Artiodactyla derives from this (Greek *artios*: even-numbered, and *daktylos*: finger). However, the decisive development for all even-toed ungulates was in the ankle joint. While in the odd-toed ungulates the ankle permits only one hinge-like turning movement, the two hinges in the even-toed ungulates enable the leverage to be altered as required. In addition to the ability to chew the cud this 'invention' serves to explain the enormous development which the even-toed ungulates have accomplished in the course of their evolution.

Even-toed ungulates were for a long time completely unknown at Messel. Only in 1980 was it possible to simultaneously recognize two different genera. One of them represents the phylogenetic root of the even-toed ungulates as such, the family Dichobunidae (Greek *dicho*: double; *bunos*: cusp; the name relates to the cusps of the lower-jaw molars which are arranged in pairs). The genus concerned is *Messelobunodon*. It is based on the finds of two skeletons (Fig. 354) recovered during the

Fig. 354: *Messelobunodon schaeferi*, a primitive even-toed ungulate, from the 1978 Senckenberg excavation. The gut contents are quite well known from two individuals. On the basis of these, the animals appear to have foraged in the surface leaf litter of the forest floor.

excavations of the Senckenberg Research Institute in 1978 and 1981, respectively (Franzen 1981, 1983). These were nimble and slimly built mammals. According to their habit, they correspond most closely, if their long tail is disregarded, to the small deer of the genus *Tragulus* (chevrotains). These animals live today in the lowland and mountain forests of South-east Asia, which are dense with undergrowth. Like chevrotains, *Messelobunodon* may also have been a fast runner and versatile mover through the scrub. Its very strong posterior extremities enabled it to jump far and the long tail acted as a balancing and steering organ (Fig. 355). Its overall habit made *Messelobunodon* obviously less fitted for defence than for flight. This is indicated by the elongated limbs, but most of all by the very weak canines and by the absence of any armour on the head. In cases of emergency the strong hind-limbs could also have served for defence.

Like the ant-eater *Eurotamandua joresi*, *Messelobunodon* is a good example of the fact that, in some instances of gut-content analysis,

the 'occasional inclusions', i.e. admixtures without nutritive value, tell us more about the life-style and foraging of a species (fossil or Recent) than the food itself (Richter 1981, 1987). For fossil species whose behaviour can no longer be observed but must be pieced together laboriously from circumstantial evidence, such indicators are particularly welcome.

In one of the two animals investigated, the very old individual of *M. schaeferi*, one can regard as food in the strict sense of the word only the numerous fragments of fungi (or lichens), massive lumps of densely packed hyphae with one bumpy and one smooth surface (Fig. 356). Everything else (individual seed capsules, sand grains, and the very numerous fragments of leaves) was certainly at the time of consumption already devoid of any nutritive value for the animal, since the seed capsules are empty and the leaves are deteriorated to a large extent by bacterial or fungal decay (Figs 357 and 359). A comparison proves that *M. schaeferi* had eaten large quantities of old leaf litter, but surely not

to nourish itself on such extremely poor-quality food. This gut contents sample only makes sense when it is assumed that the animal foraged (in this case for fungi) with its long extended, lightly built snout by rummaging in the leaf litter of the forest floor. Thus it inevitably consumed, while eating food hidden in the leaves, the worthless dead leaves.

The gut of the second animal contains not only individual leaf fragments and pieces of fungus, but, in addition, large quantities of seeds which all come, apparently, from only one plant species (Figs 358, 360, and 361). Other pieces of thick plant tissue derive, presumably, from the fossilized flesh of the fruit from which the seeds probably originated. These apparently very significant differences in the diet of the two animals are, however, if studied more closely, at least ecologically negligible. One animal consumed fungi for which it foraged on the forest floor; the second ate small fruits but probably also gathered these from the ground. We know this because the gut not only contains the firm flesh of fresh fruit but also numerous vestiges of obviously

Fig. 355: The primitive even-toed ungulate, *Messelobunodon*, was approximately the size of a dachshund. It was a quick runner, which nimbly negotiated the undergrowth and could leap considerable distances.

Fig. 356: Gut contents of *Messelobunodon schaeferi* showing fungal hyphae with a bumpy surface on one side. Such fragments are particularly frequent in the gut of this animal and were obviously the only proper food that it had eaten just prior to its death. SEM: magnification, 165×.

Fig. 357: *Messelobunodon schaeferi*: part of gut contents showing a largely decayed leaf with numerous fungal hyphae criss-crossing the damaged cuticle. These are fungi which live on dead tissues (saprophytes). Similar leaf remains can be found in any half-rotted leaf litter of extant forests. SEM: magnification 400×.

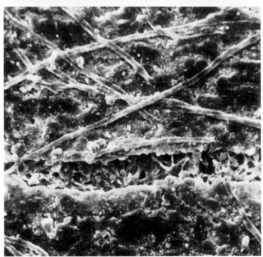

Fig. 358: Part of gut contents of *Messelobunodon* sp. Pericarp of an unknown fruit. It probably corresponds with the seed shown in Figs 360 and 361. More or less decomposed fruit flesh, permeated by fungal hyphae, was found in considerable quantity in this gut. SEM: magnification, 75×.

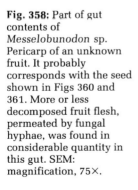
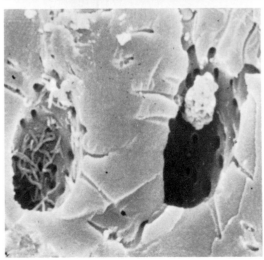

Fig. 359: Part of gut contents of *Messelobunodon schaeferi*. Cuticle of a broad leaf with decomposition by bacteria. SEM: magnification, 925×.

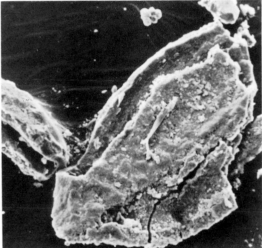

Fig. 360: Part of gut contents of *Messelobunodon* sp. An oval seed which cannot be precisely identified is very frequent in the gut of this animal. SEM: magnification, 65×.

Fig. 361: Part of gut contents of *Messelobunodon* sp. The same seed as in Fig. 360. Sclerenchyma of the seed coat (testa). SEM: magnification, 3000×.

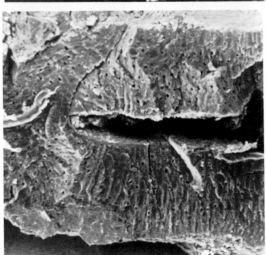

253

decayed fruits whose disintegrated tissue was permeated by the hyphae of saprophile fungi. Both animals foraged for food (fungi, fruit, perhaps also worms or slugs, which we cannot yet identify) presumably in the same way and in the same ecosystem, on the floor of the Messel rain forest.

Messelobunodon is phylogenetically intermediate between the oldest and most primitive even-toed ungulates known to date, in the form of the genus *Diacodexis* from the Lower Eocene of North America (Rose 1982, 1985), Europe (Godinot 1978, 1981), and Pakistan (Thewissen *et al.* 1983), and the more highly developed forms like *Dichobune* from the European Upper Eocene (Fig. 363).

The second genus of even-toed ungulates from Messel was documented in print in 1980 by Professor Tobien from Mainz. It is based on a complete skeleton discovered and prepared in 1974 by private collectors. In their honour and in that of Messel the new genus and species was named *Masillabune martini* (Fig. 362).

Masillabune, in contrast to *Messelobunodon*, displays relatively strong upper canines and a first lower premolar that has assumed the shape of a canine (Tobien 1985). *Masillabune* belongs to the family Haplobunodontidae (Greek *haploos*: simple; *bunos*: cusp; the name hints at the one-cusped lower premolar) and, hence, the superfamily Anthracotherioidea (Greek *anthrakos*: coal; *therion*: animal; as a reminder of the first finds of this animal group in the brown coal). These fossil mammals were residents of moist and boggy habitats and are regarded as being, phylogenetically, distantly related to extant hippos (Fig. 363). It was, however, not possible until now to document a direct phylogenetic connection. Analyses of gut contents resulted,

as for the primitive horses, in finds of remains of laurel leaves as well as less readily identified plant tissues (Schaarschmidt in Tobien 1980). This suggests that these animals, unlike *Messelobunodon*, probably foraged in a tier of vegetation above the forest floor (Tobien 1985). Compared with *Messelobunodon*, *Masillabune* had significantly longer thighs and shorter, thick-set, shanks (Tobien 1985). This suggests powerful rather than fast locomotion, which is certainly useful in a boggy area.

A third genus of even-toed ungulates from Messel was announced only recently (Franzen 1987). This is the genus *Aumelasia*. Like *Messelobunodon* it is judged to belong to the ancestral group of all even-toed ungulates, namely the family *Dichobunidae*. The finds from Messel—a complete skeleton from the Senckenberg excavations of 1985 (Fig. 364), another one from a private collection, and a

Fig. 362: The rare even-toed ungulate *Masillabune martini* was rather more sturdily built than *Messelobunodon*. It had stronger canines and a shorter tail. The animal probably foraged in the undergrowth of the forest floor. Total length, *c.*45 cm.

partial skeleton from the excavations of the Regional Natural History Collection, Karlsruhe in 1985—illustrate particularly well the bulk of information that this extraordinary site holds in store for science and for the interested public. Until these finds at Messel, *Aumelasia* was known only from several jaw fragments as well as from some isolated teeth from the Paris Basin (Sudre 1980; Sudre *et al.* 1983). With the complete skeletons at hand, the genus suddenly takes shape. For the first time we have a well-founded idea not only of the appearance of these animals, but also of their life-style, as two finds have also furnished gut contents.

In the body proportions and thus in their general aspect, *Aumelasia* and *Messelobunodon* were largely similar. The life styles of these two genera also should have been largely congruent, since G. Richter's investigations of the gut contents of one animal have furnished seed grains and a lot of sand in *Aumelasia* as well. In spite of this, *Aumelasia* was certainly a valid genus as evidenced by the characteristic differences in the structure of the dentition, the morphology of the lower jaw, and that of the thigh (Franzen 1987). Phylogenetically, *Aumelasia* is traced back to the genus *Protodichobune* which evolved in the European early Eocene from *Diacodexis*. Towards the close of the Middle Eocene *Aumelasia* obviously already had become extinct (Sudre *et al.* 1983).

In their entirety the new finds from Messel are evidence that the even-toed ungulates in the European Middle Eocene were already far more highly differentiated than was originally assumed. However, in terms of the number of individuals found, these primitive, even-toed

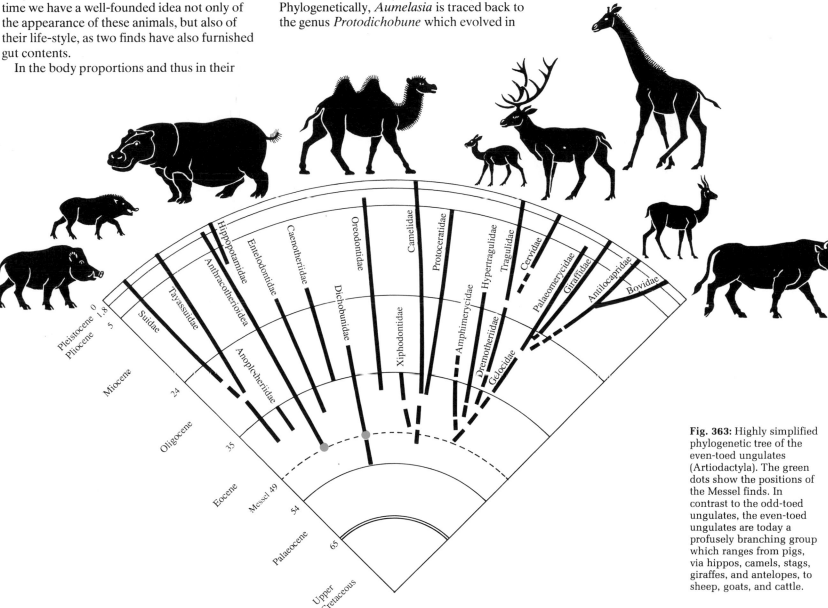

Fig. 363: Highly simplified phylogenetic tree of the even-toed ungulates (Artiodactyla). The green dots show the positions of the Messel finds. In contrast to the odd-toed ungulates, the even-toed ungulates are today a profusely branching group which ranges from pigs, via hippos, camels, stags, giraffes, and antelopes, to sheep, goats, and cattle.

255

ungulates are overshadowed entirely by the numerous primitive horses. This leads to the conclusion that they, in contrast to the horses, were loners rather than members of groups or herds. In their size the small- or, at most, medium-sized primitive even-toed ungulates of the Lower and the Middle Eocene obviously occupied the niches below the generally larger odd-toed ungulates of that time (Tobien 1985).

Fig. 364: *Aumelasia* is a close relative of *Messelobunodon*, but the skull and dentition are more powerful. Along with unidentified fruits, the gut contents comprise abundant sand, as for *Messelobunodon*. Hence, these animals also foraged on the ground. Find from the 1985 Senckenberg excavation. Prepared and entirely freed from the matrix.

23

Death and burial of the vertebrates

Taphonomy is a distinct discipline of palaeontology that deals with all processes occurring between the death and final burial of fossil organisms. It began with the work of J. Weigelt (1927). It represents an attempt to establish the conditions that prevailed at the time of burial, on the one hand, from the mode of burial and the degree of preservation of the fossils, and, on the other hand, from the fate of the organisms before being buried. Hence the questions raised for the Messel vertebrates: did the animals live in the vicinity of the place of burial or were they washed in via waterways? Did crocodiles feed off the corpses, or did these sink undamaged to the bottom of the lake? Were there currents at the bottom of the lake? and so on.

Death and burial of the vertebrates

MICHAEL WUTTKE

From the photographs in this book it is apparent that the vertebrate skeletons have disintegrated to varying degrees. In most of them the limbs are all preserved in a completely relaxed position, lying on their side (as in Fig. 364), but other skeletons are incomplete and individual joints are contorted in ways that would not have been possible during life (Fig. 347). An analysis and interpretation of these biostratinomic results must be based on the rules that govern the processes of decomposition today, as they can be studied by observation and by tests. This palaeontology of current processes, called 'real-time palaeontology' (Richter 1928), studies the living creature, or, respectively, its corpse in terms of how it behaves as a future deposit in the sediment, and how this deposit can be interpreted in relation to the character of its past environment (Schäfer 1962).

The basic processes of decay are, of course, similar in all vertebrates, but the life-style can have a decisive influence on the biostratinomic processes (see below).

Immediately after the death of an animal the processes of decomposition begin, instigated in the initial stages by the body's own enzymes (autolysis). The microbial decomposition of the organic constituents of the corpses occurs simultaneously; in vertebrates it mostly begins in the abdominal or the oral cavity.

The individual tissues of the body respond at different rates to decomposition by bacteria, so that a certain hierarchy of decomposition can be established. The skin of mammals, for example, and particularly that of reptiles, is very resistant to autolysis and putrefaction. In these animals, therefore, sizeable volumes of putrefaction gases can collect inside the skin bag. In corpses of drowning victims they prevent the body from sinking until the skin bag splits open and the gases disperse. The resulting long period of drifting can, however, be deduced from the skeleton of a sunken corpse. At the time of sinking many bone connections have already loosened inside the skin bag, and, in consequence, many skeletal elements do not retain their original position when the corpse is embedded.

Mammals

These few criteria already permit a first interpretation of the biostratinomic history of some of the Messel vertebrates described here, for example, the mammals. The majority of the mammals are embedded in a stable posture on their side. The limbs lie in a totally relaxed position parallel to or above one another (Fig. 340). Weigelt (1927) described this 'passive' embedding position as the typical position for corpses in water. A precondition for such an embedding of the complete animal is that the animals were brought to the place of deposition shortly after their demise. At the temperatures assumed for Messel this must have been approximately 4–6 hours (see below). Otherwise, so many gases of putrefaction would have collected in the abdominal cavity that the sinking of the animal corpse at this early stage of decay would have become impossible. Such completely preserved skeletons, in which all the connections of the joints are still intact, also give information about currents at the bottom of Lake Messel; the animals must have been buried under quiet-water conditions.

Examples of corpses that drifted, distended by gases of putrefaction, for a lengthy period at the water surface are the primitive lemuroids of

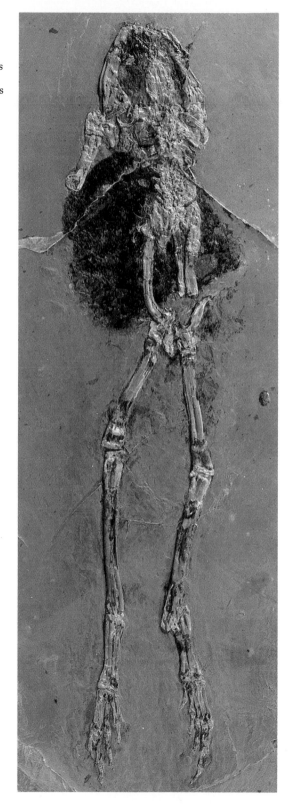

Fig. 365: *Eopelobates wagneri*, female with spawn preserved in the abdominal area. All limbs have been aligned longitudinally by currents at the bottom of the lake. Length of fossil, 19 cm.

Messel (Koenigswald 1979; Franzen 1987; Koenigswald and Wuttke 1987; see Figs 287 and 291). Both skeletons show characteristic features of decomposition. The soft-part connections between vertebral column and pelvis were completely disconnected prior to the embedding process, and in *Europolemur* the left arm and the vertebral column of the tail also became detached (Fig. 291).

As experiments testing putrefaction in extant frogs have shown (Wuttke 1983), bacterial processes of decay in the abdominal cavity can occur so violently and quickly that all connections between vertebral column and pelvis are dissolved even when the decomposition in the hind-legs has not progressed very far. The development of putrefaction gases can become so strong at higher water temperatures (*c.* 25°C) that the abdominal cavity tears open and the anterior and posterior parts of the body thus become disconnected. Similar processes may have applied to the adapid of Messel (Fig. 287): all joints of the posterior extremities are still connected firmly: only the vertebral column and the pelvis have become completely separated from one another. Evidence of decomposition, traceable to a strong development of putrefaction gases in the abdominal cavity after the corpses had already settled at the bottom of the lake, can be detected in larger mammals of the approximate size of an early horse. The gases here escaped regularly from the area of the loins on the side closer to the water surface (Fig. 345). Such occurrences can be recognized by an arc-shaped arrangement of the ribs as well as by missing parts of the skeleton in the loin region in skeletons which otherwise still retain their anatomically correct articulation.

Reptiles

Crocodiles are among the most frequently found reptiles in the Messel oil shales. As a rule their complete skeletons were buried. This can only be explained, as already discussed for mammals, if the time taken to transport the crocodile corpses from their place of death to that of deposition was short. From this point of view the animals must have lived in the lake itself or in the mouths of its tributaries. Quite a number of large specimens of the genus *Diplocynodon* show evidence of disintegration of the osteodermal armour resulting from the development of putrefaction gases in the body cavity. This disintegration is most apparent in

Fig. 366: The system of causally related factors acting together which led to the individual states of preservation of the Messel vertebrates.

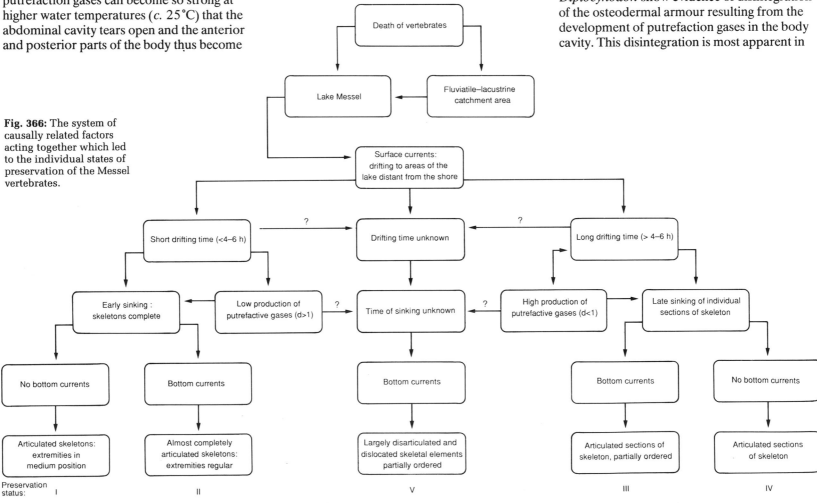

the transitional region leading from the trunk into the tail where the bony plates of the skin armour and of individual vertebrae are displaced. Because the remaining skeletal elements of these specimens are still in their original positions, these processes must have occurred inside the corpses after they were deposited at the bottom of the lake. These gases must have escaped explosively via the cloaca, thus displacing all juxtaposed skeletal elements from their original positions; the soft-part connections had already been destroyed by the processes of decay.

Finds of turtles occur with similar frequency to those of crocodiles. Interestingly, the finds are concentrated within the earlier strata (approximately in the vicinity of key horizon beta) in an area which is closest to the north-eastern tributary (Franzen *et al.* 1982). This is probably due primarily to the processes which caused the turtle corpses to stop floating and sink to the bottom. The turtle corpses were probably picked up by water currents flowing into the lake. Due to the strong retarding action of a larger body of water such as a lake on the velocity of an incoming tributary, the turtle corpses sank to the lake bottom in exactly that area in which the velocity was no longer sufficient to keep them floating. Messel is one of the few fossil sites from which it was possible to recover complete skeletons of turtles; here too this can only be explained by assuming that the period of transport between death and burial was short. There are, however, quite a few finds in which only the carapace was buried. In these cases, skull and extremities decayed on corpses drifting at the water surface, fell off, and were buried separately.

Among the rarer finds are the lizard-like reptiles (lacertiliads) of which, for some species, only a single specimen has as yet been recovered. As in the other groups of vertebrates these are mostly completely preserved specimens, whose elements are still in their original articulation. Apart from these lizards, which were deposited quickly after their deaths, specimens that drifted for an extended period on the surface of the lake are also found occasionally. Even these, however, are more or less complete, if disarticulated,

skeletons. This can be attributed to the high resistance to decay of the reptilian skin which, like a sack, kept the skeletal elements together even when the other soft parts had already suffered microbial decomposition.

A biostratinomic analysis of the fossil snake finds is somewhat more difficult than that of other reptiles. It cannot easily be decided whether the snake corpses sank quickly to the bottom of the lake or if they drifted in the water for a prolonged interval.

As our own experiments on decomposition have shown, skeletons of snakes can be buried complete even when snake corpses, distended by putrefaction gases, have drifted for a considerable time at the water surface. This can be attributed, on the one hand, to the extraordinarily firm interconnections of the vertebrae and, on the other hand, to the great resistance to decay of the skin (up to nine months).

As a rule the vertebrae are still held together by ligaments even when the other soft parts have already been bacterially decomposed. Indications of prolonged drifting times can be detected in a large number of the Messel snakes. Here the most significant indications are the loss of the skull as well as the disintegration of the vertebral column into sections of 5 to 20 vertebrae each. In drifting corpses of snakes the skull, together with some of the neck vertebrae, always hangs down, due to its larger specific gravity. The mouth, as a rule, is open. During the process of decay, the individual bones of the skull fall out of the skin bag and are embedded separately. The rest of the continuously drifting snake corpses sink only when the soft parts are almost completely decomposed microbially and when the putrefaction gases have escaped. During this period of compression and extension forces caused by currents and gravity act on the body and lead to a rupturing of the vertebral interconnections. Nevertheless, the anatomical assemblage of the vertebral column remains largely intact due to the enclosing skin bag.

Birds

The number of bird skeletons that are still preserved with their anatomical connections intact is astonishingly high. This is particularly striking because bird corpses, helped to float by the air in the quills between the down feathers and in the bones of the extremities, can drift for weeks at the water surface. For birds, therefore, the conditions are different from those prevailing for the mammals, which decompose significantly more quickly. No valid explanations for this can be given to date. In no other group of Messel vertebrates, however, is the proportion of partial skeletons (as, for example, skulls with some vertebrae of the neck, and wings with attached pectoral girdle) so high as for the birds. They all originate from bird corpses that drifted for long periods at the lake surface and which disintegrated gradually by putrefaction.

Amphibia

Investigations of the biostratinomy of an entire group of vertebrates are available for the frogs of Messel (Wuttke 1988). Through individual analyses of the biostratinomy of every anurid (partial) skeleton, it was possible to reconstruct the factors that acted in a causal way on one another, acted together, and to which the anurid cadavers were exposed until the final burial (Fig. 366). Although most anurid corpses were exposed to complex influences such as long drifting times caused by a quick development of putrefaction gases, currents at the bottom, etc. (Fig. 365), it was possible to reconstruct five states of preservation that can be generalized and into which the individual skeletons can be classified. These results can be easily transferred to the other groups of animals. The skeleton of the giant ant-eater *Eurotamandua* (Fig. 305) could, for example, be classified into preservation state I, and the above-mentioned partial skeleton of a lemuroid (Fig. 287) into preservation state IV.

The investigations into the relationships of the skeletal elements with the embedding sediment also produced insights into the suitability of Lake Messel as a habitat for an aquatic anurid fauna, as well as into the temporal and spatial distribution of currents at the lake bottom. The aquatic frogs of the species *Messelobatrachus tobieni* (Fig. 157), for example, settled only sporadically in the lake. All finds are in conjunction with a certain type of oil shale (facies type 3). The reason for this must lie in a generally poor water chemistry, which altered only at certain times, but, if so, then for several years. In this oil-shale type, strangely, it is possible to document the influence of bottom currents in only two of 20 skeletons of frogs. Both of these skeletons come from excavation sites which lie closest to the reconstructed tributaries. In the 'normal' type of oil shale (facies type 1), by contrast, only skeletons of the toad *Eopelobates wagneri*, which visited the lake solely for spawning, have been found. Of these, 75 per cent of all skeletons show the influence of bottom currents irrespective of the distances from the tributaries at which they were found.

In addition to the results of sedimentological and geochemical studies, further investigations into the biostratinomy of the Messel vertebrates will certainly lead to a constantly changing and ever more complex view of the conditions at the origin of the Messel fossil site.

24

Conservation—dissolution—transformation. On the behaviour of biogenic materials during fossilization

In the course of evolution, living creatures have managed to synthesize numerous organic and inorganic compounds which assume different functions in the organism. Thus we find structural albumins as collagen in bones, or as keratin in the hair and claws of animals. In arthropods such as insects, spiders, and crabs, the exoskeleton is constructed mainly from an organic compound, chitin. Under normal conditions such organic constituents do not survive fossilization. Under favourable circumstances, however, and herein lies the special importance of Messel, even such perishable compounds can persist through geological time.

Fig. 367: Viviparous pond snail (*Viviparus* sp.) with periostracum preserved. Following an earlier dissolution of the calcium carbonate, only the organic membrane that surrounded the calcium carbonate shell has been preserved. Height of snail, *c.* 17 mm.

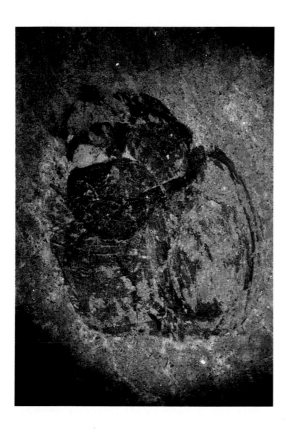

Fig. 368: Viviparous pond snail (*Viviparus* sp.) with the shell 'preserved'. The exact shape of the original calcium carbonate shell was, however, replaced by iron carbonate (siderite). Height of snail, *c.* 14 mm.

Conservation—dissolution—transformation. On the behaviour of biogenic materials during fossilization

MICHAEL WUTTKE

Many organisms produce mineral substances (biominerals) in their metabolism which serve mainly as skeletal building blocks (e.g. bones or the silica skeleton in diatoms). These mineral substances are often well preserved as fossils as regards both their mineral content and their structure. Minerals which derive from chemical reactions that took place outside the cells, but under the influence of their metabolism, can also be preserved as fossils. An example of this is the formation of pyrite (FeS_2) through the exudation of bacterial metabolic products.

Following the death of organisms the individual organic and inorganic compounds are exposed to many, often no longer comprehensible, influences (diagenesis), which can profoundly change them either immediately or in the course of geological time. Organic compounds such as albumins and sugars are altered by external chemical influences, by microbial decomposition and metabolism, and by pressure and heat. Inorganic substances such as the calcitic shells of molluscs can, while still retaining their original structure, be completely replaced by other minerals.

The research into these complex processes is still in its infancy in Messel. Their elucidation, however, should give essential information about the former biochemical and geochemical environment at the bottom of Lake Messel and, respectively, in its sediments.

Inorganic hard parts

For the palaeontologist the mineral skeleton or parts of it are frequently the only evidence of former organisms. Among skeletal biominerals, certain essential elements can be recognized in multicellular animals. Two-thirds of all skeletal elements consist of calcium minerals. The most widely distributed biominerals in the animal kingdom are, at present, the carbonates, for example, calcium carbonate ($CaCO_3$) in the shells of snails and the otoliths of fishes. Calcium phosphates are far less evident; they occur to a greater extent in higher animals, that is, primarily in the vertebrates.

The second most widely distributed biomineral is opal, that is, hydrated, amorphous silica ($SiO_2 \cdot nH_2O$); in Messel it is contained in the skeletal spicules of siliceous sponges.

Following the death of the organisms and their deposition or, respectively, burial at the bottom of Lake Messel, the individual biominerals reacted in different ways to the chemical conditions (frequently changing during the lake's history) at the floor of the lake or in its sediments.

The calcium carbonates were generally

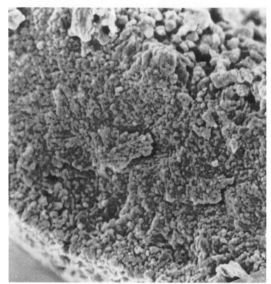

Fig. 369: Scanning electron micrograph of the sideritized snail shell of Fig. 368. The former cross-lamellar structure of the calcium carbonate shell has been retained. SEM: magnification, 1200×.

dissolved; thus, for instance, the otoliths (which are important for taxonomic research) have not yet been preserved for any fishes, and neither has the calcium content of the endolymphatic sacs in frogs been retained. From the systematic excavations which have been carried out since 1975 by museums and institutes, no fossilized snail shells were known until 1986; only the periostracum (Fig. 367), the organic membrane that once enveloped the shell, had been preserved. In the 1986 finds the original substance of the calcium carbonate was also not preserved (Fig. 368). Chemical investigations and scanning electron microscopy (SEM) showed that the original substance of the shell had been replaced by iron carbonate (siderite) in such detail, that the original cross-lamellar structure of the shell can still be recognized (Fig. 369, Richter and Wuttke unpublished).

The diagenesis of vertebrate skeletons, which consist of calcium phosphate, has not yet been studied in detail, although its preservation within the varying types of shale (see Chapter 3, 'The genesis of the Messel oil shale') shows distinct differences. In particular, in the shale of facies type 3 (Franzen *et al.* 1982), renowned for excellent soft-tissue preservation, there are all gradations from—admittedly rare—good bone preservation to the complete dissolution of this biomineral (Fig. 370). Not only did the chemical conditions at the lake bottom or, respectively, in the sediment, play a significant role in dissolving the phosphate (even the coprolites (fossilized dung) are somewhat less phosphatized in this type of facies), but chemical changes in the corpses during microbial decomposition of the organic matter must also have played a fundamental role. This is clearly demonstrated, for example, in those bats in which the bone was mainly dissolved in the area of the trunk, that is, in the region where the most substantial soft parts are found, while the skull and the bones of the extremities remained preserved (Fig. 261).

The multitude of organic and inorganic acids given off during the microbial metabolism and insufficiently buffered by the water contained in the sediment were probably responsible for this. An all-inclusive model for these processes is not yet available.

Predominantly microbial dissolution processes in inorganic skeletal substances have been demonstrated most recently by Kott and Wuttke (unpublished) on the silicic spicules of the Messel silicic sponge, *Spongilla gutenbergiana* Müller *et al.* 1984 (Fig. 371). In freshwater sponges (Demospongia) these spicules of silica also form a skeletal system in which the individual needles are connected by the structural albumin, spongin. The attraction of these spicules for micro-organisms is based on their structure. They consist of an organic material as well as of opal precipitated in layers of variable water content. These siliceous spicules are deposited by the sponge in concentric layers around an axial filament of organic matter which occupies the longitudinal axis of the spicule. In certain species of fresh-water sponge with spiny spicules (as in the Messel freshwater sponge) it has been possible to demonstrate the extension of the axial filaments into the thorns. The external covering of the spicule is composed of an organic membrane.

The death of a sponge is quickly followed by its disintegration (Schäfer 1962) and the spicules liberated in this process can then be quickly occupied by bacteria. This bacterial invasion and the dissolution of the opal of the spicule that accompanies it or is caused by it, respectively, leaves characteristic traces of dissolution; these were observed at Messel by Heil (1964) and Martini and Rietschel (1978) but were interpreted differently by these authors.

In the course of decomposition of the organic material of the siliceous spicules the bacteria settled first on the spicule surface and then progressed along the axial filament and its branches, leading to the spines, and ever further into the spicule. The actual dissolution of the spicule opal did not proceed actively in order to produce silicic acid for the bacterial metabolism, but passively via the exudation of bacterial metabolic products such as acids and bases. By this means the opal dissolved preferentially in those layers of the spine with a higher water content (Fig. 373). Because such layers are concentrated in the spines of the spicules, dissolution occurred preferentially, so that now the spicules appear to be pierced by circular holes. On the surface of the spicules the solutions have left pit-like scars which correspond in their shape and size with the former bacteria (Fig. 372). Even within the central canal of the siliceous spicules, considerably widened by the processes of dissolution (Fig. 374), there are still indirect traces of these bacteria, namely, in the form of pyrite. This was precipitated from the capillary water of the sediment in several steps during the bacterial metabolism. This process will be described in more detail later in this chapter under the heading 'Soft-part preservation at Messel'.

The part played by capillary waters in the solution of the spicule opal can no longer be elucidated because as yet there are no known distinguishing features of this process that could be identified in the fossil material. Most probably, both factors, indirect bacterial as well as purely inorganic solution processes, are responsible, although the bacterial metabolism would have had a larger share in dissolution.

Fig. 370:
Messelobatrachus tobieni in pure soft-tissue preservation. The bones are almost completely dissolved. In the skull the eyes are still clearly recognizable, as are both parts of the liver in the trunk and two veins in the thigh. Head and body length, *c*.21 mm.

Organic hard parts

Hard parts of organic material (to be precise, chitin), which encase the body surface like armour, have primarily been developed by arthropods such as insects, spiders, and crabs.

At Messel, however, not only the chitinous hard parts but in many cases even the original structural colours are preserved to such a degree of brilliance as is known otherwise only from the Geiseltal (Fig. 91). The colour scale concerned extends from blue and green through gold to copper red. The proof that these are original, structural colours, and not secondary hues developed *post mortem*, stems from the fact that, in all cases in which several specimens of a species are available, they show the same colour pattern. In addition to the preservation of structural colours, particularly in many species of Buprestidae (jewel beetles) and Chrysomelidae (leaf beetles), one even finds examples of a preservation of patterns from pigment-based colours (Fig. 95).

What then are the characteristics of the building block, chitin, which enable it, in spite of being an organic compound, to be preserved over geologically long epochs? The material, chitin (chemically polyacetylglucoseamine), is formed from metabolic exudates of the skin tissue. In the exoskeleton of arthropods it is built from stratified layers. A covering of

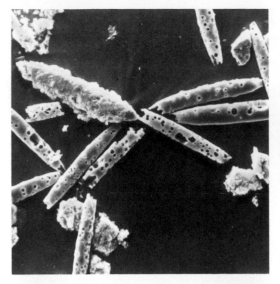

Fig. 371: Siliceous spicules with circular perforations of the spicule body from the skeletons of freshwater sponges. SEM: magnification, *c.* 140×.

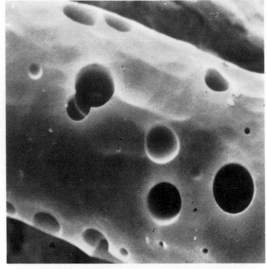

Fig. 373: Detail from Fig. 371. Channels and extensive dissolution scars on a siliceous spicule. SEM: magnification, *c.* 4200×.

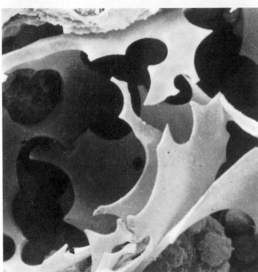

Fig. 372: An almost completely dissolved siliceous spicule. By means of selective dissolution of the individual siliceous spicules the stratified construction becomes visible. SEM: magnification, *c.* 4500×.

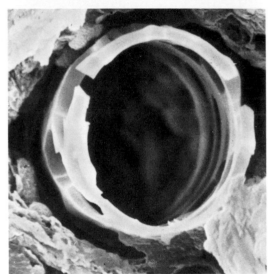

Fig. 374: Transverse section of a siliceous spicule; the diameter of the central canal is considerably enlarged by dissolution. SEM: magnification, 4000×.

specific, water-repellant substances (mainly waxes and lipoproteins; Pflug 1984) serves to protect the surface. The most important factor in the fossil preservation of chitin is the impregnation by quinones, which make chitin chemically more resistant. Here too it is again the micro-organisms, mainly bacteria but also fungi, which are able to decompose chitin skeletons in a short time. In the course of this process bacteria exude chitinases from their surface. These exoenzymes dissolve the chitin so that it can be absorbed by bacteria.

Richter (1985) was able to demonstrate such bacteria at Messel in the stomach of the hairy hedgehog *Pholidocercus hassiacus* Koenigswald and Storch 1983 (Fig. 255); here the bacteria were still attached to the chitin fragments when they were fossilized.

The dissolving of the chitin via chitinases always takes place in the immediate surroundings of the micro-organisms so that, as described above for the silicic spicules, pit-like scars are formed in the chitin material (Fig. 254) that match the bacteria in shape and size.

How can such processes of decomposition be reconciled with the excellent preservation of chitin that has been described for Messel (see Chapter 5, 'Giant ants and other rarities: the insect fauna')? According to Richter, the attack of micro-organisms did not take place after the deposition of the fossil on the lake floor, but rather the insect would already have been exposed to decomposition on land for a prolonged period before it was taken up by *Pholidocercus* with its food.

This as yet single proven example of traces

of dissolution by chitinolytic bacteria leads to the conclusion that chemical conditions that predominated on the lake floor, but also in the water column (before their sinking the insects must have drifted for quite some time on the lake surface), prevented chitin-decomposing micro-organisms from developing.

The preservation of the chitin also shows, yet again, the special conditions which led to the formation of the Messel fossil site; although geochemical and morphological evidence of bacteria has been presented, special conditions have resulted in a specific inhibition of chitinolytic micro-organisms. Perhaps comparisons with Recent lakes will help to elucidate this question.

Soft tissues

The Messel fossil site became famous early in its history for the preservation of soft tissues (Haupt 1925; Weitzel 1932) even though its real importance, for example in relation to ecological evidence (as elaborated in the other chapters of this book), became evident only in the 1970s. In addition to this evidence it is also possible to reconstruct pictures of the life of the fossil vertebrates that are not based solely on comparisons with the skeletal anatomy of Recent vertebrates; thus, contours of pelts are preserved which even provide information about the length and density of the tuft on the tail of the early horse from Messel.

As far as the preservation of the most minute soft-tissue structures in Tertiary vertebrates is concerned, the Geiseltal rather than Messel has been pre-eminent until now. Voigt (1934) was successful with a first report of remains of fossilized musculature and of connective tissue in fishes. In a series of further articles (1935, 1936, 1937, 1938) he was able not only to document similar tissues with cartilage and lipid cells, but also the original colour pigments in the skin of fishes and frogs. As he asserted in a later article (1937), the original organic content of these tissues is not present, but has been replaced completely by silica (SiO_2) (Fig. 375).

In contrast to plant tissues, animal tissues are fossilized only in very rare instances. Normally, after an animal's death the tissues are prone to

Fig. 375: Soft-tissue preservation in the Geiseltal. (1) Connective tissue from the dermis of a frog's skin, with oblique transverse sections of blood-vessels. Magnification, 403×. (2) Muscle fibres of the artiodactyl *Anthracobunodon weigelti*. Magnification, 43×. (3) Muscle fibres of a beetle. Magnification, 970×. (4) Silicified muscle fibres of a beetle. Magnification, 970×. (5) Striped wing muscle of *Eopyrophorus* sp. J, isotropic layers; Q, anisotropic layers; Qh, Hensen stripes; Z, light line in J. Magnification, 500×. (6) Contractile thickening of a flight muscle. Fine anastomoses of a trachea occur above the perimysium (arrow). Magnification, 500×. (7) Encapsulated nematode larva in the muscle of a beetle. Magnification, 1000×. (8) Muscle fibres between the neural processes of the fish *Anthracoperca siebergi* Voigt. Magnification, 18×. (9) Pigment cell (xantholocophore) from a frog skin, filled with granules of guanin. Magnification, 1300×. Originals from the brown-coal mine 'Cecilie' in the Geiseltal near Halle/Saale.

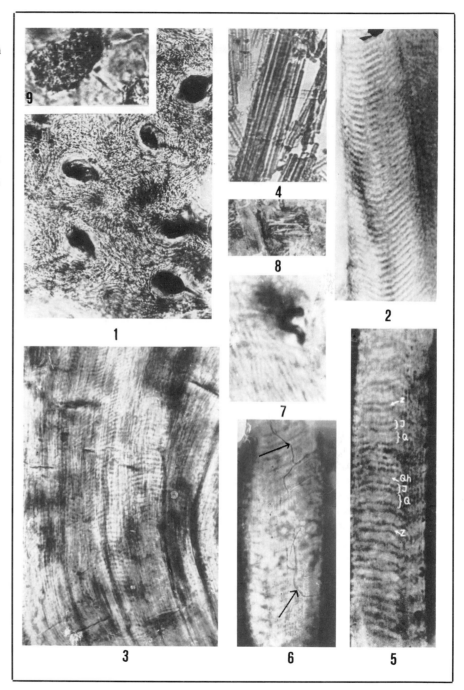

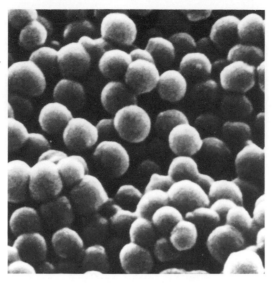

Fig. 376: Cocci-shaped bacteria preserved in iron carbonate (siderite) from the skin of the frog *Messelobatrachus tobieni*. Diameter of the bacteria, *c.*2 μm.

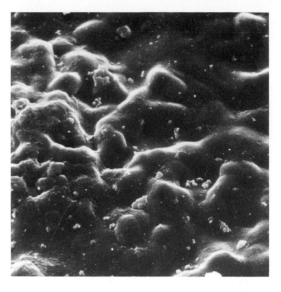

Fig. 377: Petrified bacteria from the rib area of *Messelobunodon schaeferi*. The bacteria are cemented together by organic material. Diameter of the bacteria, *c.*1 μm.

quick disintegration, particularly through decomposition by micro-organisms. A replacement of the organic substance by silicic acid, in which the shape of the original tissue is preserved requires an extended period of time and can only take place when the disintegration of the organic material is retarded. For this initial preservation Voigt postulated a form of tanning by means of the tannin present in the acidic, humus-filled waters of the Geiseltal peat bog. Similar processes are recorded, for example, as having taken place in human corpses, which have lasted, partly unchanged, in Recent peat bogs for more than 8000 years. A series of natural tannins, liberated from the bog vegetation during peat bog formation, are responsible for this preservation.

Lignin (the actual substance of wood), liberated during the microbial decomposition of plant materials, combines chemically with albumins which make the latter most resistant to microbial decomposition (Waksman and Iyer 1933). Of further significance is the permeability of animal tissues to tannin, so that, for example, the horny bills of birds or the claws of frogs, although also composed of albumins, have not been fossilized. An essential piece of circumstantial evidence for such processes is the rule established by Voigt that almost always only the external tissues—

essentially the skin and the musculature immediately below it—were fossilized.

During experimental investigations into the silicification of woody tissues, Leo and Barghorn (1976) were able to elucidate the fundamental processes which occur during the silicification of organic material. With minor modifications their model can be used for animal tissues. This was demonstrated by the investigations of Wuttke and Kott (unpublished) on silicified soft tissues of sponges from Devonian slate in the Hunsruck.

Free silicic acid occurs in natural waters, primarily in true solution in the form of orthosilicic acid $Si(OH_4)$. It was brought to the Geiseltal peat bog from neighbouring areas at higher altitude, probably principally from the Mesozoic red sandstone from which it was liberated by weathering processes. It must have been present in high concentration in the water of the peat, as numerous silica nodules are found in the coal, as well as small dumb-bell-shaped and spherical lumps of silica.

Owing to their small diameter the molecules of orthosilicic acid were easily able to penetrate the tanned animal tissue and to completely impregnate it. The precipitation of the silicic acid on to and inside the animal cells occurred via chemical surface bonds (hydrogen bonds) between the biomolecules and the molecules of

silicic acid; the latter covered all cellular surfaces as a thin film. In the course of the gradual disintegration of the organic matter, further functional groups were liberated constantly, which were able to react with the orthosilicic acid until all biological details were completely replicated. In the course of further diagenesis the molecules of silicic acid polymerized, while giving off water, to form amorphous, opaline silica which, under the microscope, reveals such minute details as cell nuclei.

Soft-part preservation at Messel

The preservation of soft-part structures such as fur shadows, feathers, and hairs in the Messel vertebrates, described so impressively in the preceding chapters, results from an entirely different phenomenon. As described in the last section, the precondition for fossilization in the Geiseltal was the slowing down or, respectively, the prevention of bacterial processes of decomposition. In the case of Messel, paradoxically, those very micro-organisms that decomposed the former soft tissues were responsible for reproducing the soft-tissue structures. Scanning electron micrographs (see Chapter 26, 'Petrified gastrointestinal contents—their analysis and interpretation') of these soft tissues showed that, with a few exceptions, the original substance has been replaced in the finest detail by dense mats of petrified micro-organisms (Fig. 376). The petrification only affected, strangely enough, those bacteria which participated in the decomposition of the soft tissues of the vertebrates embedded in the mud at the bottom of Lake Messel, but not those bacteria which inhabited that mud.

Chemical analyses of these bacteria showed that an iron carbonate, namely siderite ($FeCO_3$), served as the means of petrification. These analyses also made it possible to replicate the steps leading to the petrification of the bacteria (Wuttke 1983).

In the mostly anoxic environment (see below) at the bottom of Lake Messel the decomposition of the organic material of animal corpses took place mainly by the action of sulphate-reducing and denitrifying bacteria which, in the course of their metabolism, transferred an entire series of organic and inorganic metabolic products (for example, carbon dioxide, hydrogen sulphide) into the water contained in the pores of the surrounding sediment. Among other moieties, this water contained iron and carbon dioxide in solution which were in chemical balance with yet other dissolved substances. The introduction of the above-mentioned bacterially produced substances into the water contained in the sediment pores changed the immediate chemical environment such that the saturation point for iron carbonates in solution was exceeded and they were, consequently, precipitated. This precipitation started, independently of the species of micro-organism, directly on the cell surface or within the micro-organisms themselves.

In the course of the microbial activity, a thin crust of siderite gradually formed on or within the bacteria, which led to their demise. With the death of the micro-organisms the precipitation of siderite ceased, so that the morphological structure of the decomposing bacteria was preserved (Fig. 376). In the course of further diagenesis additional organic substances of plant origin migrated between

Fig. 378: Bird with preserved feathers. Partial skeleton of an as yet unidentified bird. The rachis as well as the barbules of the tail feathers can be recognized. Height of the picture detail, 12 cm.

Fig. 379: The frog *Eopelobates wagneri* with skin preserved, which is very rare in this species. Even the webbing between the toes, which leaves no visible marks on the skeleton, can be reconstructed. Head and body length, *c.* 45 mm.

the petrified micro-organisms and partly cemented them to one another (Fig. 377). The dark coloration of the skin and feather shadows of the Messel vertebrates, which can be recognized in the colour photographs of the previous chapters (see also Fig. 379), can probably be traced to these organic substances.

The chemical environment inside the animal corpses was also subject to variations and, as a consequence, varying amounts of bacteria became petrified. Thus, in exceptional cases, even internal organs like the liver or the course of individual veins in a specimen of the frog *Messelobatrachus tobieni* can be recognized within the confines of the skin shadow (Fig. 370).

Pyrite preservation of fur shadows or, respectively, of parts of it has, to date, only been observed in one Messel mammal, the species *Macrocranion tupaiodon*, recovered in 1985 (Figs 380 and 381). The majority of the fur shadow consists of the already familiar sideritized bacteria cemented together by dark, organic substances. Individual light-coloured spots of a few millimetres in diameter, which at first led the excavators to believe that the original fur pattern had been preserved, proved to be, on closer inspection, accumulations of microscopic, spherically arranged pyrite crystals. They are designated pyrite framboids (Fig. 382). At first sight these aggregates appear to have little to do with micro-organisms, but it has just recently been possible to document their bacterial origin using the fossils of the Lower Devonian Hunsruck slate of Bundenbach, whose preservation is mostly pyrite-based (Kott and Wuttke 1987). It has been found that the formation of pyrite can be traced back primarily to the metabolic activities of the anaerobic, sulphate-reducing bacteria and the aerobic/anaerobic bacterial producers of sulphur dioxide, which decomposed the organic material of the animal corpses in Lake Messel. The formation of pyrite takes place in two steps and proceeds only when bacteria with the qualities discussed above exist in close proximity to one another.

The first step of pyrite formation is initiated by bacteria which live in a wholly anoxic environment and which, to gain energy, reduce the sulphate in the water trapped in the sediment pores. The metabolic product that they give off via the cell surface to the external environment is the chemically highly reactive hydrogen sulphide (H_2S). This combines immediately with the iron dissolved in the pore water which, in turn, precipitates as iron monosulphide (FeS) directly on to the surface of the bacteria. At this point in time the bacterial shape can still be clearly recognized because the iron monosulphide is present in non-crystalline form. The actual formation of pyrite, however, can only be achieved when the iron monosulphide can combined, in a second step, with free sulphur. This free sulphur is supplied mainly by bacteria from the oxygenated environment, but in isolated cases even anaerobes can achieve this. To gain energy both forms split the hydrogen sulphide (H_2S) produced by the anaerobes and give off free (elemental) sulphur to the external environment as the final product. This quickly reacting sulphur then combines with the iron monosulphide secreted around the first-

Fig. 380: The mammal *Macrocranion tupaiodon* with fur preserved. The black areas consist of siderite, the light, rounded ones of pyrite. Head and body length, c.15 cm.

Fig. 381: Detail from Fig. 380. The form and arrangement of the hairs can only be reconstructed when preserved in siderite. During pyritization the original shape is destroyed by crystal growth.

mentioned bacteria, to form pyrite (FeS_2). The crystallization that accompanies this process destroys the external shape (which initially was that of the bacteria), and so leads to the pyrite framboids that, at Messel, have remained intact to the present.

As yet it has not been possible to clarify what part the anaerobic H_2S-splitting bacteria have in the formation of pyrite. Thus it appears possible that the pyrite formation in the fur shadow of *Macrocranion* proves that, at least from time to time, Lake Messel was aerated throughout (even though weakly) down to its floor.

The Messel fossil site does not, however, only permit 'indirect' statements concerning the soft-part anatomy of fossil vertebrates, but, in some exceptional cases, even the original albumins of the hair and scales of mammals (Richter and Storch 1980; Koenigswald *et al.* 1981) have been preserved.

In all these cases special conditions that prevented decomposition by micro-organisms or by a chemical solution must have predominated in the sediments. An identical preservation was observed in the fossils of the Geiseltal, where Voigt (1936–39) was able to demonstrate horny scales from lizards and hair from mammals microscopically and by a flame test (smell of burnt keratin).

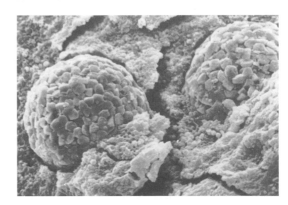

Fig. 382: Scanning electron micrograph of the pyritized part of the fur. It consists predominantly of pyrite crystalloids (framboids) which are arranged spherically. Diameter of spheres, c.0.01 mm.

In the case of both fossil sites it is, however, still not clear what special conditions contributed to this preservation; perhaps the tissues were impregnated with bactericidal substances or underwent a chemical transformation, which did not alter their external structure but which prevented their bacterial decomposition.

Apart from hard and soft tissues from animals, their excrements have also been preserved in the oil shale. The elucidation of their origin and history of fossilization form, at present, the topic for a dissertation (Schmitz). In addition to the excrement of land herbivores, found preserved in a coal-like form, fossilized dung (coprolites) of the aquatic fauna, primarily of fish and crocodiles, is prevalent at Messel. These coprolites, which have mostly been preserved in their original form, present in the form of phosphorite concretions surrounded by a crust of siderite (Figs 383 and 384). The chemical composition of the phosphorite concretions corresponds to a large extent with that of bones (hydroxyapatite). It has not yet been possible to determine if the phosphate in the concretions derives from the excrement (meat contains up to 30 per cent apatite; bone, which is digested almost completely in crocodiles, c. 95 per cent) or if it has migrated in from the sediment.

Present-day areas of phosphorite concretions are located predominantly on the west coasts of continents. Sedimentological or geochemical investigations have shown that such concretions form on the surface or within the uppermost metre of the sea-bed. Preconditions for the formation of concretions are a high phosphate content in the water contained in the sediment pores, a low to non-existent oxygen content, and a pH value of 7–9.

On the basis of these results we can establish a provisional model for the phosphatizing and sideritizing processes in Messel as far as the origin of phosphorus and the biogeochemical conditions on the floor of Lake Messel are concerned. Weber and Zimmerle (1985) traced the origin of the inorganic portion of the Messel oil shale to the weathering products of basaltic lavas and tuffs. The majority of this material was imported into the lake via tributaries as argillaceous silt.

Fig. 383: Fossilized dung (coprolites) of fish (above) and crocodiles (below) preserved in phosphate. All coprolites are surrounded by a siderite crust. Height of photograph, c. 25 cm.

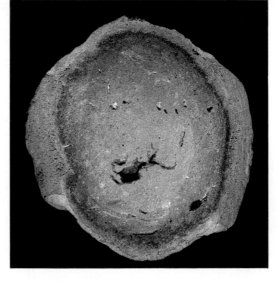

Fig. 384: Transverse section through a crocodile coprolite. The dark centre consists of calcium phosphate; the pale crust of siderite. Diameter, c. 4 cm.

In connection with this there would have been an increased transport of iron and phosphorus (in the form of apatite) into the lake. These elements would have gone into solution in the anoxic zone on the lake bed and in the sediment, respectively. A further (secondary) source of phosphorus would have been the algae which drifted to the lake bed; their remains make up the major part of the organic portion of the oil shale (see Chapter 3, 'The genesis of the Messel oil shale'). Their organic material was decomposed by anaerobic micro-organisms in the topmost centimetres of the sediment. The phosphorus fixed in the algae was transferred in the course of the microbial metabolic processes to the surrounding water in the sedimentary pores, and there contributed to a further increase in concentration.

Animal excrement sinking to the lake floor was so quickly hardened by the onset of phosphate and siderite precipitation, which began either on or in the topmost centimetres of the lake bed, that it could no longer be deformed by overlying sediments (Fig. 385). The precise working of this process, however, still requires an explanation. It is not certain at present to what extent a primary phosphate content in the excrements served as seed crystals for the phosphates contained in the water in the sedimentary pores. An increase in concentration through micro-organisms in the excrement itself appears to be a possibility. The formation of the siderite crust which surrounds the phosphorite concretion may have taken place inorganically, but may also have

Fig. 385: The string-like faeces of a fish preserved in siderite. The fossilization of this rather insubstantial fish faeces is rare at Messel and suggests fossilization over a short period of time. Diameter of excrement, c.4 mm.

occurred, as described above, through the influence of micro-organisms, in the course of soft-tissue preservation.

The biogeochemical processes in Lake Messel and their influence on the fossilization processes for plant and animal debris are understood, if at all, only to a small degree. Messel, however, offers the chance to investigate, for a single fossil site, the changes undergone by all groups of substances and by this means, perhaps, to arrive at a general understanding of the conditions for fossilization in a fossil site. These results would then, in turn, have the greatest importance for the investigation of other fossil material or sites.

25

From excavation to exhibition piece

The fossils of Messel are renowned for the completeness of their preservation and for their beauty. The site is marked, furthermore, by the presence of the most diverse animal and plant groups—yet visitors who see the site for the first time are often disappointed by the unprepossessing specimens that are found. The excavation and preservation of all this material requires various kinds of time-consuming preparation techniques without which the beauty of the specimens is not realized. The reasons for this lie in the character of the oil shale—it disintegrates as it dries out—and the variable composition of the fossils. Thus, the skeletons of the vertebrates must be mounted in resin, while plants and insects, which are still comprised of organic materials, must be freed from their embedding matrix under water and stored in glycerine. Their true information content, however, can only be obtained by means of special techniques of photography and microscopy. Here we shall describe polarization photography, fluorescence microscopy, and scanning electron microscopy (SEM) and X-ray techniques that allow hidden structures to be seen.

From excavation to exhibition piece

MICHAEL ACKERMANN, JÖRG
HABERSETZER, AND
FRIEDEMANN SCHAARSCHMIDT

Under normal circumstances fossils are found at Messel by serendipity during the splitting of the oil shale. Only once has it been possible to dig for a fossil deliberately and this turned into a nerve-racking activity.

When, in the autumn of 1980, scientific cores commissioned by the Minister of Culture in Hessen, were drilled in the pit, the neck vertebra of an obviously large crocodile was discovered in the centre of one core from just below the surface. Eager fossil collectors had apparently read the publication concerning the drilling, as a short time after the publication, and secretly gained entrance to the pit, which was barred to the unauthorized. They had found the relevant drill site (using the published drilling map), and had already removed several layers and almost reached the skeleton. In order to avert the danger of an inept 'pirate' excavation, the Senckenberg group decided to rescue the block containing the crocodile skeleton in a large-scale operation. First, the block containing the skeleton was exposed. Then, with the aid of splitting wedges and a chain-saw, broad channels were dug around it until the block lay in a large pit, standing free on a natural pedestal. Since, owing to the unfavourable terrain, no heavy lifting gear could be used, an entry corridor had to be dug on the slope inclined towards the basin, through which the skeleton, surrounded by a solid wooden frame, lifted on to a litter, and embedded in polyurethane foam, could be transported. In this way, after 2 days of hard physical labour, the excavation team secured a crocodile which is, with a length of *c.* 4 metres, one of the largest that has ever been found in Messel. During the preparation the vertebra which had

been cut out of the skeleton during the drilling was refitted in its original position.

Not every fossil recovery takes place under such dramatic circumstances. Since there are no concentrations of fossils in the Messel oil shale, the main work of the excavation teams lies in the splitting of blocks, with, however, the constant expectation of coming upon a valuable fossil at any moment (Figs 386 and 387). At the same time, scientific observations of the most varied kind must be made constantly, and all small remains such as insects, leaves, and fruits must be recovered.

Fossil recovery and preparation

For more than 10 years now, only scientific institutions and museums have been permitted to excavate at Messel. The permission for such work is given by the Ministry for Science and Art in Wiesbaden in consultation with the Office for Mines, Weilburg. The law of the Protection of Monuments in Hessen forms the basis for permission to excavate. Since 1975 admission of the public, including private fossil collectors, to the pit has been prohibited, as the full significance of the flora and fauna can only be understood in connection with the site of deposition in the lake, and only when recovered by specialists. When the private treasure hunter searches feverishly for large fossils, the unity of the findings is destroyed. Among the debris in excavation holes, finds of remains of fossil mammals were frequent, a witness to the fact that important fragments lie far removed from these in unknown private collections. The fossil itself has thus lost much of its value. It must be considered that fossils are rescued, in the scientific and cultural context, only when they are prepared and accessible in a research institute or in a public exhibition.

The excavations are executed differently by the individual institutes. But the use of smaller or larger groups, in which employees of the institutions are supported by students of geology doing their field studies or by honorary helpers, has been effective. By this means it is possible to search a sizeable amount of oil shale for fossils in a short space of time. This is

Fig. 386: (above). Excavation assistants split the massive blocks of Messel oil shale and examine it for fossils.

Fig. 387: A find is cut out of a block. To prevent drying out, the fossil and the surrounding oil shale must be moistened.

necessary because there are no concentrations of fossils in the sediment and because the frequency of finds depends in general on the amount of rock investigated. In addition, work must progress at a good pace because the exposed oil shale quickly dries in the open air, splits, and soon disintegrates into small pieces.

In a normal excavation two steps follow, once the oil shale has been laid bare by construction machinery: (1) the working of bulky blocks of oil shale is accomplished by aluminium wedges, sledge-hammers, crowbars, and shovels; (2) large butcher's knives are used for the splitting of blocks. In this process smaller fossils are mostly uncovered fortuitously; larger ones are often indicated in advance by a curvature of the layers. Fossils are frequently broken open by the splitting. Care must be taken to gather all the fragments. When a fossil lies at the edge of a block, its continuation must be sought in a different block or in the natural deposit. As long as they are exposed, all finds must be kept moist and must be packed, watertight, as quickly as possible. In order to avoid their sticking to the paper, the exposed parts of the fossil are first covered with a thin sheet of plastic and then wrapped in moist newspaper. Finally, the whole pack is put into a plastic bag which is stuck down watertight with sticky tape or even soldered closed. Very thin pieces can be supported by a board when packing, very large ones are surrounded by a wooden frame and covered in polyurethane foam. During this entire period (which can last several hours) the pieces must always be kept quite moist and the excavation sites must, if at all possible, be protected by a tarpaulin from the sun's radiation. Before packing, all larger pieces are recorded in a register, and delicate material, such as gut contents, is removed and conserved for scientific research in a 2 per cent formalin solution.

Besides these main excavations, which must be carried out quickly, excavations directed towards the collection of insects and small remains of plants are also often necessary. For these, small kitchen knives are the most suitable. These remains and also leaves and other large remains of plants are collected at first in pails of water or plastic boxes and are checked in the laboratory under the stereo- and fluorescence microscopes.

The temporarily wrapped pieces can later be prepared at leisure. The methods employed in this task depend on the composition of the material and on the size of the specimen.

Preparation of vertebrates

The method best suited for vertebrates is the transfer method first recommended for the Messel fossils by Kühne (1961), but which had

Fig. 388: The coal-like soft-body outline of this small lizard (?*Eolacerta* sp.) is exceptionally well preserved and easy to trace in the moist oil shale. Length, 15 cm.

Fig. 389: The most frequently used procedure is the preliminary preparation of only one side of the fossil before transferring it to an artificial support.

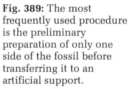

already been employed 50 years earlier in a comparable form in palaeobotany. In this technique the fossil is freed from the oil shale and transferred to a permanent support. Formerly, waxes or a chalk/wax mixture were used. But these had the disadvantage that the bones themselves were not hardened and that skin shadows could not be transferred. This, however, is not the case if one uses suitable artificial resins (for example, epoxy resins). One proceeds as follows: half a side of the fossil is freed from the oil shale, using needles

Fig. 390: After transferring the find (Fig. 388) on to a support of artificial resin, the lower side can now be prepared. In this picture the specimen is already entirely freed from the oil shale.

be investigated from all sides. Some fine details such as 'skin shadows' are not retained, however. If these are present, the transfer method is again resorted to.

As with the transfer method, half a side of the fossil is first freed from the oil shale. Now one can decide whether to prepare the specimen as a unit or as individual bones. Usually one will choose the first procedure with a view to obtaining a more attractive show-piece. The liberated side is now carefully surface-dried and hardened with a cyanacrylate-glue ('superglue'). Loose parts are fitted into place using a two-component stone cement. Then the exposed side is covered completely with cling-film and uneven areas are evened out with clay or plasticine. After a supporting mantle of gypsum has been made, the piece can be carefully inverted and the second side can be freed from the matrix. This side too must now be hardened with cyanacrylate glue. Finally, the piece is taken from the supporting mantle and treated with laquer. If the dismantling method is to be used, the bones must be, individually, completely freed from the matrix and hardened all around. Photographs or sketches, respectively, are made in advance of the preparation to document the position of the find.

Fig. 391: When preparing a skeleton so as to be free from any matrix, every bone is conserved individually and then returned to its original position.

and scrapers; harder crusts of siderite or pyrite must be removed, by contrast, with a sand-blasting apparatus using fine grains of sand (grains 50 μm or less). After this a frame of clay is modelled at some distance around the fossil which is supposed, later, to prevent the escaping of the artificial resin. Then the surface is dried with pressurized air and the oil shale is painted thinly with artificial resin. The fossil is carefully dried further, with a hair-dryer, so that the artificial resin can later penetrate the bones. Following this, the artificial resin is poured over the specimen in several thin layers, to avoid the trapping of air bubbles. When the block has hardened it can be inverted and the oil shale can be removed from the back and, thus, from the fossil. While the 'split' fossil exhibited a broken internal surface when it was recovered, the more beautiful outer surface is now visible (Fig. 390). The finished preparation is painted with a colourless laquer (such as Zaponlack).

For some time now fossils have been successfully prepared free from the matrix without a support of artificial resin. The advantage of this method is that the objects can

Preparation of insects

The preparation of insects takes place at medium magnification (20–40 ×) under a stereo microscope. The tools used in this process are very thin steel micropins bent into a hook at their tip ('needles'). Using these, the remains of sediment that cling to the insects can be scraped off, a process that can take several days for larger finds and which demands the greatest concentration, as the chitin is frequently no harder than the surrounding sediment. In order that the finds do not tear during the preparation, due to drying out, the work must always take place under water. When the work is finished, the specimens are stored in pure glycerine where, as a rule, even the delicate structural colours remain preserved. A preparation procedure, first developed for this application at Messel and which is, in principle, comparable with the vertebrate preparation procedure, opens up entirely new vistas for the researcher. If larger insects are transferred to artificial resin and freed from all vestiges of sediment, it is then possible to investigate under the light microscope the most minute details of the chitinized internal organs. For example, the information about the stinging apparatus of the 'giant ants', for example, was gained in this way. The use of the fluorescence microscope is also possible. Although chitin itself does not fluoresce, its structures, frequently in complete detail, appear black against the general background fluorescence.

Preparation of plant remains

Just as with insects, plant remains are also preserved as organic matter and are present as thin layers. As a rule, therefore, they cannot be recovered easily by simple mechanical or chemical techniques, but must be uncovered painstakingly and must remain in the oil shale.

Because the usual preparation with knives and needles normally used for other sediments is not possible, a time-consuming underwater procedure had to be developed (Schaarschmidt 1982). Normally the exposure of less delicate objects like leaves and large fruits is achieved with ultrasonic apparatus, the 'Cavitron', also used by dentists in removing tartar. Although it is certainly possible to work quickly using this, in fragile objects, damage cannot always be prevented. In these cases, for example, in blossoms and small fruits, it is in, the end, best to use bent micropins in a needle holder. With these the oil shale can be very cleanly removed from the specimens submerged in water.

The transfer method can only be used in rare cases for easily detached leaves or robust specimens. It is most useful when the specimen is 'upside-down', for example, a blossom below the calyx. After a light surface drying, the specimens are transferred to a support of hydrophilic artificial substrate and the oil shale is then removed under water. Since, unlike in the vertebrates, the artificial substrate does not penetrate into the plant specimens, the pieces must afterwards also be stored in glycerine. Leaves that are easily detached can also easily be transferred in glycerine to a Plexiglass plate, covered with an acetate sheet, and ringed with Plexiglass glue.

Normally all prepared plant remains are stored in polystyrene boxes in which they are covered by glycerine, to which thymol is added to prevent the growth of algae, fungi, and bacteria. From this substrate the pieces can at any time be transferred to water for observation and then replaced. Smaller pieces can also be dried when they have first been immersed in glycerine for 1–2 days. But the rock then becomes very dark, individual features are scarcely visible, and the pieces can also bend or tear. Specimens stored in glycerine, however, keep almost as well as in the fresh state.

In photographing these specimens one can use polarized light to obtain increased contrast (one polarized filter in front of the lamp, another in front of the objective). The colour in photographs can be distinctly enhanced by this method.

The fluorescence microscope has proved useful for investigating the most intricate structural detail. Here short-wave light (violet or ultraviolet) is used for illumination. By this means some substances, such as cutin or sporopollenin, are excited to emit light in the long-wave spectrum, for example, yellow light. Thus, without previous preparation, the structure of the epidermis and the structure of pollen grains can be made visible. The best results are obtained when the pieces are placed in a bowl of water and 'immersion objectives' are dipped into the water. This process is helpful in a first identification, it makes it possible to judge the state of preservation, and some details can only be recognized using this procedure. On the other hand, for a detailed investigation, additional microscope slides must be produced by maceration. This is achieved by the action of chemicals (for Messel specimens, for example, hydrogen peroxide) on the leaf fragments, by clearing with ammonia, and staining with safranin. Following this, the investigation can proceed with specialized light microscopes (for example, the phase-contrast microscope) or with the scanning electron microscope (SEM). In a similar fashion the pollen can be obtained from oil-shale samples (Thiele-Pfeiffer 1988; Schaarschmidt 1982).

How to render the invisible visible

Although the fossils of Messel are preserved superbly and are frequently complete, details can often not be investigated because they are overlain by other structures. Using X-ray photography the covered bone structures can be revealed without damaging the often unique fossils. In the Senckenberg Research Institute several X-ray units are used for different purposes.

One special X-ray image intensifier is used to X-ray thick plates of rock in the search for fossils. Another conventional X-ray apparatus (Siemens Naodor 2) is used at Messel to make direct 'on-site' exposures which can give important information during preparation concerning dispersed or deeper-lying parts of the fossil. Overall views serve to select parts of the specimen for more involved, detailed analyses. With a special micro-X-ray apparatus (Microx 60/1; Figs 392 and 393) and a comprehensive palette of different film emulsions and recording and development

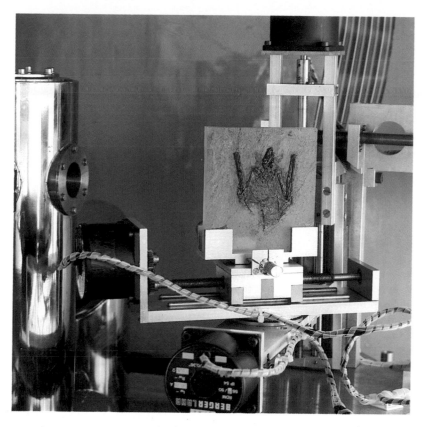

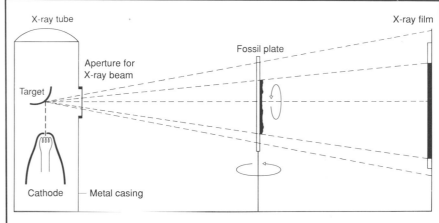

Fig. 392: In front of the micro-X-ray tube on the left is a specimen manipulator with a fossil plate. The stepper motors of this manipulator are steered by remote control. In this way any chosen rotations can be made, within the path of the beam.

Fig. 393: At the anode of the X-ray cylinder an X-ray beam is produced by the impact of accelerated electrons. The X-rays penetrate the fossil, leading to a corresponding blackening of the film. Adjusting the distances between the anode, fossil plate, and X-ray film results in an enlargement of the picture. With this technique a better-focused reproduction of detail is possible when the focal spot of the anode is very small. In this X-ray apparatus the spot diameter is only 0.01 mm.

X-ray tube

X-ray film

Aperture for X-ray beam

Fossil plate

Target

Cathode — Metal casing

techniques, high-resolution radiographs are produced according to the information required. Instead of describing the many other areas of application using this non-destructive technique, we would like to mention only a few applications.

Formerly it was assumed that, in principle, the inner ears (cochleae) of all mammals have a similar structure. Various investigations have, however, recently shown that echolocating bats not only have larger inner ears than the flying foxes, which don't use echolocation, but that their morphological details are also entirely different. These big morphological differences possibly correlate directly with various echolocation strategies. An extensive radiological–morphological investigation on the inner ears of fossil and Recent microbats and Recent flying foxes tested whether this hypothesis could be proved. An X-ray method

was developed which made it possible to measure the smallest bony structures of the cochlea. In collaboration with engineer Küber (VDO Adolf Schindling) a remote-control sample manipulator was developed, with whose help preparations could be tilted with an accuracy of 0.2° in all spatial planes (Fig. 392). Among other things, this makes it possible to obtain an image of the bony suspension apparatus of the basal membrane which is only 0.02 mm thick. This, however, is only successful when the bone lamella is in an exactly defined alignment spatially (orthogradient) in the path of the rays (Fig. 393). Apart from its non-destructive testing of valuable specimens, the advantage of the radiological method lies principally in its applicability to both Recent and fossil specimens thus allowing a direct comparison of data. Radiological studies on selected extant bats serve, on the one hand, to

test hypotheses on the functional significance of the anatomical structures, and, on the other hand, as a necessary prerequisite for the interpretation of palaeontological data. The investigation was able to clarify the question as to the morphology of the inner ear of the Messel fossil bats (see also Chapter 14, 'Bats—already highly specialized insect predators'). With this micro-X-ray method it is possible, furthermore, to investigate entirely different details as, for example, the structure of teeth and germinal teeth. For this purpose the fossil is examined at different angles of X-ray penetration and the view with the clearest reproduction of details is chosen (Fig. 394).

The adaptation of the flight apparatus of the Messel bats to distinct niches can be corroborated by exact measurements and the comparison of wing skeletons using radiographs. An appropriate enlargement of the special films as well as of the stereo-exposures makes it possible to pinpoint joint areas which are covered in the fossil and to measure telescoped bone fragments individually. From the complete reconstruction of the flight apparatus that is possible using such data, parameters important for the flight characteristics can then be ascertained (see

also Chapter 14, 'Bats—already highly specialized insect predators'). While, in some species, we can choose between many well preserved fossil slabs, in other cases we must resort to elaborate procedures involving radiological reconstruction from two slabs each containing part of the split specimen (Figs 395–397).

In contrast to a photograph, a radiograph is a summation to which all structures contribute which produce a shadow throughout the entire volume of the irradiated object. We have already referred to the great importance of the spatial projection relationships between X-ray tube, object, and film. The possibilities arising from the immense amount of information contained in a radiograph have, however, by no means been exhausted. Since even high-quality printing techniques can only inadequately reproduce this complex information, the Senckenberg Research Institute has developed a special preparation technique for the NDT-films with which the fossils are X-rayed. The negatives, produced by micro-X-ray technique, are developed so as to produce an extremely enhanced black tone with maximal contrast. There then follows a manual preparation of

several steps which reduces the range of the blackness so as to enable further work with a standard computer scanner. By a combination of these individual steps we obtain extraordinarily detailed radiographs which convey the impression of a three-dimensional print. In this way, even the untutored eye can easily become accustomed to the radiograph. In future, the radiological method will, through the additional use of X-ray video techniques and digital image processing, assume an even more fundamental role in palaeontological research.

From the excavation to the exhibition

It is the aim of every excavation to obtain fossils not only for research but also for exhibitions. The complete skeletons are particularly suitable for this. In addition, for smaller objects, for example, insects and plants, enlarged photographs will be used. Further development of preparation and investigation techniques will provide valuable acquisitions for the exhibitions.

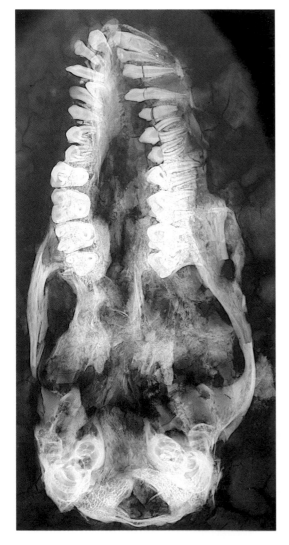

Fig. 394: Only an X-ray photograph of the skull of *Pholidocercus hassiacus* (see Chapter 13, 'Primitive insectivores—extraordinary hedgehogs and long-fingers') renders visible the vestibular organ and the inner ear (invisible in the original specimen). In the upper and lower jaws the roots of the teeth are also visible. At the front left, two erupting teeth have not yet emerged.

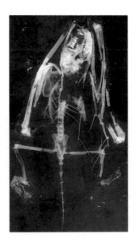

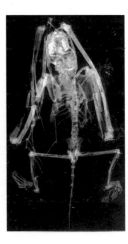

Figs 395–397: During splitting of the oil shale the bat (*Hassianycteris messelensis*) was split in half (Figs 395 and 396), so that the skeleton is not completely preserved on either slab. Exact measurement of the bones is impossible. Only by superimposing both X-ray photographs (Fig. 397; picture on the right) can the entire preserved skeleton be shown.

26

Fossilized gut contents: analysis and interpretation

The following compilation concerns the ways in which food remains are preserved in fossil mammals, but it also takes into account the general conditions of fossilization of the Messel formation. As so often happens, the preservation of plant tissues is far better and more complete than those of animals. At Messel this is true not only for pollen, which is proverbially indestructible, but also for leaf cuticle, corky or woody tissue, vascular bundles, or stone cells (from the walls of seeds with a hard coat).

Fossilized gut contents: analysis and interpretation

GOTTHARD RICHTER

The preservation of soft tissues is usually the exception in palaeontology. Yet at Messel soft plant tissues (parenchyma) are preserved not only in the form of collapsed remains of cell walls, but also as fossilized cell contents preserved in fragments of intact tissues, in which the original 'casts', the cell walls, are present only as rudiments (Fig. 398). This form of preservation of plant tissues which has, according to our information, been described only at Messel, appears to be particularly frequent in stomach contents and to be distinctly more frequent in meat/insect-eaters than in plant-eaters.

The reasons for this are obvious. This type of preservation presupposes the embedding, and at least the start of fossilization of fresh leaves whose tissues retain fully preserved cell contents and cell pressure (turgor). In dry or moist rotting leaf litter the cell walls would be either collapsed or corroded by fungal or bacterial action. The cell content would have trickled away, dried up, or decomposed before it could have become fossilized. Obviously, one would expect living, fresh leaves to be present at a higher concentration in stomach contents than in normal sediment. Carnivores, insectivores, or omnivores generally have only limited facilities to mechanically or chemically decompose and digest plant parts that were eaten intentionally or fortuitously, whereas herbivores must optimize the use of their single source of nourishment. In the stomach of a herbivore, therefore, a far smaller percentage of undamaged plant tissue will be 'available' for fossilization than in that of an animal feeding on mixed nutrients or where meat intake predominates.

The preservation of animal soft tissues is rare at Messel, although as yet only a few investigations have been carried out

concerning this (Richter 1987). Of the mineralized skeletal parts, bones are excellently preserved in stomach contents as long as digestion had not yet begun. However, it is just on the basis of such traces of digestion that 'primary' breaks (caused by the prey being crushed during feeding) can be differentiated at first sight from 'secondary' ones (sustained while the food remains were being cleaned and isolated), because the fractures of bones with initial traces of digestion have distinctly rounded edges (Fig. 399). In this connection, a second mineral skeletal element, the siliceous spicules of fresh-water sponges (never represented in stomach contents) should be mentioned. They show more or less extensive, very characteristic forms of corrosion (see Chapter 24, 'Conservation—dissolution—transformation. On the behaviour of biogenic materials during fossilization').

Horn (keratin), perhaps chemically altered by diagenetic processes, is found only rarely in

the Messel fossils. Vestiges of skin, hair, and feathers, on closer inspection, usually turn out to be an accumulation of fossilized bacteria, although these often reproduce the shape of the original structure with amazing accuracy (Wuttke 1983). Only in the stomach contents of bats do we find bat hair (swallowed when grooming) and in one pangolin some horn scales (Richter and Storch 1980; Koenigswald et al. 1981).

Chitin is generally well preserved. But even this very resistant substance must have altered in consistency after deposition in the sediment and before the actual fossilization; it must have become, to some extent, 'soft' at the surface, because we frequently find the impressions of chitinous hair pressed deeply into the surface of cuticle fragments (Fig. 400). Yet the original surface sculpture of the chitin armour is retained.

The food remains recovered from mammal fossils are, as a rule, very small particles which

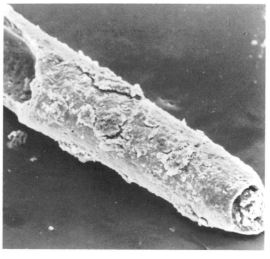

Fig. 399: (Below). Gut contents of *Leptictidium nasutum*. Hollow bone. Original break (by biting) at lower right—fresh break (by preparation) at top left. The edge of the original break shows traces of digestion and is therefore rounded; the fresh break has a sharp edge. SEM: magnification, 110×.

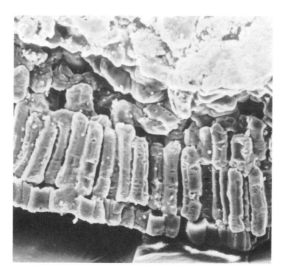

Fig. 398: (above). Gut contents of *Eurotamandua joresi* (ant-eater). Transverse section of a leaf (fossilized cell casts). From the bottom up: epidermis, palisade mesophyll from the parenchyma, spongy mesophyll from parenchyma, epidermis (not easily recognized), vestiges of the cuticle. SEM: magnification, 375×.

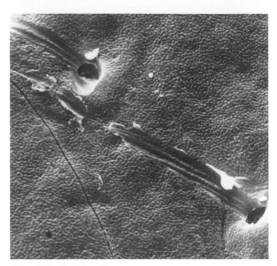

Fig. 400: Cuticle of an unidentified beetle from the Messel oil shale showing hair follicles and surface sculpting. Through the pressure of the subsiding sediment, the hairs (no longer present) have impressed themselves deeply on to the softened surface of the chitin. SEM: magnification, 1850×.

occur dispersed or more or less concentrated in the oil shale. Macroscopically, they generally cannot be distinguished from other particles or inclusions of the sediment and, then be identified. A whole series of techniques are available for the isolation of the smaller and smallest fossils, fossil remains, or for their microscopic examination in the substrate. Thus thin sections of rocks with fossil inclusions can be produced, made transparent using appropriate embedding substances, and examined under the ordinary light microscope. The soft Messel oil shales can even be sectioned with a microtome. Resistant fossils can be isolated chemically or mechanically by dissolving or by shattering the surrounding rock. In addition, the humous and bituminous cementing substances of the Messel shale can be softened or bleached, but it must be said

that only plant fossils can endure the relatively 'severe' methods used in these processes, whereas fossils of animal origin, even the chemically and mechanically resistant chitin, are, in our experience, destroyed more quickly than the oil shale surrounding them.

Superb success in the investigation of fossil animal tissue was achieved on fossils from the Geiseltal using the lacquer-film method developed by Voigt (1933). For this, a dry or surface-dried fossil slab is covered with a thin layer of transparent lacquer which is carefully detached after drying. In this process thin strata (for example, single layers of cells) remain on the lacquer film. After their embedding as a transparent microscope specimen, cell walls, nuclei, and cell organelles, in so far as they have been preserved, become visible under the microscope. This method enabled us to identify the only animal soft tissue known, to date, from Messel—a single cell layer (plate epithelium) from the inside of a beetle cuticle.

Of the methods that have been mentioned hitherto, the chemical or mechanical isolation of food remains can be used only to a limited extent on our material. A second method, the thin-polish or thin-section technique, has the disadvantage that it gives satisfactory results only when applied directly to fossil particles that are already known in principle; it is useless for routinely checking many doubtful objects, because it is too time-consuming. The lacquer-film technique, on the other hand, is only useful for transparent structures (for example, cells) and, in our experience these are far less well preserved at Messel than in the Geiseltal. For these reasons and especially because of the relatively simple and time-saving preparation techniques and the possibility of high magnification, we prefer to use scanning electron microscopy for the examination of putative stomach contents.

The advantage of the scanning electron microscope (SEM) lies in its high resolution and its great depth of focus, as well as in the fact that the magnification can be varied quickly and without problems between 20 × and 50 000 ×. Almost more important is the fact that the SEM images the natural surfaces

with their original surface sculptures. Internal structures, even if they lie closely below the surface, are not reproduced. Consequently, the objects to be investigated must be split into the thinnest possible lamellae and the new surfaces obtained in this manner must be scanned.

In the stomach contents of herbivores and omnivores, the cuticles of leaves, frequently present in great quantity and good preservation, often permit an at least approximate identification of the food plants based on the characteristic patterns of the cuticular cells and stomata (Wilde 1989). But, because the fragments are small, only a few mm^2, the identification is rather difficult. The same is true, to an even greater extent, of the remains of fruits and seeds, because the preserved fragments only rarely permit an inference as to the form, size, and surface pattern of the whole seeds and because there is still a dearth of reliably identified comparative material. It is, however, to be hoped that, through intensive research on the rich plant material, such gaps in knowledge will be closed in time.

In the stomach contents of the Tertiary vertebrates of Messel, food organisms of animal origin are represented by hard tissues, predominantly chitin or bone. The superb preservation of the surface sculptures of chitin has already been mentioned, but contributes little to the possibility of identification of the insect food. Firstly, the preserved chitin fragments are always very small, because the insects are partly chewed and partly ground down by stones in the stomach. Larger pieces that are still connected can, moreover, only rarely be detached from the relatively solid stomach contents. What is worse, almost all hollow parts—including the mouth-parts and the extremities that are so important in identification—have been squashed and flattened by the subsiding sediments and form, in the main, a collection of uncharacteristic single fragments. The extent to which the actual appearance of a hollow organ, and thus the possibility of identifying it, is impaired by this particular type of deformation, is shown in the figures which compare a trachea from the breathing system of a Recent insect (dried,

Fig. 401: Trachea with its original shape preserved intact, from the abdomen of a large moth (Sphingidae); dry preparation. SEM: magnification, 950×.

broken open; Fig. 401) with a collapsed trachea from the stomach contents of a Recent shrew (Fig. 403) and, finally, with a trachea fragment from a fossil beetle from the Messel fossil site (Fig. 402).

But while certain characteristics of the chitin exoskeleton of insects (the form of hair or scales, the thickness and surface sculpture of the cuticle) still permit at least tentative conclusions as to the identity of the food organisms, the determination of prey animals from only a few bone fragments is hardly possible. One has to be content here if certain skeletal elements, such as, for example, vertebrae, or hand- and foot bones, at least permit a distinction between reptiles, birds, and mammals.

The rather fragmentary representation of the food spectrum of a fossil animal species *per se* would not be a satisfying outcome of such long and elaborate technical investigations. Only the concomitant clues as to the behaviour, habit, and habitat of the species in question make

such work interesting and worthwhile. The pieces of information here are meagre. They consist, in the case of Messel, of often complete skeletons which, in the most favourable cases, still exhibit the so-called 'skin shadow' which

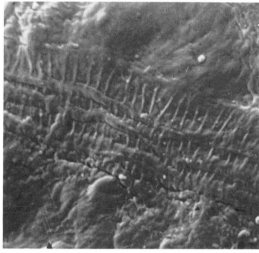

Fig. 402: Longitudinally folded and collapsed trachea of an unidentified insect from the Messel oil shale. SEM: magnification, 9500×.

reproduces the body contours and the remains of the food intake. Additional, mostly vague information is available concerning the fossil landscape (relief, forms of vegetation) and the climate (temperatures, height, and distribution of rainfall, etc.), but information as to the performance of the senses, energy balance, mobility, and behaviour, obtained in Recent species by observation of the living animal, is missing.

We may, however, presuppose that fossil and Recent organs of the same or similar form of construction also have the same or similar function and produce a comparable performance. It follows that in fossil and

Recent representatives of a related group (genus, family, order), given the same or similar morphological equipment, we may suppose the habitat and habit to be at least similar. The Eocene *Palaeochiropteryx tupaiodon* (see Chapter 14, 'Bats—already highly specialized insect predators') is an impressive example of how far the hypothetical reconstruction of a specific life habit can lead when conditions are favourable.

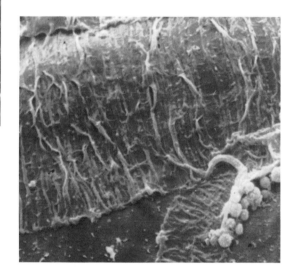

Fig. 403: Collapsed trachea of an unknown insect from the stomach content of a Recent shrew (*Crocidura russula*). SEM: magnification, 4000×.

27

The Messel fauna and flora:
a biogeographical puzzle

When the Messel flora and fauna were alive, Europe was separated by oceans from all other continents. Furthermore, shallow ocean inlets had spread across the European land-mass and transformed it into an archipelago. The fauna mirrors this geographical situation in numerous endemic phylogenetic developments which were limited to Europe, or even only to parts of Europe. Regional differences in faunal and floral composition are even detectable between such spatially and contemporaneously adjacent fossil sites as Messel and the Lower Coal of the Geiseltal near Halle/Saale.

The Messel fauna and flora: a biogeographical puzzle

GERHARD STORCH AND
FRIEDEMANN SCHAARSCHMIDT

The Messel fauna and flora have a variety of origins. Very ancient forms once had a wide distribution on the ancient supercontinents before these broke up. Their presence is therefore based solely on the breakup of an originally united area of distribution, and the organisms were able, on now separated continents, to survive for varying lengths of time and to embark, to a certain extent, on new evolutionary paths (Fig. 404).

Many other evolutionary lineages had, however, first begun on other continents, and the species of Messel or their ancestors had, at some time, migrated to Europe. We can frequently reconstruct their phylogenetic and geographical origin. It becomes apparent that, in contrast to all 'provincialism' during the European middle Eocene, our research area must be extensively widened, since all continents, with the exception of Antarctica, are involved. There are various reasons for using the mammals as a first illustration of this intercontinental puzzle. The level of scientific research on mammals is relatively advanced; their phylogenetic lineages generally change quickly; the early Tertiary was a period of important mammalian radiations world-wide. In the delimitation of the 'when' and 'where' of biogeographical relations mammals are, therefore, particularly helpful (Storch 1984, 1986).

Our considerations have to start, of course, with a mobile picture of the Earth. The framework for this is based on plate tectonics, which traces continental drift and the formation of the present-day oceans in the geological past, using geophysical and oceanographical methods. In this science, large-scale movements of continental plates—their breakup, drifting, and collision—are in the forefront. Details, however, such as small-scale or only temporarily traversable intercontinental land bridges, can often be better assessed in terms of the former distribution of animals.

It is, first of all, surprising that only a few Messel species belong to families or orders that evolved, with some certainty, in Europe. In the case of mammals, the search for such pre-Eocene groups meets with certain basic difficulties which arise from gaps in the fossil record. All that we know from the latest Cretaceous, approximately 70–65 million years before the present, is a single small molar from southern France and five tooth fragments from Portugal (Antunes *et al.* 1986). This situation does not improve when, 65 million years ago, the Cretaceous ends with the extinction of the dinosaurs, and the Cainozoic—with the Palaeocene—begins. At first we are left in the dark. Only from a period possibly 60 million years ago do we have some information from a single fossil site, Hainin in Belgium. We are better informed about the middle to late Palaeocene up to approximately 54 million years ago, thanks to mammal faunas from the Paris Basin and from Walbeck in Germany (Savage and Russell 1983; Storch 1986). The faunas show marked endemic characteristics as well as some relationships to North America. Primitive 'insectivores' of the family Pantolestidae are evidence for such scarce Palaeocene connections between Europe and North America. The finds, otherwise, are hardly an aid to the better understanding of the origin of the Messel species. The mammals from the beginning of the Tertiary were primitive and none belonged to an evolutionary lineage leading towards the modern groups—they represent the end of a Cretaceous radiation rather than the 'dawn' of modern mammals (Russell 1975). It is, however, not impossible that three extinct families of primitive mammals from Messel have their roots in the European Palaeocene: hedgehog relatives (Amphilemuridae), primitive 'insectivores' (Pseudorhyncocyonidae), and early ungulates (Paroxyclaenidae).

At the start of the Eocene, approximately 54 million years ago, a radical shift occurred in the European fauna. There followed a mass invasion by modern types and the local 'old-timers' for the most part disappeared quickly. This wave of immigration has marked the Messel fauna and has also laid the foundations for today's fauna. Thus, for the first time ever, the even- and odd-toed ungulates appeared, and also the hitherto unknown prosimians. But where did all these new forms come from, and within what geographical framework did their migrations take place?

The northern North Atlantic (the Greenland–Norwegian Sea) had first begun to split open and spread in the late Palaeocene. In the early Eocene there existed—most probably—two land bridges between the North American and European land-masses. The Thule bridge connected Europe, via the Faroes, Iceland, and Greenland, with Ellesmere and Baffin Islands in Canada. Further north the DeGeer route led from Norway via Spitsbergen and Greenland again to Ellesmere Island and Canada. Favourable climatic conditions enabled a largely unhindered exchange of land animals between both continents via the Thule bridge before this was partially flooded later in the early Eocene. The DeGeer route could have existed longer but certainly with severe climatic filters. Furthermore, during the early Tertiary, Scandinavia was separated from the rest of Europe by a sea inlet. Unfortunately, we lack contemporary mammal-bearing fossil sites in that area. In any case, in the early Eocene the mammal faunas of North America and Europe had, with almost 60 per cent of genera in common, a similarity which they would not again attain in their entire history (Koenigswald 1981; McKenna 1975, 1983; Savage and Russell 1983). It is also worth mentioning that the closest relatives of both Messel giant-ant species of the genus *Formicium*, are known from the Middle Eocene of southern England and from the early Eocene of Tennessee in North America.

Ellesmere Island, the northernmost part of

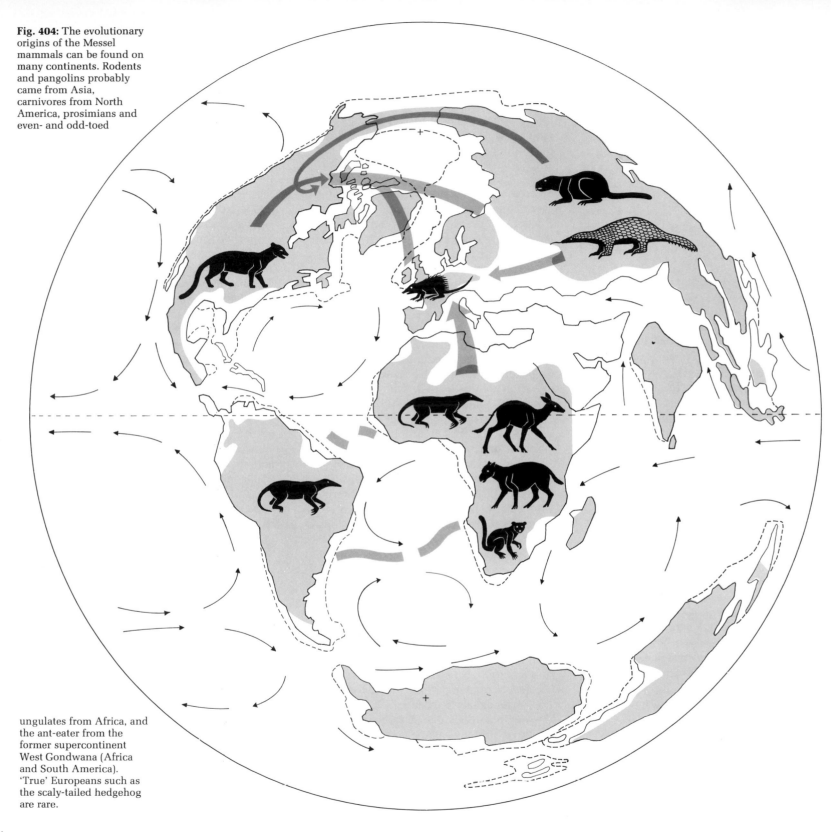

Fig. 404: The evolutionary origins of the Messel mammals can be found on many continents. Rodents and pangolins probably came from Asia, carnivores from North America, prosimians and even- and odd-toed ungulates from Africa, and the ant-eater from the former supercontinent West Gondwana (Africa and South America). 'True' Europeans such as the scaly-tailed hedgehog are rare.

North America in the high Arctic, is the pivot of the postulated migration routes. Since 1975 vertebrate fossils from the early Middle Eocene, which were evidence for the expected bilateral relationships have, indeed, been found here (McKenna 1983; West and Dawson 1978). Finds of crocodiles and monitor lizards are witness to a warm-temperate to warm climate and plant fossils lead to the conclusion that the high Arctic was once densely forested. The forests consisted of deciduous trees and there is, therefore, no difficulty in explaining their existence in terms of warm summers. There remains a mystery, however, as to how a varied fauna could have survived a polar night that lasted for several months, since the latitude of Ellesmere Island has remained approximately the same according to palaeomagnetic results.

There were similarities only between the mammal fauna between Europe and North America, as, unlike now, Asia was separated from Europe. In the early Eocene and during the existence of the animals at Messel, an epicontinental sea (Ob Sea and Turgai Strait) along the eastern margin of the Urals connected the Arctic Ocean with the Tethys Sea in the south. Although, with our present state of knowledge, we cannot trace the direct migration of any of our Messel mammals from Asia, there are indications that at least pangolins could have taken such a route. Among the reptiles, the terrestrial monitor-like necrosaurids, so well documented at Messel, may have already migrated directly from Asia before the Eocene.

It is surprising that the, contrary to expectations 'modernization' of the European mammal fauna in the early Eocene and the sudden rise of 'Euroamerican' congruity can be explained only to a small degree by the migration of North American mammals via North Atlantic routes. This supposition is, however, certainly correct for the ancestral group of modern carnivores, the family Miacidae. Rodents too had taken this route to Europe, but they themselves had probably reached North America, from Asia via the Bering strait, only a short time before, in the latest Palaeocene. The majority of 'Euro-

American' genera, however, were new to both continents and without plausible direct ancestors. They must, therefore, have immigrated to Europe and America from outside and used the North Atlantic corridors in either direction as routes of dispersal.

The origin of most newcomers is still unknown, but there are distinct indications that Africa must have played an important role as an area of departure. The oldest known genera of horses (*Hyracotherium*, which is ancestral to the early horse of Messel) and of the even-toed ungulates (*Diacodexis* from which *Messelobunodon* is descended), obviously appeared somewhat earlier in Europe than in North America (Godinot 1978, 1981, 1982; Hooker 1980; Estravis and Russell 1989). Moreover, it appears as if they occurred, within Europe, earlier in the south than in the north (Antunes and Russell 1981). This suggests Africa–Europe–North America as the sequence of dispersal. The oldest of the modern primates, the prosimians, could also have migrated from Africa. Most recently, Sigé *et al.* (1990) corroborated this idea with the discovery in Morocco of the oldest modern primate from the late Palaeocene. Unfortunately, the knowledge of Palaeocene and early Eocene African mammals is deficient and limited, up to now, to findings from the north of the continent.

Coming from Africa, a group of most 'exotic' vertebrates must also have crossed the Tethys Sea to Europe: 'South Americans' in Europe. Horned frogs (Ceratophryinae) are today restricted to South America; as fossils we know them from the Tertiary of Argentina and France. A group of ziphodont mesosuchians (extinct terrestrial crocodiles) are documented from the late Cretaceous onward in South America, and in the Eocene from Messel and the Iberian Peninsula. An extinct family of birds, the Phorusrhacidae, has been documented for the early Tertiary only in Uruguay and Argentina apart from Messel and the French fossil sites of the Quercy; their presence in Messel is the earliest evidence. Ant-eaters, finally, live today exclusively in South and Central America. Here prior to the discovery of *Eurotamandua*

at Messel, their only fossils were recorded. Remarkably, all these groups are missing in the late Cretaceous and the early Tertiary of North America, the possible bridging continent between South America and Europe.

Africa, consequently, must have played this role, as hinted at by the youngest discovery of a ziphodont mesosuchian in the Middle Eocene of Algeria (Buffetaut 1982). However, the above-mentioned poor knowledge of African vertebrates from the early Cretaceous and the late Tertiary still largely prevents actual proofs.

Africa could have been the stepping-stone for groups of animals which once originated from South America. The Southern Atlantic between northern Brazil and West Africa was still narrow and intermittent land connections existed until the late Cretaceous and perhaps even the early Palaeocene. Beyond that, several ridges of the spreading ocean floor were above water; until the Oligocene they could have provided land vertebrates with migration routes that were interrupted by marine barriers and so could have imposed strong ecological filters. But it is more probable that Africa and South America were a common area of distribution for the groups of vertebrates under consideration before the two continents finally separated and drifted apart, approximately 90 million years ago (Storch 1986). The 'South Americans' in Europe, in any case, belong to phylogenetically very ancient groups. Thus, the ant-eaters probably split off from the placental mammals back in the Cretaceous and should, therefore, be older than the final separation of Africa and South America. There are vertebrates, such as the tongueless frogs (pipids), which, even today, have an African–South American distribution as an expression of a once undivided area. Similarly, parasites from the body cavity of freshwater fish from the Amazon Basin show the closest relationships with those from south-west Nigeria.

If we trace a significant proportion of the mammals, but also other vertebrates from Messel, to migrations from Africa, a question arises concerning the reverse route of dispersal. Do forms exist in Messel which left their traces, perhaps up to the present, on southern continents? It was recognized only recently that the crane-sized, flightless bird *Palaeotis weigelti*, which is known from the Geiseltal and from Messel, has a close relationship with the ostriches (Houde 1986). The opinion, held by P. Houde, that the African ostrich could have very well evolved on northern continents and that it could have emigrated from Europe to Africa at some time in the Tertiary, has to be modified, according to recent findings, in so far as the same could also be valid for the South American rhea. *Palaeotis* actually shares more similarities with the rhea than with the African ostrich (Peters 1988).

The spread of plants conforms to rules other than those for animals, because they cannot, on a whim, move to new areas that are more favourable to them, but have to rely on the passive transport of seeds and fruits. However, this can take place, aided by wind, water, or animals, over large distances and even over barriers that are insurmountable for animals, such as small expanses of water; island chains, too, can allow migrations of plants. By contrast, large expanses of water and mountain ranges can be absolute barriers to plant species. In this way it can be explained that some families, such as the cacti and bromelias, are restricted to America and, furthermore, that many species which became extinct in Europe during the glacial epoch have not returned.

Because plants are tied to one place, their distribution mirrors the world's climatic zones. Thus we know today that a belt of tropical rain forests joined, in the north and south, on to two belts of arid zones. These are succeeded by two warm-temperate belts with evergreen forests of subtropical character. Only in the Northern Hemisphere is this succeeded by a belt with deciduous trees, or conifers, respectively, in a climate of distinct seasons. Finally, before the onset of the polar ice deserts, two belts of tundra are established.

We have already seen from the early Tertiary fauna and flora of Ellesmere Island that the climatic zonation cannot always have been the same in the geological past as it is today. At that time there grew, where we now have arctic

conditions, deciduous forests like those of present-day Central Europe. During the Eocene, at the time of Lake Messel, polar ice caps did not yet exist and in Antarctica forests of the southern beech *Nothofagus* flourished. At that period a more even, warmer temperature than today must have prevailed over the entire earth and, presumably, as is documented by the crocodiles of Ellesmere Island, also during the polar night. We have other evidence that the middle Eocene, in particular, was the climatically most favourable period of the Tertiary.

It is thus no surprise that the Messel flora resembles extant equatorial floras. The finds of palms are reliable evidence, but other families also have their main distribution in the tropics at present: such as the Lauraceae (laurels), the Menispermaceae (moonseed family), and the Icacinaceae. Even among the Juglandaceae (walnut family), whose main distribution is now in the temperate zone, there are genera which occur in warm areas of Asia and America, such as *Engelhardia* (Jähnichen *et al*. 1977), also recorded in Messel. However, some other, today purely tropical families are absent from Messel and from other, contemporary floras of Europe. The reason for this is, presumably, the seasons which existed in Europe during the Eocene, in spite of the warm climate. But this is not contrary to expectation. At that palaeolatitude (only a little further south than today) the hours of sunshine must have been fewer in winter than in summer. But, in view of the example of Ellesmere Island, we must doubt that this also led to a lowering of temperature. There, in still higher latitudes, the temperature can hardly have fallen below 10 °C even during the polar night. In an Eocene archipelago-like Central Europe, the geographical situation, with regard to the temperature, probably had an ameliorating effect as well. There is also no botanical evidence (deciduous trees) for seasonal dry periods. Only growth rings in wood from the contemporary brown coal of Helmstedt allow one to conclude that there was a seasonal rhythm. But could the decrease in sunshine during the winter have led to a cessation of growth? A favourite procedure for the

interpretation of palaeoclimate transfers the present-day climatic diagram of a locality with similar vegetation to the fossil flora. That this is possible neither for Messel nor for Ellesmere Island becomes apparent when it is considered that there are, today, no comparably warm localities in such high latitudes. Obviously, at present, we have completely different climatic conditions with much higher geographical and seasonal variations. Summing up, one can say that in the middle Eocene an equably warm climate prevailed at Messel, just as now in the tropics and subtropics (designated as paratropical by Mai 1981), and that seasonal variations in rainfall are not documented. However, according to the latitude, there must certainly have been seasonal differences in the daily hours of sunshine.

We have based these conclusions on the flora preserved in the oil shale. But is this flora at all representative of the middle Eocene of Central Europe? One has to consider that the large remains (fruits, seeds, leaves) come, for the most part, from the immediate vicinity of the lake, that is, from plants which grew either as water plants in the lake or in a relatively moist area close to its borders. Does this, perhaps, create a distorted picture? This question, indeed, touches on a basic problem in palaeobotany. Sites of fossilization are, as a rule, moist areas such as lakes and bogs—floras from dry localities are almost completely absent. Only one fossil plant group at Messel could have been carried by the wind over greater distances—the spores and pollen grains. We can, indeed, note remarkable differences in the composition of both fossil groups. These are:

1. The ferns. Seventeen species of spores contrast with only four types of fronds, which are rare.

2. Five species of conifer pollen occur regularly, while cone and branch remains are great rarities.

3. While the nine species from the Fagaceae (beech family) are among the most frequently found forms of pollen, no fruit or leaves have been found up to the present.

These differences allow us to draw a conclusion as to the ecological sequence of the forest. Besides the moist lowland forest at the border of the lake, wooded areas in drier, perhaps more elevated sites must also have existed. Conifers and Fagaceae could have had a more significant role here. The ferns too did not grow in large numbers at the border of the lake, but, presumably, in moist places (perhaps along small brooks) in the interior of the forest. There are no indications of an open, steppe-like vegetation.

The comparison with other contemporary sites, undertaken by Wilde (1989), is also informative. In Central Europe this applies in the first instance to the rich floras from the Geiseltal near Halle and from Helmstedt. It is evident here that both of these floras have many species in common, but that both floras exhibit large differences from the Messel flora. This, at first disconcerting, result can be explained easily. The Messel oil shale is a lake deposit, but both other sites are brown-coal deposits which have evolved from bogs. From observations on Recent tropical-forest bogs as occur, for example, in Sumatra, we know, moreover, that the vegetation there is impoverished and shows a different composition from the 'normal' vegetation of the surrounding area. Transferred to the Eocene this means that we observe a normal vegetation at Messel, while the floras from Helmstedt and the Geiseltal are special cases of bog vegetation. The vagaries of fossilization are responsible for the fact that, particularly in the Tertiary, such bog vegetation, in connection with brown coal, has been so well preserved in terms of both quantity and quality. In understanding the normal vegetation and climate, Messel is therefore of exceptional importance.

Unfortunately, we still know too little today about the general distribution of plant groups during the Eocene or even about the migration routes of individual species. Here there are only a few relevant good pieces of research, which concern anatomical structure and which therefore allow reliable conclusions. Nevertheless, Dilcher (1987) was recently able to show that two laurel leaves from Messel and south-eastern North America, respectively, differ only insignificantly in the structure of the upper leaf epidermis. Collinson (1986, 1987, 1988) has even established that a fruit form, which belongs, presumably, to the Lythraceae (loose-strife family) and which is frequent in Messel, also occurs in the Middle Eocene of western North America.

Better statements, concerning the further fate of individual plant families, are possible today if we include their Recent distribution in our considerations. If one disregards the now-cosmopolitan families Nymphaeaceae (water-lilies) and Leguminosae (pea family), then most of the families that occur in Messel have a tropical–subtropical distribution. The Menispermaceae (moon-seed family) and Icacinaceae are purely tropical families: the Vitaceae (grape family), palms, and Araceae (aroids) are tropical–subtropical families with their main distribution in the Old- and New World tropics. The Rutaceae (citrus fruit family) also occur throughout the tropics, but extend into the temperate zone of the Southern Hemisphere. Although the Lauraceae (laurels) and Theaceae (tea family) occur world-wide in the tropics and subtropics, they have their main distribution in America and Asia. The Pandanaceae (screw pines) are limited to the Old World tropics and the Mastixiaceae to South-east Asia alone. Three further families deviate in their pattern of distribution from those families with a tropical–subtropical range. The Juglandaceae (walnut family) are distributed today predominantly throughout temperate to subtropical climes, but extend with some species into tropical Asia and to South America. The Taxodiaceae (swamp cypresses) today have a scattered distribution, which includes smaller occurrences in North America, East Asia, and Tasmania. With the exception of one Vietnamese species, the Restionaceae are now restricted to the South Hemisphere, with centres of distribution in Australia and South Africa.

The predominance of tropical–subtropical families in Messel is evidence that the Eocene climate was warm well into European latitudes. Through other fossil sites we know that the belt of tropical vegetation was then far broader over the entire earth and we also know that Central European floras had close relationships to present-day floras in North America but most of all to South-east Asia. Evidence of a connection to African floras, in contrast, cannot be proved. Of particular interest are families with residual areas of distribution in South-east Asia or with centres in Asia and America. Fossil deposits like that of Messel show that the areas of now purely South-east Asian families spread, in the Tertiary, into Europe and that scattered areas of present distribution (East Asia/North America) were once connected to one another by deposits in Europe.

Appendix 1

Details of illustrations

Sources of illustrations

Private collections: Dr T. Bastelberger; Dr G. Jores; M. Keller; T. Martin; H. P. Schierning; R. Wald, M. Wuttke; O. Feist; E. Pohl; A. Kessler.

Photographs: J. F. Bornhardt, E. Haupt; Prof. Dr W. v. Koenigswald; G. Krebs; W. Kumpf; T. Martin; Dr N. Micklich; Dr S. Schaal; Prof. Dr F. Schaarschmidt; Dr G. Storch; Dr H. Thiele-Pfeiffer; N. Wolz.

SEM—(scanning electron micrographs): Dr G. Richter; M. Starck.

X-ray photographs: Dr J. Habersetzer; Prof. Dr W. Stürmer; Dr S. Tuengerthal.

Line drawings: K. Albrecht; G. Eder; Dr J. L. Franzen; A. Helfricht; M. Hinkel; E. Junqueira; T. Keller; I. Lehnen; Dr N. Micklich; M. Möller; H. Schäfer; M. Wuttke.

Abbreviations of institution names

HLMD = Hessisches Landesmuseum, Darmstadt (Hessen Regional Museum, Darmstadt);

LNK = Landessammlungen für Naturkunde Karlsruhe (Natural History Collection, Karlsruhe);

NhMM = Naturhistorisches Museum Mainz (Natural History Museum, Mainz);

IRScNB = Institut Royal des Sciences Naturelles de Belgique, Bruxelles (Royal Institute of Natural Sciences, Brussels);

SM & SMF = Senckenberg Museum, Frankfurt am Main.

The figures

1. Photograph: Haupt
2. Design: Schaal, Franzen; diagram: Schäfer
3. From Storch (1987)
4. Redrawn after Matthess (1966); illustration: Schäfer
5. Thin section: Keuerleber; photograph: Haupt
6–7. Photographs: Schaal
8. Redrawn from Möller, a 1986 aerial photograph commissioned by the Ministry for Science and Art, Wiesbaden
9. Redrawn after Franzen et al. (1982); montage: Möller
10. From Franzen et al. (1982)
11. Illustration: Junqueira
12. Illustration: Eder
13. SMF, specimen no. ME 1799; photograph: Haupt
14. SMF, specimen no. ME 1800; illustration: Junqueira
15. SM, specimen no. B Me 1266; photograph: Schaarschmidt
16. SM, specimen no. B o. Nr.; photograph: Thiele-Pfeiffer
17. Redrawn after Thenius (1981); illustration: Eder
18. SM, specimen no. B o. Nr.; photograph Thiele-Pfeiffer
19. SM, specimen no. B Me 1303; photograph: Schaarschmidt
20. SM, specimen no. B Me 2731; photograph: Schaarschmidt
21. Redrawn after Heywood (1979); illustration: Eder
22. Photograph: Schaarschmidt
23. SM, specimen no. B Me 2107; photograph: Schaarschmidt
24. SM, specimen no. B o. Nr.; photograph: Thiele-Pfeiffer
25. Redrawn after Heywood (1979), illustration: Eder
26. SM, specimen no. B Me 3681; photograph: Haupt
27. SM, specimen no. B Me 2738; photograph: Schaarschmidt
28. SM, specimen no. B o. Nr.; photograph: Schaarschmidt
29. SM, specimen no. B Me 4003; photograph: Schaarschmidt
30. SM, specimen no. B o. Nr.; photograph: Schaarschmidt
31. Redrawn after Heywood 1979; illustration: Eder
32. SM, specimen no. B Me 3706; photograph: Haupt
33. Redrawn after Heywood 1979; illustration: Eder
34. SM, specimen no. B Me 825; photograph: Schaarschmidt
35. SM, specimen no. B o. Nr.; photograph: Schaarschmidt
36. SM, specimen no. B Me 2739; photograph: Schaarschmidt
37. SM, specimen no. B Me 874; photograph: Schaarschmidt
38. SM, specimen no. B Me 4125; photograph: Schaarschmidt
39. Redrawn after Heywood 1979; illustration: Eder
40. Photograph: Schaarschmidt
41. SM, specimen no. B Me 2734; photograph: Schaarschmidt
42. Illustration: Eder
43. SM, specimen no. B Me 2736; photograph: Schaarschmidt
44, 45. SM, specimen no. B o. Nr.; photograph: Schaarschmidt
46. SM, specimen no. B Me 2735; photograph: Schaarschmidt
47. Redrawn after Heywood; illustration: Eder
48. SM, specimen no. B Me 4728; photograph: Schaarschmidt
49. SM, specimen no. B Me 911; photograph: Schaarschmidt
50. Redrawn after Heywood (1979); illustration: Eder
51. SM, specimen no. B Me 3701; photograph: Schaarschmidt
52. SM, specimen no. B Me 1267; photograph: Schaarschmidt
53. Redrawn after Heywood (1979); illustration: Eder
54. SM, specimen no. B Me 2733; photograph: Haupt
55. Redrawn after Heywood (1979); illustration: Eder
56. SM, specimen no. B Me 4921; photograph: Schaarschmidt
57. Redrawn after von Koenigswald and Schaarschmidt 1983; illustration: Eder
58. SM, specimen no. B Me 4936; photograph: Schaarschmidt
59. Photograph: Schaarschmidt
60. SM, specimen no. B Me 1402; photograph: Schaarschmidt
61. SM, specimen no. Me o. Nr.; photograph: Thiele-Pfeiffer
62. Redrawn after Heywood (1979); illustration: Eder
63. Photograph: Schaarschmidt
64. SM, specimen no. B Me 3212; photograph: Schaarschmidt
65. SM, specimen no. B Me 2740; photograph: Schaarschmidt
66. SM, specimen no. B Me 4763; photograph: Schaarschmidt
67. Redrawn after Heywood (1979); illustration: Eder
68. SM, specimen no. B Me 3700; photograph: Haupt
69. SM, specimen no. B o. Nr.; photograph: Thiele-Pfeiffer
70. SM, specimen no. B Me 2741; photograph: Schaarschmidt
71. SM, specimen no. B Me 2732; photograph: Schaarschmidt
72. SM, specimen no. B Me 4988; photograph: Schaarschmidt
73. SM, specimen no. B o. Nr.; photograph: Thiele-Pfeiffer
74. Redrawn after Kirchheimer (1943); illustration: Eder
75. Redrawn after Heywood (1979); illustration: Eder
76. Redrawn after Heywood (1979); illustration: Eder
77. SM, specimen no. B Me 2743; photograph: Schaarschmidt
78. SM, specimen no. B Me 2742; photograph: Schaarschmidt
79. SM, specimen no. B o. Nr.; photograph: Schaarschmidt
80, 81. HLMD, specimen no. Me 7582; photographs: von Koenigswald
82. SM, specimen no. B Me 1758; photograph: Haupt
83. SM, specimen no. B Me 2737; photograph: Schaarschmidt
84. SM, specimen no. B Me 2737; photograph: Schaarschmidt
85. Illustration: Helfricht
86. Table; data from Thiele-Pfeiffer (1988), Collinson (1986), and Wilde (1989).
87. SMF, specimen no. Mel 455; photograph: Haupt
88. SMF, specimen no. Mel 432; photograph: Haupt
89. SMF, specimen no. Mel 308; photograph: Haupt
90. SMF, specimen no. C. 16016; photograph: Haupt
91. SMF, specimen no. Mel 119; photograph: Haupt
92. SMF, specimen no. Mel 1514; photograph: Haupt
93. SMF, specimen no. Mel 1257; photograph: Haupt
94. SMF, specimen no. Mel 1549; photograph: Haupt
95. SMF, specimen no. Mel 685; photograph: Haupt
96. SMF, specimen no. Mel 432; photograph: Haupt
97. SMF, specimen no. Mel 1671; photograph: Haupt
98. SMF, specimen no. Mel 1017; photograph: Haupt
99. SMF, specimen no. Mel 1548; photograph: Haupt
100. SMF, specimen no. Mel 1515; photograph: Haupt
101. SMF, specimen no. Mel 1509; photograph: Haupt
102. SMF, specimen no. Mel 1513; photograph: Haupt
103. SMF, specimen no. Mel 1512; photograph: Haupt
104. SMF, specimen no. Mel 1510; photograph: Haupt
105. SMF, specimen no. Mel 1511; photograph: Haupt
106. SMF, specimen no. Mel 1014; photograph: Haupt
107. LNK; photograph: Haupt
108. SMF, specimen no. Mel 1670; photograph: Haupt
109. SMF, specimen no. Mel 952; photograph: Haupt
110. SMF, specimen no. Mel 1550; photograph: Haupt
111. SMF, specimen no. Mel 975; photograph: Haupt
112. Private collection: Keller; photograph: Micklich
113. SMF, specimen no. ME 719; photograph: Haupt
114. SMF, specimen no. ME 719; photograph: Haupt
115. LNK, specimen no. Me 442; photograph: Micklich

116. LNK, specimen no. Me 634; photograph: Micklich
117. LNK, specimen no. Me 37a; photograph: Micklich
118. LNK, specimen no. Me 24a; photograph: Micklich
119. SMF, specimen no. ME 1618; photograph: Haupt
120. SMF, specimen no. ME 1655; photograph: Haupt
121. SMF, specimen no. ME 464; photograph: Haupt
122. SMF, specimen no. ME 561; photograph: Haupt
123. Private collection: Keller; photograph: Micklich
124. SMF, specimen no. ME 552; photograph: Haupt
125, 126. SMF, specimen no. ME 1599; photograph: Haupt
127. SMF, specimen no. ME 1600; photograph: Haupt
128. NhMM, specimen no. PW 1981/4, (holotype); photograph: Haupt
129. NhMM, specimen no. PW 1981/4; X-ray photographs: Stürmer; log$_e$ editing: Röntgen-Schnitzer, Frankfurt am Main
130. From Micklich (1985); illustration: Micklich
131. SMF, specimen no. ME 622; photograph: Haupt
132. SMF, specimen no. ME 658b; photograph: Haupt
133. SMF, specimen no. ME 625; photograph: Haupt
134. SMF, specimen no. ME 737; photograph: Haupt
135. SMF, specimen no. ME 622; photograph: Haupt
136. From Micklich 1985; illustration: Micklich
137. SMF, specimen no. ME 261; photograph: Haupt
138. Private collection, Keller; photograph: Haupt
139. SMF, specimen no. ME-A 91 (cast); photograph: Haupt
140. SMF, specimen no. ME 1536; photograph: Haupt
141. SMF, specimen no. ME 998; photograph: Haupt
142. SMF, specimen no. ME 1720; photograph: Haupt
143. SMF, specimen no. ME 432; X-ray photograph: Habersetzer
144. Private collection, Keller; photograph: Haupt
145. SMF, specimen no. ME 719; photograph: Haupt
146. SMF, specimen no. ME 840; photograph: Haupt
147. SMF, specimen no. ME 1600; photograph: Richter
148. SMF, specimen no. P 6076 (counterpart of type specimen, donated by E. Bettag, Dudenhofen); photograph: Richter
149. SMF, specimen no. ME 737; photograph: Micklich
150. SMF, specimen no. P 4286; photograph: Micklich
151. Table
152. SMF, specimen no. ME 483; photograph: Haupt
153. SMF (loan); private collection, Wuttke; photograph: Haupt
154. SMF, specimen no. ME 1301b; photograph: Haupt
155. SMF, specimen no. ME 1301a; photograph: Haupt
156. IRScNB, no number; photograph: Haupt
157. SMF, specimen no. ME 752a; photograph: Haupt
158. SMF, specimen no. ME 1824; photograph: Haupt
159. SMF, specimen no. ME 476; photograph: Haupt
160. SMF, specimen no. ME 1340; photograph: Haupt
161. SMF, specimen no. ME 1340; photograph: Haupt
162. SMF, specimen no. ME 611; photograph: Haupt
163. SMF, specimen no. ME 715; photograph: Haupt
164. SMF, specimen no. ME 1211; photograph: Haupt
165. SMF, specimen no. ME 1003; photograph: Haupt
166. SMF, specimen no. ME 722a; photograph: Haupt
167. SMF, specimen no. ME 1341; photograph: Haupt
168. SMF, specimen no. ME 776; photograph: Haupt
169. SMF, specimen no. ME 1137; photograph: Haupt
170–172. SMF, specimen no. ME 1801; photograph: Haupt
173. SMF, specimen no. ME 1338; photograph: Haupt
174. Drawing modified after Berg (1966); photograph: Haupt
175. HLMD; photograph: Haupt
176. SMF, specimen no. ME 899; photograph: Haupt
177. HLMD, specimen no. Me 5346; photograph: Haupt
178. SMF, specimen no. R 4126; photograph: Haupt
179, 180. HLMD, specimen no. ME 7003; photograph: Haupt
181. SMF, specimen no. ME 1802; photograph: Haupt
182. SMF, specimen no. ME 546; photograph: Haupt
183. Drawing after finds from the private collections of Feist and Pohl, as well as of SMF; design: Keller; illustration: Eder
184. Private collection, Feist; illustration: Eder
185, 186. IRScNB; illustration: Keller
187. SMF, specimen no. ME 2a; illustration: Keller
188. SMF, specimen no. ME 1802; illustration: Keller
189. HLMD, specimen no. ME 3950; photograph: Haupt
190. SMF, specimen no. ME 901; photograph: Haupt
191. SMF, specimen no. ME 2a; photograph: Haupt
192. SMF, specimen no. ME 1249a; photograph: Haupt
193. HLMD, specimen no. ME 7915; photograph: Haupt
194. SMF, specimen no. ME 1607; photograph: Haupt
195. SMF, specimen no. Me 958b; photograph: Haupt
196, 197. SMF, specimen no. ME 1566; photograph: Haupt
198. SMF, specimen no. ME 1578; photograph: Haupt
199. Illustration: Peters
200. Private collection, Bastelberger; photograph: Haupt
201. SMF, specimen no. ME 1818; photograph: Haupt
202. LNK, specimen no. ME 555; photograph: Haupt
203. SMF, specimen no. Av 406; photograph: Haupt
204. SMF, specimen no. ME 610; photograph: Haupt
205. SMF, specimen no. ME 216; photograph: Haupt
206. SMF, cast, original in private collection, Kessler; photograph: Haupt
207. SMF, specimen no. ME 1144a; photograph: Haupt
208. SMF, specimen no. ME 1635a; photograph: Schaal
209. LNK, specimen no. Me 301; X-ray photograph: Habersetzer
210. HLMD, specimen no. Me 7598; X-ray photograph: Habersetzer
211. SMF, specimen no. ME 599; X-ray photograph: Habersetzer
212. SMF, specimen no. ME 606a; photograph: Haupt
213. SMF, specimen no. ME 8a; photograph: Haupt
214. SMF, specimen no. ME 1452a; photograph: Haupt
215. SMF, specimen no. ME 1233b; photograph: Haupt
216. SMF, specimen no. ME 1268; photograph: Haupt
217. SMF, specimen no. ME 1758a; photograph: Haupt
218. SMF, specimen no. ME 1121b; photograph: Haupt
219. HLMD, specimen no. ME 8035; photograph: Haupt
220. LNK, loan from private collection; photograph: Haupt
221. Illustration: Helfricht
222. LNK, specimen no. Me 983; X-ray photograph: Habersetzer
223, 224. Illustrations: Helfricht
225. LNK, specimen no. Me 983a; photograph: Haupt
226. LNK, specimen no. Me 983b; photograph: Haupt
227. From Storch and Lister (1985); illustration: Helfricht
228. SMF, specimen no. ME 1143; X-ray photograph: Habersetzer
229. SMF, specimen no. ME 1143; photograph: Storch
230. From Maier, Richter, and Storch (1986); illustration: Helfricht
231. HLMD, specimen no. Me 7582; photograph: Kumpf
232–235. SEM—micrographs: Richter
236. LNK, specimen no. ME 464; photograph: Haupt
237. SEM—micrograph: Richter
238. Photograph: von Koenigswald
239. HLMD, specimen no. Me 7431; from Heil *et al.* (1987)
240. IRScNB; photograph: Haupt
241. SMF, specimen no. Me 977b; photograph: Haupt
242–244. SEM: Richter
245. SMF, specimen no. ME 758a; photograph: Haupt
246. HLMD, specimen no. Me 7577; radiograph: Habersetzer
247. SMF, specimen no. ME 758a; photograph: Haupt
248. HLMD, specimen no. Me 7577; photograph: Haupt
249. From von Koenigswald and Storch (1983); illustration: Helfricht
250–255. SEM: Richter
256. Illustration: Schäfer
257. HLMD, specimen no. Me 8850; photograph: Haupt
258. Illustration: Junqueira
259. Private collection, Schierning; photograph: von Koenigswald
260. SMF, specimen no. ME 963; photograph: Haupt
261. HLMD, specimen no. Me 8057; photograph: Haupt
262. SMF, specimen no. ME 963; radiograph: Habersetzer
263. SMF, specimen no. ME 1518; photograph: Haupt
264. SMF, specimen no. ME 1414a; photograph: Haupt
265. SMF, specimen no. ME 1024a; photograph: Haupt
266. SMF, specimen no. ME 1469a; photograph: Haupt
267. (Top) SMF, specimen no. ME 1789, (centre) SMF, specimen no. ME 1089, (lower) SMF, specimen no. ME 1499; illustration: Helfricht
268. Illustration: Helfricht
269–276. SEM: Richter
277, 278. SMF, specimen no. 59567; radiographs: Habersetzer
279–281. SMF, specimen no. 59562; radiographs: Habersetzer
282, 283. LNK, specimen no. Me 414; radiographs: Habersetzer
284. LNK, specimen no. Me VIII 189; radiographs: Habersetzer
285. LNK, specimen no. Me VIII Gb; radiographs: Habersetzer
286. SMF, specimen no. ME 1477b; radiographs: Habersetzer
287, 288. HLMD, specimen no. Me 7430; photograph: Haupt
289. From Franzen (1987); illustration: Franzen
290. SMF, specimen no. ME 1228a; photograph: Haupt
291. SMF, specimen no. ME 1228b; photograph: Haupt
292. SMF mammal collection
293. Design: Franzen; graphics: Schäfer
294. SMF, specimen no. ME 1683; photograph: Haupt
295. Cast: specimen no. SMF 83/3; photograph: Storch; original in IRScNB
296. Private collection, Wald; photograph: Storch
297. From Storch (1978); illustration: Storch
298. From Storch (1978); illustration: Helfricht
299. From Storch (1978); illustration: Helfricht
300. Redrawn after von Koenigswald *et al.* (1981); illustration: Helfricht
301. HLMD, specimen no. Me 4247c; photograph: von Koenigswald *et al.* (1981)
302, 303. SEM: Richter
304. Illustration: Albrecht
305. Private collection, Jores; photograph: Storch
306. From Storch (1984); illustration: Storch, Helfricht

307. From Storch (1984); illustration: Helfricht
308–310. From Storch (1981); illustrations: Helfricht
311–318. SEM: Richter
319. SMF, specimen no. ME 1528; photograph: Haupt
320. Illustration: Helfricht
321. HLMD, specimen no. Me 7596; photograph: Haupt
322–324. SEM: Richter
325. Illustration: Junqueira
326. SMF, specimen no. ME 1287; photograph: Haupt
327. SMF, specimen no. ME 1287; X-ray photograph: Habersetzer
328. Graphics: Schäfer
329. Illustration: Schäfer
330. From Springhorn (1980), private collection: Feist; photograph: Bornhardt
331. Illustration: Schäfer
332. From Springhorn 1980, private collection: Feist; photograph: Bornhardt
333. SMF, specimen no. ME 1284; radiograph: Tuengerthal
334, 335. SMF, specimen no. ME 1284; photograph: Haupt
336. From Springhorn (1985); illustration: Hinkel
337, 338. SMF, specimen no. ME 1465b; photographs: Haupt
339. IRScNB, specimen no. I. G.26533; radiograph: Tuengerthal
340. SMF, specimen no. ME 2401; photograph: Oleschinski, Bonn
341. From von Koenigswald (1983); illustration: Lehnen
342. Illustration: Schäfer
343. Redrawn after von Koenigswald (1983); illustration: Junqueira
344. SMF, specimen no. ME 1285; photograph: Haupt
345. SMF, specimen no. ME 1397; photograph: Haupt
346. SMF, specimen no. ME 12; SEM–micrograph: Richter
347, 348. HLMD, specimen no. Me 8989; photographs: von Koenigswald
349. From von Koenigswald
350. Redrawn after Franzen (1986); illustration: Schäfer
351. SMF, specimen no. ME-A116 (cast); original: private collection: Jores
352. Design: Franzen; illustration: Schäfer
353. Design: Franzen; graphics: Schäfer
354. SMF, specimen no. ME 510 (holotype); photograph: Haupt
355. After Franzen (1983); illustration: Schäfer
356–361. SEM: Richter
362. Private collection, Martin; photograph: Martin
363. Design: Franzen; graphics: Schäfer
364. SMF, specimen no. ME 1527a; photograph: Haupt
365. SMF, specimen no. ME 1229; photograph: Haupt
366. From Wuttke (1988)
367. SMF, specimen no. Me M 3; photograph: Krebs
368. SMF, specimen no. Me M 30; photograph: Krebs
369. SEM: Richter
370. SMF, specimen no. ME 978; photograph: Haupt
371–374. SEM: Richter
375. From Voigt (1988), originals: Geiseltal Museum, Halle/Saale
376. SMF, specimen no. ME 1226; SEM: Starck
377. SEM: Richter
378. SMF, specimen no. ME 1074; photograph: Haupt
379. HLMD, specimen no. Me 9724; photograph: W. Kumpf
380, 381. IRScNB; photograph: Haupt
382. SEM: Richter
383. (Top left) SMF, specimen no. MeK 91, (top centre) SMF, specimen no. MEK 163, (top right) SMF, specimen no. MEK 210, (bottom left) SMF, specimen no. MEK 53, (bottom centre) SMF, specimen no. MEK 23, (bottom right) SMF, specimen no. MEK 512; photograph: Haupt
384. SMF, specimen no. MEK 52; photograph: Haupt
385. SMF, specimen no. MEK 315; photograph: Haupt
386, 387. Photographs: Haupt
388. Find of 14 July 1986, from site 21; photograph: Haupt
389. Photograph: Wolz, with kind permission: DuPont Frankfurt am Main
390. SMF, specimen no. ME 1661a; photograph: Haupt
391. Photograph: Haupt
392. Photograph: Wolz, with kind permission: DuPont Frankfurt am Main
393. Illustration: Eder
394. SMF, specimen no. 81/619; radiograph: Habersetzer
395. SMF, specimen no. ME 1469a; radiograph: Habersetzer
396. SMF, specimen no. ME 1469b; radiograph: Habersetzer
397. SMF, specimen no. ME 1469 a + b; radiograph: Habersetzer
398–403. SEM—microphotographs: Richter
404. Redrawn after Storch (1986); graphics: Schäfer

Appendix 2

Biographical profiles of the authors

Michael Ackermann

Born in 1956, since 1984 he has been workshop superintendent of the Senckenberg Research Institute at the Messel field site. He trained as a technician in natural history museum curation and research, and took an apprenticeship in preparing zoological specimens. The focal point of his work is the preparation of the Messel fossils.

Jens Lorenz Franzen

Born in 1937, he has been head of the department of palaeo-anthropology at the Senckenberg Research Institute since 1982. He studied geology, palaeontology, and anthropology at Kiel and Freiburg im Breisgau. In 1968 he obtained his doctorate in natural sciences, working on palaeotheres, early relatives of the horses. He initiated and then led the Senckenberg excavations at the Messel pit from 1975 to 1984. His particular interest, besides primates and ungulates, is the taphonomy and the biostratigraphical assessment of the fossil site.

Jörg Habersetzer

Born in 1952, he has worked at the Senckenberg Research Institute since 1985. After studying biology, human biology, and medicine in Göttingen and Frankfurt, he obtained his Ph.D. in zoology and has since done work on behaviour, echolocation, and flight biology of bats. In addition, he has developed new X-ray applications to fossils.

Thomas Keller

Born in 1947, from 1986 to 1990 he was spokesman for the Preservation of Palaeontological Monuments in the Hessian Ministry of Science and Art (now in the Hessian Administration for Monument Conservation). Since 1987 he has been co-ordinator of all scientific excavations at the Messel pit. Following an extended career in the preparation of geological–palaeontological specimens, he obtained his degree in geology and palaeontology at Frankfurt am Main. He is currently preparing a dissertation about a group of Messel lizards.

Wighart von Koenigswald

Born in 1941, he is professor of palaeontology at the University of Bonn. After studying in Bonn and Munich he worked as a palaeontologist in Tübingen and Darmstadt. In Darmstadt, as curator at the Hessen Regional Museum, he led the excavations at the Messel pit for several years and initiated the permanent exhibition of the Messel finds. In addition to more general articles, he has published scientific papers on several mammals of the Messel fauna, putting special emphasis on the palaeobiological aspects.

Herbert Lutz

Born in 1953, he is a palaeontologist at the Museum of Natural History in Mainz. Until 1986 he was scientific assistant to Professor R. Kinzelbach at the Institute for Zoology at the Technical University, Darmstadt. He studied biology and chemistry at Darmstadt and obtained his Ph.D. in zoology in 1988, working on the fossil insects of the Messel pit. His main interest is in the insect faunas of the Tertiary and the relevant problems of systematics, taxonomy, and palaeo-ecology.

Norbert Micklich

Born in 1951, he has been concerned with the fish of Messel, both in the framework of a research project for the Deutsche Forschungsgemeinschaft at the Senckenberg Research Institute and later as a scientific volunteer at the State Museum of Natural History (Karlsruhe). He studied biology at Darmstadt and there obtained his Ph.D. in zoology. His interest is in Tertiary fish in general, focusing on the perciformes (perch-like fishes) due to their multitude of forms and their various ecological adaptations.

Dieter Stefan Peters

Born in 1932, he is keeper of the vertebrate department and of the ornithological section at the Senckenberg Research Institute. He studied biology, chemistry, and politics in Wroclaw and Frankfurt am Main (first and second degree and Ph.D.) and became a supernumerary professor at the Johann Wolfgang Goethe University in Frankfurt am Main. At present he works on the fossil birds of the Messel pit, with an emphasis on the systematics and phylogeny of birds in general; he is also concerned with structural morphology and the theoretical basis of evolutionary biology.

Gotthard Richter

Born in 1924, he is curator of the invertebrate department and the Marine Vertebrates II section at the Senckenberg Research Institute in Frankfurt am Main. He studied biology in Giessen and obtained a Ph.D. in zoology. His research areas are taxonomy, morphology, and ecology of pelagic molluscs. Within the research programme of the Messel fossil site he investigates the food remains found inside many of the fossils, with a view to understanding the behaviour and habitats of the Tertiary organisms.

Stephan Schaal

Born in 1955, he has been curator of the Messel section at the Senckenberg Research Institute in Frankfurt am Main since 1984. He studied palaeontology and geology in Berlin and obtained his Ph.D. in 1984. A focal point of his research is the investigation of the microstratigraphy and the sedimentation of the Messel oil shale. An additional interest is the snakes of the fossil site.

Friedemann Schaarschmidt

Born in 1934, he is custodian and head of the palaeobotanical section at the Senckenberg Research Institute. He studied biology at Jena and Frankfurt, obtained his Ph.D. in botany in 1962, and became Assistant Professor of Palaeontology at the University of Frankfurt in 1984. He is interested in all the Messel plant fossils. He has a particular interest in the flowering plants and in the interrelationships of the organisms.

Rainer Springhorn

Born in 1948, he is director of the Landesmuseum of Lippe, Detmold, and university lecturer in palaeontology at the University of Freiburg im Breisgau. He studied geology and palaeontology in Freiburg in Breisgau and Innsbruck, and obtained his Ph.D. in 1976, his lectureship in 1983, and his professorship in 1990. Since 1980 he has worked on primitive carnivores and on extant carnivores within the scientific framework of the fossil excavations at the Messel pit.

Gerhard Storch

Born in 1939, he is curator and head of the fossil mammal section at the Senckenberg Research Institute. He studied biology in Darmstadt, Vienna, and Frankfurt and obtained his Ph.D. in zoology. He is fascinated by the possibility of reconstructing the biology of the Messel fossil mammals, at the same time as studying their phylogeny and zoogeography.

Michael Wuttke

Born in 1950, from 1987 to 1989 he was entrusted with the new arrangement of the geological–palaeontological display in the Castlepark Museum in Bad Kreuznach. He studied geology and palaeontology at Frankfurt am Main and Mainz. His dissertation was on the Messel anurids, with emphasis on fossil diagenesis and biostratinomy. Since 1989 he has been curator of the section of geological history for the protection of geological/palaeontological monuments at the Administration for Monument Conservation Rhineland-Palatinate.

References

Preface

Behnke, C., Eikamp, H., and Zollweg, M. (1986). *Die Grube Messel*. Goldschneck Verlag Weidert, Korb.

Heil, R., Koenigswald, W. von, Lippmann, H. G., Graner, D., and Heunisch, C. (ed.) (1987). *Fossilien der Messel-Formation*. Hessisches Landesmuseum, Darmstadt.

Preface to the English edition

Franzen, J. L. (in press). The oldest primate hands—additional remarks and observations. In *The hands of primates* (ed. H. Preuschoft and D. Chwers), 'Il Sedicesimo', Florence.

Franzen, J. L. (1990). *Hallensia* (Mammalia, Perissodactyla) aus Messel und dem Pariser Becken sowie Nachträge aus dem Geiseltal. *Bull. Inst. R. Sci. Natur. Belgique* (in press).

Franzen, J. L. and Michaelis, W. (ed.) (1988). Der eozäne Messelsee—Eocene Lake Messel. *Cour. Forsch.-Inst. Senckenberg*, **107**, 1–452.

Goth, K. (1990). Der Messeler Ölschiefer—ein Algenlaminit. *Cour. Forsch.-Inst. Senckenberg*, **131**, 1–143.

Habersetzer, J. and Storch, G. (1988). Grube Messel: akustische Orientierung der ältesten Fledermäuse. *Spektrum der Wissenschaft*, **7**, 12–14.

Habersetzer, J. and Storch, G. (1990). Ecology and echolocation of the Eocene Messel bats. In *European bat research 1987* (ed. V. Hanak, T. Horacek, and J. Gaisler), pp. 213–33. Charles University Press, Prague.

Habersetzer, J., Richter, G., and Storch, G. (1989). Paleoecology of the Middle Eocene Messel bats. *Abstracts of the 5th*

International Theriol. Congress, Rome, Vol. 2, pp. 629–30.

Habersetzer, J., Richter, G., and Storch, G. (in press). Paleoecology of the Middle Eocene Messel bats. *Histor. biol.*

Hesse, A. (1989). Taxonomie der ordnung Gruiformes (Aves) nach osteologischen morphologischen Kriterien unter besonderer Berücksichtigung der Messel-ornithidae Hesse 1988. *Cour. Forsch.-Inst. Senckenberg*, **107**, 235–47.

Koenigswald, W. von (1990). Die Paläobiologie der Apatemyiden (Insectivora s.l.) und die Ausdeutung der Skelettfunde von *Heterohyus nanus* aus dem Mittel-eozän von Messelbei Darmstadt. *Palaeontographica*, **A210** (1–3), 41–77.

Lutz, H. (1990). Systematische und palökologische Untersuchungen an Insekten aus dem Mittel-eozän der Grube Messel bei Darmstadt. *Cour. Forsch.Inst. Senckenberg*, **124**, 1–165.

MacPhee, R. D. E., Novacek, M. J., and Storch, G. (1988). Basicranial morphology of early Tertiary Erinaceomorphs and the origin of primates. *Am. Mus. Novitates*, **2921**, 1–42.

Peters, D. S. (1989*a*). Ein vollständiges Exemplar von *Palaeotis weigelti. Cour. Forsch.-Inst. Senckenberg*, **107**, 223–33.

Peters, D. S. (1989*b*). Fossil birds from the oil shale of Messel (Lower Middle Eocene, Lutetian). In *Acta XIX Congressus Internation. Ornithologici*, Vol. 2, pp. 2056–64. University of Ottawa Press.

Peters, D. S. (in press). A new species of owl (Aves: Strigiformes) from the Middle Eocene Messel oil shale.

Schmitz, M. (1991). Die Koprolithen mitteleozäner Vertebraten aus der Grube Messel bei Darmstadt. *Cour. Forsch.-Inst. Senckenberg*, **137**, 1–199.

Storch, G. (1989*a*). Die eozänen Fledermäuse von Messel—frühe Zeugen der Stammesgeschichte. *Laichinger Höhlenfreund*, **24**, 21–30.

Storch, G. (1989*b*). The Eocene mammalian

fauna from Messel—palaeobiographical jigsaw puzzle. *Int. Sympos. Vertebr. Biogeogr. Syst. Tropics, Bonn, Abstracts*, p. 49.

Storch, G. (1990). The Eocene mammalian fauna from Messel—palaeobiographical jigsaw puzzle. In *Proceedings of the Int. Sympos. Vertebr. Biogeogr. Syst. Tropics, Bonn* (ed. G. Peters and R. Hutterer), pp. 23–32.

Storch, G. (in press). 'Grube Messel' and African–South African faunal connections. In *The Africa–South America connection* (ed. W. George and R. Lavocat). Oxford University Press.

Storch, G. and Habersetzer, J. (1988). *Archaeonycteris pollex* (Mammalia, Chiroptera), eine neue Fledermaus aus dem Eozän der Grube Messel bei Darmstadt. In *Der eozäne Messelsee—Eocene Lake Messel* (ed. J. L. Franzen and M. Michaelis), Cour. Forsch.-Inst. Senckenberg, **107**, 263–73.

Storch, G. and Haubold, H. (1989). Additions to the Geiseltal mammalian faunas, Middle Eocene: Didelphidae, Nyctitheriidae, Myrmecophagidae. *Palaeovertebrata*, **19**(3), 95–114.

Thiele-Pfeiffer, H. (1988). Die Mikroflora aus dem mitteleozänen Ölschiefer von Messel bei Darmstadt. *Palaeontographica*, **B211** (1–3), 1–86.

Wilde, V. (1989). Untersuchungen zur Systematik der Blattreste aus dem Mitteleozän der Grube Messel bei Darmstadt (Hessen, Bundesrepublik Deutschland). *Cour. Forsch.-Inst. Senckenberg*, **115**, 1–213.

1 Curriculum vitae of the Messel pit

Beeger, W. (1970). *Chronik der Grube Messel*. YTONG AG, Munich.

2 Europe in the Eocene: Messel in time and space

Barron, E. J., Harrison, C. G. A., Sloan II, J. L., and Hay, W. W. (1981). Paleogeography,

180 million years ago to the present. *Eclogae geol. Helv.*, **74**(2), 443–70.

Franzen, J. L. (1968). Revision der Gattung *Palaeotherium* Cuvier 1804 (Palaeotheriidae, Perissodactyla, Mammalia). Unpublished PhD dissertation, University of Freiburg in Breisgau.

Franzen, J. L. (1981). Das erste Skelett eines Dichobuniden (Mammalia, Artiodactyla), geborgen aus mittel-eozänen Ölschiefern der 'Grube Messel' bei Darmstadt (Deutschland, S. Hessen). *Senckenbergiana Lethaea*, **61** (3/6), 299–353.

Franzen, J. L. (1987). Ein neuer Primate aus dem Mitteleozän der Grube Messel (Deutschland, S. Hessen). In *Forschungsergebnisse zu Grabungen in der Grube Messel bei Darmstadt* (ed. S. Schaal), Cour. Forsch.-Inst. Senckenberg, **91**, 151–87.

Franzen, J. L. and Haubold, H. (1985). The European Middle Eocene of mammalian stratigraphy. *Terra cognita*, **5** (2–3), 134.

Franzen, J. L. and Haubold, H. (1986a). Revision der Equoidea aus den eozänen Braunkohlen des Geiseltales bei Halle (DDR). *Palaeovertebrata*, **16** (1), 1–34.

Franzen, J. L. and Haubold, H. (1986b). The Middle Eocene of European mammalian stratigraphy. *Modern Geology*, **10** (2/3), 159–70.

Franzen, J. L. and Haubold, H. (1987). The biostratigraphic and palaeoecologic significance of the Middle Eocene locality Geiseltal near Halle (German Democratic Republic). In *International Symposium on Mammalian Biostratigraphy and Paleoecology of the European Paleogene, Mainz, February 18th–21st 1987* (ed. N. Schmidt-Kittler), Münchner Geowiss. Abh., **A10**, 93–100.

Franzen, J. L. and Krumbiegel, G. (1980). *Messelobunodon ceciliensis* n. sp.—ein neuer Dichobunide aus der mitteleozänen Fauna des Geiseltales bei Halle (DDR). *Z. geol. Wiss.*, **8** (12), 1585–92.

Hartenberger, J.-L. (1970). Les Mammifères d'Egerkingen et l'histoire des faunes de l'Eocène d'Europe. *Bull. Soc. géol. France* (7) **12** (5), 886–93.

Krumbiegel, G., Rüffle, L., and Haubold, H. (1983). *Das eozäne Geiseltal, ein mitteleuropäisches Braunkohlenvorkommen und seine Tier- und Pflanzenwelt*. A. Ziemsen, Wittenberg Lutherstadt.

Krutzsch, W. (1976). Die Mikroflora des Geiseltales. Part IV. Die stratigraphische Stellung des Geiseltalprofils im Eozän und die sporeenstratigraphische Untergliederung des mittleren Eozäns. *Abh. Zentr. Geol. Inst. Paläontol.* **H26** (Eozäne Floren des Geiseltales), 47–92.

Kubanek, F., Nöltner, T., Weber, J., and Zimmerle, W. (1988). On the lithogenesis of the Messel oil shale. In *Der eozäne Messelsee—Eocene Lake Messel* (ed. J. L. Franzen and W. Michaelis). Cour. Forsch.-Inst. Senckenberg, **107**, 13–28.

Lippolt, H. J., Baranyi, I., and Todt, W. (1975). Die Kalium–Argon-Alter der postpermischen Vulkanite des nord-östlichen Oberrheingrabens. *Der Aufschluss* (special volume) **27** (Odenwald), 205–12.

Negendank, J. F. W., Irion, G., and Linden, J. (1982). Ein eozänes Maar bei Eckfeld nordöstlich Manderscheid (SW-Eifel). *Mainzer geowiss Mitt.*, **11**, 157–72.

Owen, H. G. (1983). *Atlas of continental displacement, 200 million years to the present*. Cambridge University Press.

Panza, G. F., Calcagnile, G., Scandone, P., and Mueller, S. (1985). Die geologische Tiefenstruktur des Mittelmeerraumes. In *Ozeane und Kontinente. Ihre Herkunft, ihre Geschichte und Struktur* (3rd edn), pp. 132–42. Spektrum der Wissenschaft, Heidelberg.

Russell, D. E. (1968). Succession en Europe, des faunes mammaliennes au début du Tertiaire. *Mém. B.R.G.M.*, **58** (Colloque sur l'Eocène, Paris 1968), 291–7.

Savage, D. E. and Russell, D. E. (1983).

Mammalian paleofaunas of the world. Addison-Wesley, London.

Schmidt-Kittler, N. and Vianey-Liaud, M. (1975). Les relations entre les faunes de rongeurs d'Allemagne du Sud et de France pendant l'Oligocène. *C.R. Acad. Sci. Paris*, **D281**, 511–14.

Smith, A. G., Hurley, A. M., and Briden, J. C. (1982). *Paläokontinentale Weltkarten des Phanerozoikums*. Enke, Stuttgart.

Storch, G. (1986). Die Säuger von Messel: Wurzeln auf vielen Kontinenten. *Spektrum Wiss.*, **1986** (6), 48–65.

Strauch, F. (1970). Die Thule-Landbrücke als Wanderweg und Faunenscheide zwischen Atlantik und Skandik im Tertiär. *Geol. Rundschau*, **60**, 381–417.

Tobien, W. (1968a). Das biostratigraphische Alter der mitteleozänen Fossilfundstätte Messel bei Darmstadt (Hessen). *Notizbl. hess. L.-Amt Bodenforsch.*, **96**, 111–19.

Tobien, W. (1968b). Mammifères éocènes du Bassin de Mayence et de la partie orientale du fossé rhénan. *Mém. B.R.G.M.*, **58** (Colloque sur l'Eocène, Paris 1968), 297–307.

Weber, J. and Zimmerle, W. (1985). Pyroclastic detritus in the lacustrine sediments of the Messel Formation. *Senckenbergiana Lethaea*, **66** (1/2), 171–6.

Wuttke, M. (1983a). 'Weichteil-Erhaltung' durch lithifizierte Mikroorganismen bei mittel-eozänen Vertebraten aus den Ölschiefern der 'Grube Messel' bei Darmstadt. *Senckenbergiana Lethaea*, **64** (5/6), 509–27.

Wuttke, M. (1983b). Aktuopaläontologische Studien über der Zerfall von Wirbeltieren. Part 1. Anura. *Senckenbergiana Lethaea*, **64** (5/6), 529–60.

3 The genesis of the Messel oil shale

Franzen, J. L., Weber, J., and Wuttke, M. (1982). Senckenberg-Grabungen in der Grube Messel bei Darmstadt. Part 3.

Ergebnisse 1979–1981. *Cour. Forsch.-Inst. Senckenberg*, **54**, 1–118.

Goth, K. (1986). Mikrofazielle Untersuchungen am Messeler Ölschiefer. In *Wissenschaftlicher Jahresbericht 1985 des Forschungsinstituts Senckenberg, Frankfurt a. M.* (ed. W. Ziegler), Cour. Forsch.-Inst. Senckenberg, **85**, 205–30.

Irion, G. (1966). Der eozäne See von Messel. *Natur u. Museum*, **107**, 213–18.

Matthess, H. (1966). Zur Geologie des Ölschiefervorkommens von Messel bei Darmstadt. *Abh. Hess. L.-Amt Bodenforsch.*, **51**, 1–87.

Reineck, H.-E. and Weber, J. (1983). Trümmer und Trübeströme im eozänen See von Messel. *Natur u. Museum*, **113**, 307–12.

Rietschel, S. (1987). Der See von Messel—eine vulkanische Falle für Urwaldtiere des Eozäns? *Frankfurter Allgemeine Zeitung*, No. 104, 6 May.

Schaal, S., Schmitz-Münker, M., and Wolf, G. (1987). Neue Korrelationsmöglichkeiten von Grabungsstellen in der eozänen Fossillagerstätte Grube Messel. *Cour. Forsch.-Inst. Senckenberg*, **91**, 203–11.

Weber, J. (1988). Die Geologie der Grube Messel. Unpublished Ph.D. dissertation, University of Frankfurt-am-Main.

Weber, J. and Hofmann, U. (1982). Kernbohrungen in der eozänen Fossillagerstätte Grube Messel bei Darmstadt. *Geol. Abh. Hessen*, **83**, 1–58.

4 The vegetation: fossil plants as witnesses of a warm climate

Barthel, M. (1976). Farne und Cycadeen. In Eozäne Floren des Geiseltales. *Abh. Zentr. Geol. Inst.*, **26**, 439–98.

Collinson, M. E. (1982). A preliminary report on the Senckenberg-Museum collection of fruits and seeds from Messel bei Darmstadt. *Cour. Forsch.-Inst. Senckenberg*, **56**, 49–57.

Collinson, M. E. (1983). *Fossil plants of the London Clay.* Palaeontological Association, London.

Collinson, M. E. (1986). Früchte und Samen aus dem Messeler Ölschiefer. *Cour. Forsch.-Inst. Senckenberg*, **85**, 217–20.

Engelhardt, H. (1922). Die alttertiäre Flora von Messel bei Darmstadt. *Abh. Hess. Geol. L.-Anst. Darmstadt*, **7**(4), 17–128.

Franzen, J. L. (1977). Urpferdchen und Krokodile—Messel vor 50 Millionen Jahren. *Kl. Senckenberg-Reihe*, **7**, 1–36.

Friis, E. M. (1983). Upper Cretaceous (Senonian) floral structures of juglandalean affinity containing Normapollespollen. *Rev. Palaeobot. Palynol.*, **39**, 161–88.

Goth, K. (1990). Der Messeler Ölschiefer—ein Algenlaminit. *Cour. Forsch.-Inst. Senckenberg*, **131**, 118 pp.

Heywood, V. H. (1979). *Flowering plants of the world.* Oxford University Press.

Kirchheimer, F. (1943). Die Mastixioideen in der Flora der Gegenwart. *Braunkohle*, **42**, 17–19.

Koenigswald, W. von and Schaarschmidt, F. (1983). Ein Urpferd aus Messel, das Weinbeeren fraß. *Natur u. Museum*, **113** (3), 79–84.

Mai, D. H. (1981). Entwicklung und klimatische Differenzierung der Laubwaldflora Mitteleuropas im Tertiär. *Flora*, **171**, 525–82.

Mai, D. H. and Walther, H. (1985). Die obereozänen Floren des Weisselster-Beckens und seiner Randgebiete. *Abh. Staatl. Mus. Mineral. Geol. Dresden*, **33**, 1–176.

Müller-Stoll, W. (1935). Palmenreste aus dem Eozän des Oberrheingebietes und ihre Erhaltung. *Paläont. Z.*, **17**, 55–73.

Pflug, H. D. (1953). In *Pollen und Sporen des mitteleuropäischen Tertiärs* (ed. P. W. Thomson and H. D. Pflug), Palaeontographica, **B94**, 1–138.

Schaarschmidt, F. and Wilde, V. (1986). Palmenblüten und -blätter aus dem Eozän von Messel. *Cour. Forsch.-Inst. Senckenberg*, **86**, 177–202.

Sturm, M. (1971). Die eozäne Flora von Messel bei Darmstadt. 1. Lauraceae. *Palaeontographica*, **B134**, 1–60.

Sturm, M. (1978). Maw contents of an Eocene horse (*Propalaeotherium*) out of the oil shale of Messel near Darmstadt. *Cour. Forsch.-Inst. Senckenberg*, **30**, 120–2.

Thenius, E. (1981). *Versteinerte Urkunden.* Springer, Berlin.

Thiele-Pfeiffer, H. (1988). Die Mikroflora aus dem mitteleozänen Ölschiefer von Messel bei Darmstadt. *Palaeontographica*, **B211** (1–3), 1–86.

Wilde, V. (1989). Untersuchungen zur Systematik der Blattreste aus dem Mitteleozän der Grube Messel bei Darmstadt (Hessen, Bundesrepublik Deutschland). *Cour. Forsch.-Inst. Senckenberg*, **115**, 1–213.

Wolfe, J. A. (1969). *Palaeogene floras from the Gulf of Alaska Region*, US Geological Survey Open-file Report. US Government Printing Office, Washington, DC.

5 Giant ants and other rarities: the insect fauna

Franzen, J. L., Weber, J., and Wuttke, M. (1982). Senckenberg-Grabungen in der Grube Messel bei Darmstadt. 3. Ergebnisse 1979–81. *Cour. Forsch.-Inst. Senckenberg*, **54**, 1–118.

Gahl, H. and Maschwitz, U. (1977). Eine Ameise aus dem Mitteleozän von Messel bei Darmstadt (Hessen). *Geol. Jb. Hessen*, **105**, 69–73.

Goth, K. (1986). Stand der paläobotanischen Untersuchungen in Messel. 1.5. Die Mikroflora des Messeler Ölschiefers. *Dt. Forsch.-Gemeinschaft, Referat Geowiss.*, II C 6. Protokoll des Rundgesprächs über die Ergebnisse der 'Forschungsvorhaben der Grube Messel', p. 23 (Appendix 5).

Kinzelbach, R. (1970a). Eine Gangmine aus dem eozänen Ölschiefer von Messel (Insecta: ?Lepidoptera). *Paläont. Z.*, **44**, 93–6.

Kinzelbach, R. (1970b). Wanzen aus dem eozänen Ölschiefer von Messel (Insecta: Heteroptera). *Notizbl. hess. L.-Amt Bodenforsch.*, **98**, 9–18.

Kinzelbach, R. and Lutz, H. (1985). Stylopid larva from the Eocene—a spotlight on the phylogeny of the Stylopids (Strepsiptera). *Ann. Entomol. Soc. America*, **78**, 600–2.

Lutz, H. (1985). Eine wasserlebende Käferlarve aus dem Mittel-Eozän der Grube Messel. *Natur u. Museum*, **115** (2), 55–60.

Lutz, H. (1986). Eine neue Unterfamilie der Formicidae (Insecta: Hymenoptera) aus dem mittel-eozänen Ölschiefer der 'Grube Messel' bei Darmstadt (Deutschland, S.-Hessen). *Senckenbergiana Lethaea*, **67** (1/4), 177–218.

Lutz, H. (1987). Die Insekten-Thanatocoenose aus dem Mittel-Eozän der 'Grube Messel' bei Darmstadt: Erste Ergebnisse. *Cour. Forsch.-Inst. Senckenberg*, **91**, 189–201.

Meunier, F. (1921). Die Insektenreste aus dem Lutetien von Messel bei Darmstadt. *Abh. hes. geol. L.-Anst.*, **7**, 1–15.

Micklich, N. (1985). Biologisch paläontologische Untersuchungen zur Fischfauna der Messeler Ölschiefer (Mittel-Eozän, Lutetium). *Andrias*, **4**, 1–171.

Richter, G. (1987). Untersuchungen zur Ernährung eozäner Säuger aus der Fossilfundstätte Messel bei Darmstadt. *Cour. Forsch-Inst. Senckenberg*, **91**, 1–33.

Richter, G. and Storch, G. (1980). Beiträge zur Ernährungsbiologie eozäner Fledermäuse aus der 'Grube Messel'. *Natur u. Museum*, **110**, 353–67.

Schaarschmidt, F. (1986). Stand der paläobotanischen Untersuchungen in Messel. 1.1. Einführung. *Dt. Forsch.-Gemeinschaft, Referat Geowiss.*, II C 6. Protokoll des Rundgesprächs über die Ergebnisse der 'Forschungsvorhaben der Grube Messel', pp. 16–17 (Appendix 5).

Schmitz-Müncker, M. (1986). Messelkoprolithen—Forschungsergebnisse und Ausblick. *Dt. Forsch.-Gemeinschaft, Referat Geowiss.*, II C 6. Protokoll des Rundgesprächs über die Ergebnisse der 'Forschungsvorhaben der Grube Messel', pp. 51–2 (Appendix 15).

Voigt, E. (1939). Weichteile an fossilen Insekten aus der eozänen Braunkohle des Geiseltales bei Halle (Saale). *Nova Acta Leopoldina, N.F.*, **6**, 1–38.

Wunderlich, J. (1986). Die ersten Spinnen aus dem Mittel-Eozän der Grube Messel. *Senckenbergiana Lethaea*, **67** (1/4), 171–6.

Wuttke, M. (1986). Untersuchungen zur Biostratinomie und Fossildiagenese der Messeler Anuren. Schlussfolgerungen für die Zielrichtung weiterer Untersuchungen zur Genese der Messeler Fossillagerstätte. *Dt. Forsch.-Gemeinschaft, Referat Geowiss.*, II

C 6. Protokoll des Rundgesprächs über die Ergebnisse der 'Forschungsvorhaben der Grube Messel', pp. 35–7 (Appendix 8).

6 Ancient knights-in-armour and modern cannibals

Andreae, A. (1893). Vorläufige Mitteilung über die Ganoiden (*Lepisosteus* und *Amia*) des Mainzer Beckens. *Verh. naturhist.-med. Ver. N.F.*, **5**, 7–15.

Andrews, S. M., Gardiner, B. G., Miles, R. S., and Patterson, C. (1967). Pisces. In *The fossil record* (ed. W. B. Harland *et al.*), pp. 637–98. Geological Society, London.

Arratia, G. (1982). A review of freshwater percoids from South America (Pisces, Osteichthyes, Percichthyidae, and Perciliidae). *Abh. senckenb. naturforsch. Ges.*, **540**, 1–52.

Balon, E. K. (1959). Die Beschuppungsentwicklung der Texas-Cichlide (*Herichthys cynoguttatus* Baird and Girard). *Zool. Anz.*, **163**, 82–9.

Bauch, G. (1961). *Die einheimischen Süsswasserfische*, (4th edn). Neumann, Radebeul.

Blot, J. (1978). Les apodes fossiles du Monte Bolca I. *Studi e ricerce sui giacimenti terzeari di Bolca*, **3** (1), 1–123.

Blot, J. (1980). La faune ichthyologique des gisements du Monte Bolca (Province de Vérone, Italie). Catalogue systématique présentant l'état actuel des recherches concernant cette faune. *Bull. Mus. nat. Hist. natur., Paris*, **2** (4), 339–96.

Boreske, J. R. (1974). A review of the North American fossil amiid fishes. *Bull. Mus. Comp. Zool., Harvard Univ.*, **146** (1), 1–87.

Colette, B. and Banarescu, P. (1977). Systematics and zoogeography of the Percidae. *J. Fisheries Res. Board Can.*, **34** (10), 1450–63.

Franzen, J. L., Weber, J., and Wuttke, M. (1982). Senckenberg-Grabungen in der Grube Messel bei Darmstadt. 3. Ergebnisse 1979–1981. *Cour. Forsch-Inst. Senckenberg*, **54**, 1–118.

Gaudant, J. (1979). Mise au point sur l'ichthyofaune paléocène de Menat (Puy-de-Dôme). *C.R. Acad. Sci. Paris*, **D288**, 1461–3.

Gaudant, J. (1980). Sur *Amia kehreri* Andreae (Poisson Amiidae du Lutétien) de Messel, Allemagne et sa signification paléogéographique. *C.R. Acad. Sci., Paris*, **D290**, 1107–10.

Gaudant, J. (1981). Sur *Thaumaturus* Reuss (poisson téléostéen), Ostéoglossomorphe fossile du Cénozoique européen. *C.R. Acad. Sci., Paris*, **293** (2), 787–90.

Gaudant, J. (1987). Mise au point sur certains Poissons Amiidae du Cénozoique européen: Le genre *Cyclurus* Agassiz (=*Kindleia* Jordan). *Paläont. Z.*, **61** (3/4), 321–30.

Gaudant, J. (1988). L'ichthyofaune éocéne de Messel et de la Geiseltal (Allemagne): Essai d'approche paléobiogéographique. *Cour. Forsch-Inst. Senckenberg*, **107**, 355–67.

Gaudant, J. and Micklich, M. (1990). *Rhenanoperca minuta* nov. gen., nov. sp., ein neuer Percoide (Pisces, Perciformes) aus der Messel-Formation (Mittel-Eozän, Unteres Geiseltalium). *Paläont. Z.*, **64** (3/4). (in press).

Gosline, W. A. (1960). Contribution towards a classification of modern isospondylous fishes. *Bull. Br. Mus. (Natur. Hist.) Zool.*, **6**, 325–65.

Gosline, W. A. (1966). The limits of the fish family Seranidae with notes on other lower Percoids. *Proc. Calif. Acad. Sci.*, **33** (4), 91–112.

Grande, L. (1980). Paleontology of the Green River Formation, with a review of the fish fauna. *Geol. Surv. Wyoming Bull.*, **63** (entire volume).

Jerzmańska, A. (1977a). The freshwater fishes from the Middle Eocene of Geiseltal. In *Eozäne Wirbeltiere des Geiseltales* (ed. H. Matthes and B. Thaler). Wiss. Beitr. Martin-Luther-Univ. Halle-Wittenberg, 1977/2/P5, 41–66.

Jerzmańska, A. (1977a). Süswasserfische des älteren Tertiärs von Europa. In *Eozäne Wirbeltiere des Geiseltales* (ed. H. Matthes and B. Thaler), Wiss. Beitr. Martin-Luther-Univ. Halle-Wittenberg, 1977/2/P5, 67–76.

Johnson, G. D. (1984). Percoidei: development and relationships. In *Ontogeny and systematics of fishes*, American Society of Ichthyology and Herpetology Special

Publication, no. 1 (ed. H. G. Moser *et al.*), pp. 464–98. American Society of Ichthyology and Herpetology, Lawrence, Kansas.

Kinkelin, F. (1884). Über Fossilien aus Braunkohlen in der Umgebung von Frankfurt a. M. *Ber. senckenberg, naturforsch. Ges.*, **1884**, 165–83.

Lee, D. S., Gilbert, C. R., Hocutt, C. H., Jenkins, R. E., McAllister, D. E., and Stauffer, J. R. (1980 *et seq.*). *Atlas of North American freshwater fishes*. North Carolina Biological Survey.

Liem, K. F. (1973). Evolutionary strategies and morphological innovations: Cichlid pharyngeal jaws. *Syst. Zool.*, **22**, 425–41.

Lutz, H. (1987). Die Insekten-Thanatocoenose aus dem Mittel-Eozän der 'Grube Messel' bei Darmstadt: Erste Ergebnisse. In *Forschungsergebnisse zu Grabungen in der Grube Messel bei Darmstadt* (ed. S. Schaal), Cour. Forsch-Inst. Senckenberg, **91**, 189–201.

McDonald, C. M. (1978). Morphological and biochemical systematics of Australian freshwater and estuarine Percichthyid fishes. *Austr. J. Mar. Freshwater Res.*, **29**, 667–98.

Micklich, N. (1978). *Palaeoperca proxima*, ein neuer Knochenfisch aus dem Mittel-Eozän von Messel bei Darmstadt. *Senckenbergiana Lethaea*, **59** (4/6), 483–501.

Micklich, N. (1983). Ein Aal aus der 'Grube Messel'. Gedanken und Probleme bei Aussagen zu Fossilfunden. *Natur u. Museum*, **113**, 211–21.

Micklich, N. (1985). Biologisch-paläontologische Untersuchungen zur Fischfauna der Messeler Ölschiefer (Mittel-Eozän, Lutetium). *Andrias*, **4**, 1–171.

Micklich, N. (1987). Neue Beiträge zur Morphologie, Ökologie und Systematik Messeler Knochenfische. I. Die Gattung *Amphiperca* Weitzel 1933 (Perciformes, Percoidei). In *Forschungsergebnisse zu Grabungen in der Grube Messel bei Darmstadt* (ed. S. Schaal), Cour. Forsch-Inst. Senckenberg, **91**, 35–106.

Micklich, N. (1989). Percoid fishes of the Messel Oilshale Formation. *Cour. Forsch-Inst. Senckenberg*, **107**, 199–210.

Micklich, N. and Gaudant, J. (1989). *Anthracoperca siebergi* Voigt 1934 (Pisces, Perciformes). Ergebnisse einer Neuuntersuchung der mitteleozänen Barsche des Geiseltales. *Z. geol. Wiss.*, **17**, 503–21.

Nelson, G. J. (1968). Gill arches of teleostean fishes of the division Osteoglossomorpha. *J. Linn. Soc. (Zool.)*, **47** (312), 261–77.

Nikolski, G. W. (1957). *Spezielle Fischkunde*. Deutscher Verlag der Wissenschaft, Berlin.

Nolf, D. (1985). Otolithi piscium. In *Handbook of paleoichthyology*, Vol. 10 (ed. H. P. Schultze), pp. 1–145. G. Fischer, Stuttgart.

Obrhelová, N. (1971). Über einen Serranid (Pisces) aus dem nordböhmischen Süsswassertertiär. *Čas. Mineral. Geol., Prague*, **16**(4), 371–87.

Obrhelová, N. (1975). Osteologischer Bau von *Thaumaturus furcatus* Reuss 1844 (Pisces) aus dem nordböhmischen Süsswassertertiär. *Čas. Mineral. Geol., Prague*, **20** (3), 272–90.

Patterson, C. (1964). A review of Mesozoic acanthopterygian fishes with special reference to those of the English Chalk. *Philos. Trans. roy. Soc. London*, **B242**, 213–482.

Patterson, C. (1973). Interrelationships of holosteans. In *Interrelationships of fishes* (ed. P. H. Greenwood, R. Miles, and C. Patterson), Zool. J. Linnean Soc., **53** (suppl. 1), 233–305.

Reighard, J. (1903). The natural history of *Amia calva* Linnaeus. In *Mark Anniversary Volume*, pp. 57–109. Holt and Co, New York.

Schulze, H. P. and Wiley, E. O. (1984). The Neopterygian *Amia* as a living fossil. In *Living fossils* (ed. N. Eldridge and S. Stanley), pp. 153–9. New York.

Scott, T. D., Glover, C. J., and Southcott, R. V. (ed.) (1980). *The marine and freshwater fishes of South Australia* (2nd edn). (Facsimile reprint)

Taverne, L. (1977). Ostéologie, phylogenèse et systématique des Téléostéens fossiles et actuels du supre-ordre Ostéoglossomorphes. Première partie: Ostéologie des genres *Hiodon, Eohiodon, Lycoptera, Osteoglossum, Scleropages, Heterotis* et *Arapaima. Acad. roy. Belg. Mém. Cl. Sci.*, 8°(2) **42** (3), 1–234.

Taverne, L. (1978). Ostéologie, phylogenèse et systématique des Téléostéens fossiles et actuels du supre-ordre Ostéoglossomorphes. Deuxième partie: Ostéologie des genres *Phareodus, Phareoides, Brychaetus, Musperia, Pantodon, Singida, Notopterus, Xenomystus* et *Papyrocranus. Acad. roy. Belg. Mém. Cl. Sci.*, 8°(2) **42** (6), 1–212.

Taverne, L. (1979). Ostéologie, phylogenèse et systématique des Téléostéens fossiles et actuels du supre-ordre Ostéoglossomorphes. Troisième partie: Evolution des structures ostéologiques et conclusions générales relatives à la phylogenèse et à la systématique du super-ordre. Addendum. *Acad. roy. Belg. Mém. Cl. Sci.*, 8°(2) **43** (3), 1–168.

Turner, S. (1982). A catalogue of fossil fish in Queensland. *Mem. Queensland Mus.*, **20** (3), 599–611.

Voigt, E. (1934). Die Fische aus der mitteleozänen Braunkohle des Geiseltales. *Nova Acta Leopoldina N.F.*, **2** (1–2), 1–146.

Weiler, W. (1963). Die Fischfauna des Tertiärs im oberrheinischen Graben, des Mainzer Beckens, des unteren Maintals und der Wetterau, unter besonderer Berücksichtigung des Untermiozäns. *Abh. Senckenb. Naturforsch. Ges.*, **504**, 1–75.

Weitzel, K. (1933). *Amphiperca multiformis* n.g. n. sp. und *Thaumaturus intermedius* n. sp., Knochenfische aus dem Mitteleozän von Messel. *Notizbl. Ver. Erdkde. u. Hess. geol. L.-Anst.*, **5** (14), 89–97.

Weitzman, S. H. (1960). The systematic position of Piton's presumed Characid fishes from the Eocene of Central France. *Stanford Ichthyol. Bull.*, **7**, 114–23.

Wiley, E. O. (1976). *The phylogeny and biogeography of fossil and recent gars (Actinopterygii: Lepisosteidae)*. University of Kansas Museum nat. Hist. Misc. Publ., no. 64, pp. 1–111. University of Kansas, Lawrence, Kansas.

Wiley, E. O. and Schultze, H. P. (1984). Family Lepisoteidae (gars) as living fossils. In *Living fossils* (ed. N. Eldredge and S. Stanley), pp. 160–5. New York.

Woolcott, W. S. (1957). Comparative osteology of serranid fishes of the genus *Roccus* (Mitchill). *Copeia*, **1957** (1), 1–10.

7 Amphibia at Lake Messel: salamanders, toads, and frogs

Weitzel, K. (1938). *Propelodytes wagneri* n.g. n. sp., ein Frosch aus dem Mitteleozän von Messel. *Notizbl. hess. geol. L.-Amt (5)*, **19**, 42–6.

Westphal, F. (1980). *Chelotriton robustus* n. sp., ein Salamandride aus dem Eozän der Grube Messel bei Darmstadt. *Senckenbergiana Lethaea*, **60** (4/6), 475–87.

8 Freshwater turtles

Brattstrom, B. H. (1965). Body temperature of reptiles. *Am. Midl. Nat.*, **73** (2), 376–422.

Broin, F. D. (1977). Contribution à l'étude des Chéloniens continentaux du Crétacé supérieur et du Tertiaire de France. *Mém. Mus. Hist. Natur., Paris* N.S., **C38.**

Harrassowitz, H. L. F. (1922a). Die Schildkrötengattung *Anosteira* von Messel bei Darmstadt und ihre stammesgeschichtliche Bedeutung. *Abh. hess. geol. L.-Anst.*, **6** (3), 133–238.

Harrassowitz, H. L. F. (1922b). Die Schild-krötengattung *Anosteira* von Messel bei Darmstadt und die Abstammung der Trionychiden (Vortrag und Diskussion). *Paläont. Z.*, **4**, 93–8.

Hummel, K. (1927). Die Schildkrötengattung *Trionyx* im Eozän von Messel bei Darmstadt und im aquitanischen Blättersandstein von Münzenberg in der Wetterau. *Abh. hess. geol. L.-Anst.*, **8** (2), 1–96.

Pritchard, P. C. H. (1979). *Encyclopedia of turtles*. T. F. H. Publications, London.

Reinach, A. von (1900). Schildkrötenreste im Mainzer Tertiärbecken und in benachbarten, ungefähr gleichalterigen Ablagerungen. *Abh. Senckenberg Naturforsch. Ges.*, **28**, 1–135.

Schleich, H. H. (1981). Jungtertiäre Schildkröten Süddeutschlands unter besonderer Berücksichtigung der Fundstelle Sandelzhausen. *Cour. Forsch-Inst. Senckenberg*, **48**, 1–372.

Staesche, K. (1928). Sumpfschildkröten aus hessischen Tertiärablagerungen. *Abh. hess. geol. L.-Anst.*, **8**, 1–72.

Weitzel, K. (1949). Neue Wirbeltiere (Rodentia, Insectivora, Testudinata) aus dem Mitteleozän von Messel bei Darmstadt. *Abh. Senckenberg Naturforsch. Ges.*, **480**, 1–24.

9 Crocodiles: large ancient reptiles

Berg, D. E. (1964). Krokodile als Klimazeugen. *Geol. Rdsch.*, **54**, 328–33.

Berg, D. E. (1966). Die Krokodile, insbesondere *Asiatosuchus* und aff. *Sebecus?*, aus dem Eozän von Messel bei Darmstadt/Hessen. *Abh. hess. L.-Amt Bodenforsch.*, **52**, 1–105.

Buffetaut, E. (1979). Evolución de los cocodrilos. *Scientific American* (Spanish edition, Barcelona), 88–99.

Buffetaut, E. (1988). The ziphodont mesosuchian crocodile from Messel: a reassessment. *Cour. Forsch-Inst. Senckenberg*, **107**, 211–21.

Estes, R. and Hutchison, J. H. (1980). Eocene lower vertebrates from Ellesmere Island, Canadian arctic archipelago. *Palaeogeogr., Palaeoclimatol., Palaeoecol.*, **30**, 325–47.

Frey, E., Laemmert, A., and Riess, J. (1987). *Baryphracta deponiae* n.g. n. sp. (Reptilia, Crocodylia), ein neues Krokodil aus der Grube Messel bei Darmstadt (Hessen, Bundesrepublik Deutschland). *N. Jb. Geol. Paläont. Mh.*, **1987** (1), 15–26.

Koenigswald, W. von and Wuttke, M. (1987). Zur Taphonomie eines unvollständigen Skelettes von *Leptictidium nasutum* aus dem Ölschiefer von Messel. *Geol. Jb. Hessen*, **115**, 65–79.

Ludwig, R. (1877). Fossile Crocodiliden aus der Tertiärformation des Mainzer Beckens. *Palaeontogr. Suppl.*, **3**, 1–52.

Tchernov, E. (1986). Evolution of the crocodiles in East and north Africa. *Cahiers de Paléontol.*, 1–64.

Weitzel, K. (1935). *Hassiacosuchus haupti* n.g. n. sp., ein durophages Krokodil aus dem Mitteleozän von Messel. *Notizbl. hess. Geol. L.-Anst.* (5) **16**, 40–9.

Weitzel, K. (1938). *Pristichampsus rollinati* (Gray) aus dem Mitteleozän von Messel. *Notizbl. hess. Geol. L.-Anst.* (5) **19**, 47–8.

10 Lizards: reptiles *en route* to success

Barnes, B. (1927). Eine eozäne Wirbeltier-Fauna aus der Braunkohle des Geiseltals. *Jb. Hallesch. Verband Erforsch, mitteldt. Bodensch. N.F.*, **6**, 5–24.

Bellairs, A. (1971). Die Reptilien. *Enzykl. d. Natur*, **11**, 394–767.

Borsuk-Bialynicka, M. (1983). The early phylogeny of Anguimorpha as implicated by craniological data. *Acta Palaeontologica Polonica*, **28** (1–2), 31–42.

Borsuk-Bialynicka, M. (1984). Anguimorphans and related lizards from the late Cretaceous of the Gobi Desert, Mongolia. *Palaeontologia polonica*, **46**, 5–105.

Estes, R. (1982). The fossil record and early distribution of lizards. In *Advances in herpetology and evolutionary biology: essays in honor of E. E. Williams*, pp. 365–98. Museum of Comparative Zoology, Harvard University, Cambridge, Massachusetts.

Estes, R. (1983). Sauria terrestria, Amphisbaenia. In *Handbuch der Paläoherpetologie*, Part 10 A. Stuttgart.

Frey, E. (1982). *Ophisaurus apodus* (Lacertilia, Anguidae)—a stemming digger? *N. Jb. f. Geol. Paläont. Abh.*, **164**, 217–21.

Gauthier, J. A. (1982). Fossil xenosaurid and anguid lizards from the early Eocene Wasatch Formation, southeast Wyoming, and a revision of the Anguioidea. *Contr. Geol., Univ. Wyoming*, **21** (1), 7–54.

Greene, H. W. (1983). Dietary correlates of the origin and radiation of snakes. *Am. Zool.*, **23**, 431–41.

Haubold, H. (1979). Zur Kenntnis der Sauria (Lacertilia) aus dem Eozän des Geiseltals. *Wiss. Beitr. Martin-Luther-Univ., Halle*, **1977/2** (P5), 107–12.

Kuhn, O. (1939). Die Schlangen aus dem Mitteleozän des Geiseltales. *Nova acta Leopoldina*, **7**, 119–33.

Meszoely, C. M. A. and Haubold, H. (1975). The status of the Middle Eocene Geiseltal limbless anguid lizards. *Copeia*, **1975** (1), 36–43.

Meszoely, C. M. A., Estes, R., and Haubold, H. (1978). Eocene anguid lizards from Europe and a revision of the genus *Xestops*. *Herpetologica*, **34** (2), 156–66.

Pregill, G. K., Gauthier, J. A., and Greene, H. W. (1986). The evolution of helodermatid squamates, with description of a new taxon and an overview of Varanoidea. *Trans. San Diego Soc. Nat. Hist.*, **21** (11), 167–202.

Schwarzbach, M. (1974). *Das Klima der Vorzeit*. Enke, Stuttgart.

Stritzke, R. (1983). *Saniwa feisti* n. sp., ein Varanide (Lacertilia, Reptilia) aus dem Mittel-Eozän von Messel bei Darmstadt. *Senckenbergiana Lethaea*, **64**, 497–508.

Sullivan, R. M. (1979). Revision of the Paleogene genus *Glyptosaurus* (Reptilia, Anguidae). *Bull. Am. Mus. Nat. Hist.*, **63** (1), 1–72.

Trutnau, L. (1981). *Schlangen*, Vol. 1. Ulmer, Stuttgart.

11 Messel birds: a land-based assemblage

Andors, A. V. (1991). Paleobiology and relationships of the giant groundbird *Diatryma* (Aves: Gastornithiformes). *Acta XX Congressus Internationalis Ornithologici*, Vol. 1, pp. 563–71. Ornithological Congress Trust Board, Wellington, New Zealand.

Berg, D. E. (1965). Nachweis des Riesenlaufvogels *Diatryma* im Eozän von Messel bei Darmstadt/Hessen. *Notizbl. Hess. Landesamt Bodenforsch.*, **93**, 68–72.

Cracraft, J. (1968). A review of the Bathornithidae (Aves, Gruiformes), with remarks on the relationships of the suborder Cariamae. *Am. Mus. Novitates*, **2326**, 1–46.

Cracraft, J. (1971). Systematics and evolution of the Gruiformes (Class Aves). 2. Additional comments on the Bathornithidae with descriptions of new species. *Am. Mus. Novitates*, **2449**, 1–14.

Fischer, K.-H. (1978). Neue Reste des Riesenlaufvogels *Diatryma* aus dem Eozän des Geiseltales bei Halle (DDR). *Mitt. zool. Museum Berlin*, **54** (suppl. Ann. Orn. 2), 133–44.

Garrod, A. H. (1876). On the anatomy of *Aramus scolopaceus*. *Proc. Zool. Soc. London*, **1876**, 275–7.

Hesse, A. (1988). Die Messelornithidae—eine neue Familie der Kranichartigen (Aves: Gruiformes: Rhynocheti) aus dem Tertiär Europas und Nordamerikas. *J. Orn.*, **129** (1), 83–95.

Hesse, A. (1989). Taxonomie der Ordnung Gruiformes (Aves) nach osteologischen morphologischen Kriterien unter besonderer Berücksichtigung der Messelornithidae Hesse 1988. *Cour. Forsch-Inst. Senckenberg*, **107**, 235–47.

Hoch, E. (1980). A new middle Eocene shorebird (Aves: Charadriiformes, Charadrii) with columboid features. *Contr. Sci. nat. Hist. Mus. Los Angeles County*, **330**, 33–49.

Houde, P. (1986). Ostrich ancestors found in the Northern Hemisphere suggest new hypothesis of ratite origins. *Nature*, **324**, 563–5.

Houde, P. and Olson, S. L. (1981). Palaeognathous carinate birds from the early Tertiary of North America. *Science*, **214**, 1236–7.

Lambrecht, K. (1928). *Palaeotis weigelti* n.g. n. sp., eine fossile Trappe aus der mitteleozänen Braunkohle des Geiseltales. *Jb. hallesch. Verb. Erforsch. mitteldt. Bodenschätze, N.F.*, **7**, 20–9.

Mourer-Chauviré, C. (1981). Première indication de la présence des Phorusrhacidés, famille d'oiseaux géants d'Amérique du Sud, dans le tertiaire européen: *Ameghinornis* nov. gen. (Aves, Ralliformes) des Phosphorites du Quercy, France. *Geobios*, **14** (5), 637–47.

Mourer-Chauviré, C. (1983). Les Gruiformes (Aves) des Phosphorites du Quercy (France). 1. Sous-Ordre Cariamae (Cariamidae et Phorusrhacidae): systématique et biostratigraphie. *Palaeovertebrata*, **13** (4), 83–143.

Mourer-Chauviré, C. (1987). Les Strigiformes (Aves) des Phosphorites du Quercy (France). Systématique, biostratigraphie et paléobiogéographie. In *L'Evolution des Oiseaux d'après le témoignage des fossiles. Table Ronde international du CNRS, Lyon-Villeurbanne, 18–21 September 1985*, Docum. Lab. géol. Lyon, no. 99 (Co-

ordinator C. Mourer-Chauviré), pp. 89–135. Lab. geol. Lyon, Lyons.

Olson, S. L. (1979). Multiple origins of the Ciconiiformes. *Proc. colonial Waterbird Group 1978*, 165–70.

Olson, S. L. (1985). The fossil record of birds. In *Avian biology*, Vol. 8 (ed. D. S. Farner *et al.*), pp. 79–256. Academic Press, Orlando, Florida.

Olson, S. L. and Feduccia, A. (1980). Relationship and evolution of flamingos (Aves: Phoenicopteridae). *Smithson. Contrib. Zool.*, **323**, 1–24.

Peters, D. S. (1983). Die 'Schnepfenralle' *Rhynchaeites messelensis* Wittich 1894 ist ein Ibis. *J. Orn.*, **124** (1), 1–27.

Peters, D. S. (1985). Ein neuer Segler aus der Grube Messel und seine Bedeutung für den Status der Aegialornithidae (Aves: Apodiformes). *Senckenbergiana Lethaea*, **66** (1–2), 143–64.

Peters, D. S. (1987a). Mechanische Unterschiede paläognather und neognather Vogelschädel. *Natur u. Museum*, **117** (6), 173–82.

Peters, D. S. (1987b). Eine 'Phorusrhacide' aus dem Mittel-Eozän von Messel (Aves: Gruiformes, Cariamae). In *L'Evolution des Oiseaux d'après le témoignage des fossiles. Table Ronde international du CNRS, Lyon-Villeurbanne, 18–21 September 1985*, Docum. Lab. géol. Lyon, no. 99 (Co-ordinator C. Mourer-Chauviré), pp. 71–87. Lab. geol. Lyon, Lyons.

Peters, D. S. (1987c). *Juncitarsus merkeli* n. sp. stützt die Ableitung der Flamingos von Regenpfeifervögeln (Aves: Charadriiformes: Phoenicopteridae). *Cour. Forsch.-Inst. Senckenberg*, **97**, 141–55.

Peters, D. S. (1989a). Ein vollständiges Exemplar von *Palaeotis weigelti*. *Cour. Forsch-Inst. Senckenberg*, **107**, 223–33.

Peters, D. S. (1989b). Fossil birds from the oil shale of Messel (lower Middle Eocene, Lutetian). *Acta XIX Congressus Internationalis Ornithologici*, Vol. 2, pp. 2056–64. University of Ottawa Press.

Peters, D. S. (1991). Zoogeographical relationships of the Eocene avifauna from Messel (Germany). *Acta XX Congressus*

Internationalis Ornithologici, Vol. 1, pp. 572–7. Ornithological Congress Trust Board, Wellington, New Zealand.

Peters, D. S. (in press). A new species of owl (Aves: Strigiformes from the Middle Eocene Messel oil shale. LACM Special Publ. Avian Paleont.

Wittich, E. (1898). Beiträge zur Kenntnis der Messeler Braunkohle und ihrer Fauna. *Abh. geol. Landesanstalt*, **3**, 79–147.

12 The marsupials: inconspicuous opossums

Crochet, J. Y. (1980). *Les marsupiaux du tertiaire d'Europe*. Editions de la Fondation Singer-Polignac, Paris.

Crochet, J. Y. (1986). *Kasserinotherium tunisiense* nov. gen., nov. sp., troisième marsupial découvert en Afrique (Eocène inférieur de Tunisie). *C.R. Acad. Sci. Paris*, **302** (Sér. II (14)), 923–6.

Crochet, J. Y. and Sigé, B. (1983). Les mammifères montiens de Hainin (Paléocène de Belgique). Part III: Marsupiaux. *Palaeovertebrata*, **13** (3), 51–64.

Gabunia, L. K., Shevyreva, N. C., and Gabunia, U. D. (1985). Über den ersten Fund eines Marsupialiers in Asien. *Dokl. Akad. Nauk SSSR*, **281** (3), 684–5. [In Russian.]

Koenigswald, W. von (1982). Die erste Beutelratte aus dem mitteleozänen Ölschiefer von Messel bei Darmstadt. *Natur u. Museum*, **112**, 41–8.

Storch, G. (1986). Die Säuger von Messel: Wurzeln auf vielen Kontinenten. *Spektrum der Wiss.*, **1986** (6), 48–65.

Woodburne, M. O. and Zinsmeister, W. J. (1984). The first land mammal from Antarctica and its biogeographic implications. *J. Paleont.*, **58** (4), 913–48.

13 Primitive insectivores, extraordinary hedgehogs, and 'long-fingers'

Buchholz, H. (1986). Die Höhle eines Spechtvogels aus dem Eozän von Arizona, USA (Aves, Piciformes). *Verh. naturwiss. Ver. Hamburg, N.F.*, **28**, 5–25.

Cartmill, M. (1974). *Daubentonia*, *Dactylopsila*, woodpeckers and

klinorhynchy. In *Prosimian biology* (ed. R. D. Martin and G. A. Doyle), pp. 655–69. Duckworth, London.

Heller, F. (1935). *Amphilemur eocaenicus* n.g. et n. sp., ein primitiver Primate aus dem Mitteleozän des Geiseltales bei Halle a. S. *Nova Acta Leopoldina N.F.*, **2** (3/4), 293–314.

Hürzeler, J. (1947). *Alsaticopithecus leemanni* nov. gen. nov. spec., ein neuer Primate aus dem unteren Lutétien von Bucksweiler im Unterelsass. *Ecl. geol. Helv.*, **40** (2), 343–56.

Jaeger, J. J. (1970). Pantolestidae nouveaux (Mammalia, Insectivora) de l'Eocène moyen de Bouxwiller (Alsace). *Palaeovertebrata*, **3**, 63–82.

Koenigswald, W. von (1980). Das Skelett eines Pantolestiden (Proteutheria, Mamm.) aus dem mittleren Eozän von Messel bei Darmstadt. *Paläont. Z.*, **54**, 267–87.

Koenigswald, W. von (1983). Der erste Pantolestide (Proteutheria, Mammalia) aus dem Eozän des Geiseltals bei Halle. *Z. geol. Wiss.*, **11**, 781–7.

Koenigswald, W. von (1987*a*). Die Fauna des Ölschiefers von Messel. In *Fossilien der Messel-Formation* (ed. R. Heil *et al.*). Hessisches Landesmuseum, Darmstadt.

Koenigswald, W. von (1987*b*). Apatemyiden Skelette aus dem Mitteleozän von Messel und ihre paläobiologische Aussage. *Carolinea*, **45**, 31–5.

Koenigswald, W. von (1987*c*). Ein zweites Skelett von *Buxolestes* (Pantolestidae, Proteutheria, Mammalia) aus dem Mitteleozän von Messel bei Darmstadt. *Carolinea*, **45**, 36–42.

Koenigswald, W. von (1987*d*). The ecological niche of early Tertiary apatemyids—an extinct group of mammals. *Nature*, **326**, 595–7.

Koenigswald, W. von (1990). Die Paläobiologie der Apatemyiden (Insectivora s.l.) und die Ausdeutung der Skelettfunde von *Heterohyus nanus* aus dem Mittel-eozän von Messel bei Darmstadt. *Palaeontographica*, **A210** (1–3), 41–77.

Koenigswald, W. von and Storch, G. (1983). *Pholidocercus hassiacus*, ein

Amphilemuride aus dem Eozän der 'Grube Messel' bei Darmstadt (Mammalia, Lipotyphla). *Senckenbergiana Lethaea*, **64**, 447–95.

Koenigswald, W. von and Storch, G. (1987). *Leptictidium tobieni* n. sp., ein dritter Pseudorhyncocyonide (Proteutheria, Mammalia) aus dem Eozän von Messel. *Cour. Forsch-Inst. Senckenberg*, **91**, 107–16.

Krishtalka, L. and Setoguchi, T. (1977). Paleontology and geology of the Badwater Creek Area, central Wyoming. *Ann. Carnegie Mus.*, **46**, 71–99.

Maier, W. (1977). *Macrocranion tupaiodon* Weitzel 1949—ein igelartiger Insektivor aus dem Eozän von Messel und seine Beziehungen zum Ursprung der Primaten. *Z. Systematik u. Evol.-Forsch.*, **15**, 311–18.

Maier, W. (1979). *Macrocranion tupaiodon*, an adapisoricid (?) insectivore from the Eocene of 'Grube Messel' (Western Germany). *Paläont. Z.*, **53**, 38–62.

Maier, W., Richter, G., and Storch, G. (1986). *Leptictidium nasutum*—ein archaisches Säugetier aus Messel mit aussergewöhnlichen biologischen Anpassungen. *Natur u. Museum*, **116**, 1–19.

Matthew, W. D. (1909). The carnivora and insectivora of the Bridger Basin, Middle Eocene. *Mem. Am. Mus. Nat. Hist.*, **9** (4), 291–567.

McKenna, M. C. (1963). Primitive Paleocene and Eocene Apatemyidae (Mammalia, Insectivora) and the primate–insectivore boundary. *Am. Mus. Novitates*, **2160**, 1–39.

Moeller, H. (1976). *Dactylopsila trivigata* (Phalangeridae) Klettern und Fressen. *Publ. Wiss. Film*, **9** (2), 129–38.

Richter, G. (1987). Untersuchungen zur Ernährung eozäner Säuger aus der Fossilfundstätte Messel bei Darmstadt. *Cour. Forsch-Inst. Senckenberg*, **91**, 1–33.

Russell, D. E., Louis, P., and Savage, D. E. (1975). Les Adapisoricidae de l'Eocène Inférieur de France. Réévaluation des formes considérées affines. *Bull. Mus. nat. Hist. natur., Science de la Terre*, **54**, 129–79.

Russell, D. E., Godinot, M., Louis, P., and Savage, D. E. (1979). Apatotheria

(Mammalia) de l'Éocène inférieur de France et de Belgique. *Bull. Mus. nat. Hist. natur.*, **3**, 203–43.

Stehlin, H. G. (1916). Die Säugetiere des schweizerischen Eocaens. Critischer Catalog der Materialien. *Abh. Schweizer paläont. Ges.*, **41** (7), 1299–552.

Storch, G. and Haubold, H. (1985). *Macrocranion tupaiodon* (Mammalia, Lipotyphla) aus dem Mittel-Eozän des Geiseltals bei Halle. *Z. geol. Wiss. Berlin*, **13**, 727–30.

Storch, G. and Lister, A. M. (1985). *Leptictidium nasutum*, ein Pseudorhyncocyonide aus dem Eozän der 'Grube Messel' bei Darmstadt. (Mammalia, Proteutheria.) *Senckenbergiana Lethaea*, **66**, 1–37.

Tobien, H. (1962). Insectivoren (Mamm.) aus dem Mitteleozän (Lutetium) von Messel bei Darmstadt. *Notizbl. hess. L.-Amt Bodenforsch.*, **90**, 7–41.

West, R. M. (1973). *Review of the North American Eocene and Oligocene Apatemyidae (Mammalia, Insectivora)*, Special Publication of the Museum of Texas Technical University, no. 3, pp. 1–42. Texas Technical University, Lubbock, Texas.

14 Bats: already highly specialized insect predators

Habersetzer, J. (1986). Vergleichende flügelmorphologische Untersuchungen an einer Fledermausgesellschaft in Madurai. In *Biona reports. 5. Bat flight—Fledermausflug* (ed. W. Nachtigall), pp. 75–104. G. Fischer, Stuttgart.

Habersetzer, J. and Storch, G. (1987). Klassifikation und funktionelle Flügelmorphologie paläogener Fledermäuse (Mammalia, Chiroptera). *Cour. Forsch-Inst. Senckenberg*, **91**, 117–50.

Habersetzer, J. and Storch, G. (1988). Grube Messel: akustische Orientierung der ältesten Fledermäuse. *Spektrum der Wissenschaft*, **7**, 12–14.

Habersetzer, J. and Storch, G. (1990). Ecology and echolocation of the Eocene Messel bats. In *European bat research 1987* (ed.

V. Hanak, T. Horacek, and J. Gaisler), pp. 213–33. Charles University Press, Prague.

Habersetzer, J., Richter, G., and Storch, G. (1989). Paleoecology of the Middle Eocene Messel bats. *Abstract, 5th International Theriol. Congr., Rome*, Vol. 2, pp. 629–30.

Habersetzer, J., Richter, G., and Storch, G. (in press). Paleoecology of the Middle Eocene Messel bats. *Histor. Biol.*

Habersetzer, J., Schuller, G., and Neuweiler, G. (1984). Foraging behaviour and Doppler shift compensation in echolocating hipposiderid bats, *Hipposideros bicolor* and *Hipposideros speoris*. *J. comp. Physiol.*, **155**, 559–67.

Jepsen, G. L. (1966). Early Eocene bat from Wyoming. *Science*, **154**, 1333–9.

Neuweiler, G., Bruns, V., and Schuller, G. (1980). Ears adapted for the detection of motion, or how echolocating bats have exploited the capacities of the mammalian auditory system. *J. Acoust. Soc. Am.*, **68**, 741–53.

Novacek, M. (1985). Evidence for echolocation in the oldest known bats. *Nature*, **315** (6015), 140–1.

Revilliod, P. (1917). Fledermäuse aus der Braunkohle von Messel bei Darmstadt. *Abh. Grossherzog. Hess. Geol. L.-Anst. Darmstadt*, **7** (2), 157–201.

Richter, G. and Storch, G. (1980). Beiträge zur Ernährungsbiologie eozäner Fledermäuse aus der 'Grube Messel'. *Natur u. Museum*, **110** (12), 353–67.

Russell, D. E. and Sigé, B. (1970). Révision des chiroptères lutetiens de Messel (Hessen, Allemagne). *Palaeovertebrata*, **3** (4), 83–182.

Smith, J. D. and Storch, G. (1981). New Middle Eocene bats from 'Grube Messel' near Darmstadt, W.-Germany. *Senckenbergiana biologica*, **61**, 153–67.

Storch, G. (1989). Die eozänen Fledermäuse von Messel—frühe Zeugen der Stammesgeschichte. *Laichinger Höhlenfreund*, **24**, 21–30.

Storch, G. and Habersetzer, J. (1988). *Archaeonycteris pollex* (Mammalia, Chiroptera), eine neue Fledermaus aus dem Eozän der Grube Messel bei Darmstadt. In

Der eozäne Messelsee—Eocene Lake Messel (ed. J. L. Franzen and M. Michaelis), Cour. Forsch-Inst. Senckenberg, **107**, 263–73.

van Valen, L. (1979). The evolution of bats. *Evol. Theory*, **4**, 103–21.

Vater, M., Feng, A. S., and Betz, M. (1985). An HRP-study of the frequency–place map of the horseshoe bat cochlea: morphological correlates of the sharp tuning to a narrow frequency band. *J. comp. Physiol.*, **A157**, 671–86.

15 Our closest relatives: the primates

Franzen, J. L. (1983). *Ein neuer Primate aus dem Eozän von Messel*, Paläont. Ges., no. 53, Jahresvers., Programm u. Kurzfassungen der Vorträge. Paläontologische Gesellschaft, Mainz.

Franzen, J. L. (1987). Ein neuer Primate aus dem Mitteleozän der Grube Messel (Deutschland, S-Hessen). In *Forschungsergebnisse zu Grabungen in der Grube Messel bei Darmstadt* (ed. S. Schaal), Cour. Forsch-Inst. Senckenberg, **91**, 151–87.

Franzen, J. L. (1988). Ein weiterer Primatefund aus der Grube Messel bei Darmstadt. In *Der eozäne Messelsee— Eocene Lake Messel* (ed. J. L. Franzen and M. Michaelis), Cour. Forsch-Inst. Senckenberg, **107**, 275–89.

Franzen, J. L. (1989). The oldest primate hands. Additional remarks and observations. In *The hands of primates* (ed. H. Preuschoft and D. Chwers). Editrice 'Il Sedicesimo', Florence.

Koenigswald, W. von (1979). Ein Lemurenrest aus dem eozänen Ölschiefer der Grube Messel bei Darmstadt. *Paläont. Z.*, **53** (1/2), 63–76.

Koenigswald, W. von (1985). Der dritte Lemurenrest aus dem mitteleozänen Ölschiefer der Grube Messel bei Darmstadt. *Carolinea*, **42**, 145–8.

Koenigswald, W. von and Wuttke, M. (1987). Zur Taphonomie eines unvollständigen Skelettes von *Lepticitidium nasutum* aus dem Ölschiefer von Messel. *Geol. Jb. Hessen*, **115**, 65–79.

Lippmann, H. G. and Wiemer, G. (1979).

Bergung und Präparation von Fossilien aus der Grube Messel unter Berücksichtigung eines Primatenfundes. *Der Präparator*, **25** (1), 3–13.

16 Pangolins: almost unchanged for 50 million years

Emry, R. J. (1970). A North American Oligocene pangolin and other additions to the Pholidota. *Bull. Amer. Mus. Nat. Hist.*, **142**(6), 455–510.

Koenigswald, W. von, Richter, G., and Storch, G. (1981). Nachweis von Hornschuppen bei *Eomanis waldi* aus der 'Grube Messel' bei Darmstadt (Mammalia, Pholidota). *Senckenbergiana Lethaea*, **61**, 291–8.

Richter, G. (1987). Untersuchungen zur Ernährung eozäner Säuger aus der Fossilfundstätte Messel bei Darmstadt. *Cour. Forsch-Inst. Senckenberg*, **91**, 1–33.

Starck, D. (1941). Zur Morphologie des Primordialkraniums von *Manis javanica* Desm. *Jb. Morphol. mikroskop. Anat.*, *Part 1*, **86**, 1–122.

Storch, G. (1968). Die Kaumuskulatur des Weissbauch-Schuppentiers, *Manis tricuspis* (Mammalia). *Senckenbergiana biologica*, **49**, 423–7.

Storch, G. (1978*a*). Ein Schuppentier aus der Grube Messel—zur Paläobiologie eines mitteleozänen Maniden. *Natur u. Museum*, **108** (10), 301–7.

Storch, G. (1978*b*). *Eomanis waldi*, ein Schuppentier aus dem Mittel-Eozän der 'Grube Messel' bei Darmstadt (Mammalia: Pholidota). *Senckenbergiana Lethaea*, **59** (4/6), 503–29.

17 The ant-eater *Eurotamandua*: a South American in Europe

De Jong, W. W., Zweers, A., Joysey, K. A., Gleaves, J. T., and Boulter, D. (1985). Protein sequence analysis applied to Xenarthran and Pholiodote phylogeny. In *The evolution and ecology of armadillos, sloths, and vermilinguas* (ed. G. G. Montgomery), pp. 65–76. Smithsonian Institute Press, Washington, DC.

Montgomery, G. G. (1985). Impact of vermilinguas (*Cyclopes, Tamandua*:

Xenarthra = Edentata) on arboreal ant population. In *The evolution and ecology of armadillos, sloths, and vermilinguas* (ed. G. G. Montgomery), pp. 351–63. Smithsonian Institute Press, Washington, DC.

Sarich, V. M. (1985). Xenarthran systematics: albumin immunological evidence. In *The evolution and ecology of armadillos, sloths, and vermilinguas* (ed. G. G. Montgomery), pp. 77–81. Smithsonian Institute Press, Washington, DC.

Storch, G. (1981). *Eurotamandua joresi*, ein Myrmecophagide aus dem Eozän der 'Grube Messel' bei Darmstadt (Mammalia: Xenarthra). *Senckenbergiana Lethaea*, **61**, 247–89.

Storch, G. (1986). Die Sänger von Messel: Wurtelm auf viden Kontinenten. *Spektrum der Wissenschaft*, **6**, 48–65.

Storch, G. and Habersetzer, J. (1991). Rückverlagerte Choanen und akzessorische Bulla tympanica bei rezenten Vermilingua und *Eurotamandua* aus dem Eozän von Messel (Mammalia: Xenarthra). *Z. Saugetierkunde*, **56**, 257–71.

Storch, G. and Haubold, H. (1989). Additions to the Geiseltal mammalian faunas, Middle Eocene: Didelphidae, Nyctitheriidae, Myrmecophagidae. *Palaeovertebrata*, **19**, 95–114.

18 Rodents: at the start of a great career

Hartenberger, J. L. (1968). Les Pseudosciuridae (Rodentia) de l'Eocène moyen et le genre *Masillamys* Tobien. *C.R. Acad. Sci. Paris*, **267D**, 1870–1920.

Haupt, O. (1911). *Propalaeotherium* cf. *rollinati* Stehlin aus der Braunkohle von Messel bei Darmstadt. *Notizbl. Ver. Erdkde.*, **32** (4), 59–70.

Reisinger, E. (1982). Ein fossiles Säugetier aus Messel, Beschreibung und funktionelle Interpretation in ihrer methodischen Problematik. Unpublished thesis, University of Frankfurt, Frankfurt-am-Main.

Schmidt-Kittler, N. and Storch, G. (1985). Ein vollständiges Theridomyiden-Skelett (Mammalia: Rodentia) mit Rennmaus-Anpassungen aus dem Oligozän von

Céreste, S-Frankreich. *Senckenbergiana Lethaea*, **66** (1/2), 89–109.

Tobien, H. (1954). Nagerreste aus dem Mitteleozän von Messel bei Darmstadt. *Notizbl. hess. L.-Amt Bodenforsch.*, **82**, 13–29.

Weitzel, K. (1949). Neue Wirbeltiere (Rodentia, Insectivora, Testudinata) aus dem Mitteleozän von Messel bei Darmstadt. *Abh. senckenberg naturforsch. Ges.*, **480**, 1–24.

Wilde, V. (1989). Untersuchungen zur Systematik der Blattreste aus dem Mitteleozän der Grube Messel bei Darmstadt (Hessen, Bundesrepublik Deutschland). *Cour. Forsch-Inst. Senckenberg*, **115**, 1–213.

Wood, A. E. (1976). The Paramyid rodent *Ailuravus* from the Middle and Late Eocene of Europe and its relationships. *Palaeovertebrata*, **7** (1–2), 117–49.

19 Carnivores: agile climbers and prey catchers

Feist, O. and Springhorn, R. (1979). *Leben vor 50 Millionen Jahren. Fossilien der Grube Messel (Sammlung Feist)*, exhibition catalogue. Lippisches Landesmuseum, Detmold.

Savage, R. J. G. and Long, M. (1986). *Mammal evolution: an illustrated guide*. British Museum of Natural History, London.

Springhorn, R. (1980). *Parodectes feisti*, der erste Miacide (Carnivora, Mammalia) aus dem Mittel-Eozän von Messel. *Paläont. Z.*, **54** (1/2), 171–98.

Springhorn, R. (1982). Neue Raubtiere (Mammalia: Creodonta et Carnivora) aus dem Lutetium der Grube Messel (Deutschland). *Palaeontographica*, **A179** (4–6), 105–41.

Springhorn, R. (1985). Zwei neue Skelette von *Miacis* ? *kessleri* (Mammalia, Carnivora) aus den lütetischen Ölschiefern der 'Grube Messel'. *Senckenbergiana Lethaea*, **66** (1/2), 121–41.

Springhorn, R. (1988). Ein weiteres Individuum von *Proviverra edingeri* (Mammalia, Creodonta) aus dem Mittel-Eozän von Messel. *Senckenbergiana Lethaea*, **68**, 371–91.

20 The arboreal *Kopidodon*, a relative of primitive hoofed animals

Koenigswald, W. von (1983). Skelettfunde von *Kopidodon* (Condylarthra, Mammalia) aus dem mittelcozänen Ölschiefer von Messel bei Darmstadt. *N. Jb. Geol.-Palaeont. Abh.*, **167**, 1–39.

Russell, D. E. (1964). Les mammifères Paléocène d'Europe. *Mém. Mus. Nat. Hist. Natur., N.S.*, **C13**, 1–324.

Russell, D. E. and Godinot, M. (1988). The Paroxyclaenidae (Mammalia) and a new form from the early Eocene of Palette, France. *Paläont. Z.*, **62**(3/4), 319–23.

Russell, D. E. and McKenna, M. C. (1961). Étude de *Paroxyclaenus*, Mammifère des phosphorites du Quercy. *C.R. Bull. Soc. Geol. France*, **3** (Ser. 7), 274–82.

Thenius, E. (1969). *Phylogenie der Mammalia, Stammesgeschichte der Säugetiere einschliesslich der Hominiden.* W. de Gruyter, Berlin.

Tobien, H. (1969). *Kopidodon* (Condylarthra, Mammalia) aus dem Mittelcozän (Lutetium) von Messel bei Darmstadt (Hessen). *Notizbl. hess. L.-Amt Bodenforsch.*, **97**, 7–37.

21 The Messel horse show, and other odd-toed ungulates

Camp, C. L. and Smith, N. (1942). Phylogeny and functions of the digital ligaments of the horse. *Mem. Univ. Calif.*, **13**, 65–122.

Fischer, K.-H. (1964). Die tapiroiden Perissodactylen aus der eozänen Braunkohle des Geiseltales. *Geologie*, **13** (Suppl. 45), 1–101.

Franzen, J. L. (1968). Revision der Gattung *Palaeotherium* Cuvier 1804 (Palaeotheriidae, Perissodactyla, Mammalia). Unpublished Ph.D. dissertation, University of Freiburg.

Franzen, J. L. (1976). Senckenbergs Grabungskampagne 1975 in Messel: Erste Ergebnisse und Ausblick. *Natur u. Museum*, **106** (7), 217–23.

Franzen, J. L. (1977). *Urpferdchen und Krokodile, Messel vor 50 Millionen Jahren*, Kleine Senckenbergreihe, no. 7. Forschungsinstitut Senckenberg, Frankfurt-am-Main.

Franzen, J. L. (1981). *Hyrachyus minimus* (Mammalia, Perissodactyla, Helaletidae) aus den mitteleozänen Ölschiefern der 'Grube Messel' (Deutschland, S-Hessen). *Senckenbergiana Lethaea*, **61** (3/6), 371–6.

Franzen, J. L. (1983). Senckenberg-Grabungen 1982 in der Grube Messel. *Natur u. Museum*, **113** (5), 148–51.

Franzen, J. L. (1984). Die Stammesgeschichte der Pferde in ihrer wissenschaftshistorischen Entwicklung. *Natur u. Museum*, **114** (6), 149–62.

Franzen, J. L. (1986). Sektion Paläoanthropologie II (Fossile Großsäuger und deren Lagerstätten). In *Wissenschaftlicher Jahresbericht 1985 des Forschungsinstituts Senckenberg, Frankfurt am Main*, (ed. W. Ziegler). Cour. Forsch-Inst. Senckenberg, **85**, 299–307.

Franzen, J. L. (1989). Origin and systematic position of the Palaeotheriidae. In *The evolution of perissodactyls* (ed. D. R. Prothero and R. M. Schoch), pp. 102–8. Oxford University Press.

Franzen, J. L. (1990). *Hallensia* (Mammalia, Perissodactyla) aus Messel und dem Pariser Becken sowie Nachträge aus dem Geiseltal. *Bull. Inst. R. Sci. Natur. Belgique* (in press).

Franzen, J. L., Weber, J., and Wuttke, M. (1982). Senckenberg-Grabungen in der Grube Messel bei Darmstadt. 3. Ergebnisse 1979–1981. *Cour. Forsch-Inst. Senckenberg*, **54**, 1–118.

Godinot, M. (1982). Aspects nouveaux des échanges entre les faunes mammaliennes d'Europe et d'Amérique du Nord à la base de l'Eocene. *Gébios, Mém. spécial*, **6**, 403–12.

Haupt, O. (1911). *Propalaeotherium* cf. *rollinati*, Stehlin, aus der Braunkohle von Messel bei Darmstadt. *Notizbl. Ver. Erdkde. u. grossh. geol. L.-Anst.*, **32** (4), 59–70.

Heil, R. and Koenigswald, W. von (1979). Funde aus den Messeler Schichten. In *Fossilien der Messeler Schichten* (ed. R. Heil, W. von Koenigswald, and H. G. Lippman), pp. 39–88. Hessisches Landesmuseum, Darmstadt.

Hooker, J. J. (1980). The succession of *Hyracotherium* (Perissodactyla, Mammalia) in the English early Eocene. *Bull. Br. Mus. Nat. Hist. (Geol.)*, **33** (2), 101–14.

Koenigswald, W. von (1987). Die Fauna des Ölschiefers von Messel. In *Fossilien der Messel-Formation* (ed. R. Heil *et al.*), pp. 71–142. Hessisches Landesmuseum, Darmstadt.

Koenigswald, W. von and Schaarschmidt, F. (1983). Ein Urpferd aus Messel, das Weinbeeren frass. *Natur u. Museum*, **113** (3), 79–84.

Kowalevsky, W. O. (1876). Monographie der Gattung *Anthracotherium* Cuv. und Versuch einer natürlichen Classifikation der fossilen Hufthiere. *Palaeontographica (N.F.)*, **22**, 133–347.

Krumbiegel, G., Rüffle, L., and Haubold, H. (1983). *Das eozäne Geiseltal, ein mitteleuropäisches Braunkohlenvorkommen und seine Tier- und Pflanzenwelt.* A. Ziemsen, Wittenberg Lutherstadt.

Matthew, W. D. (1926). The evolution of the horse. A record and its interpretation. *Quart. Rev. Biol.*, **1** (2), 139–85.

Prothero, D. R., Manning, E., and Hanson, C. B. (1986). The phylogeny of the Rhinocerotoidea (Mammalia, Perissodactyla). *Zool. J. Linnean Soc.*, **87**, 341–66.

Richter, G. (1987). Untersuchungen zur Ernährung eozäner Säuger aus der Fossilfundstätte Messel bei Darmstadt. In *Forschungsergebnisse zu Grabungen in der Grube Messel bei Darmstadt* (ed. S. Schaal), Cour. Forsch-Inst. Senckenberg, **91**, 1–33.

Simpson, G. G. (1951). *Horses: the story of the Horse Family in the modern world and through sixty million years of history.* Oxford University Press.

Strübel, G. (1974). Vollständig erhaltenes Urpferd aus dem Mitteleozän. *Umschau Wiss. Techn.*, **74** (14), 434.

Strübel, G. (1975). Über ein Urpferd der Gattung *Propalaeotherium* aus der Fossilfundstelle Messel, Hessen. *Oberhess. naturwiss. Z.*, **42**, 5–12.

Sturm, M. (1978). Maw contents of an Eocene horse (*Propalaeotherium*) out of the oil shale of Messel near Darmstadt. *Cour. Forsch-Inst. Senckenberg*, **30**, 120–2.

Thenius, E. (1979). *Die Evolution der Säugetiere*, University of Tübinen, no. 865. G. Fischer, Stuttgart.

Tobien, H. (1988). A *Lophiodon* premolar (Mammalia, Perissodactyla, Tapiromorpha) from the middle Eocene of Messel. In *Der eozäne Messelsee—Eocene Lake Messel*, (ed. J. L. Franzen and W. Michaelis). *Cour. Forsch.-Inst. Senckenberg*, **107**, 299–307.

22 Primitive even-toed ungulates: loners in the undergrowth

Franzen, J. L. (1981). Das erste Skelett eines Dichobuniden (Mammalia, Artiodactyla), geborgen aus mitteleozänen Ölschiefern der 'Grube Messel' bei Darmstadt (Deutschland, S-Hessen). *Senckenbergiana Lethaea*, **61** (3/6), 299–353.

Franzen, J. L. (1983). Ein zweites Skelett von *Messelobunodon* (Mammalia, Artiodactyla, Dichobunidae) aus der 'Grube Messel' bei Darmstadt (Deutschland, S-Hessen). *Senckenbergiana Lethaea*, **64** (5/6), 381–402.

Franzen, J. L. (1988). Skeletons of *Aumelasia* (Mammalia, Artiodactyla, Dichobunidae) from Messel (M. Eocene, West Germany). In *Der eozäne Messelsee—Eocene Lake Messel*, (ed. J. L. Franzen and W. Michaelis). *Cour. Forsch.-Inst. Senckenberg*, **107**, 309–21.

Franzen, J. L. and Krumbiegel, G. (1980). *Messelobunodon ceciliensis* n. sp.—ein neuer Dichobunide aus der mitteleozänen Fauna des Geiseltales bei Halle (DDR). *Z. geol. Wiss.*, **8** (12), 1585–92.

Godinot, M. (1978). Diagnose de trois nouvelles espèces de mammifères du Sparnacien de Provence. *C.R. somm. Soc. géol. France*, **1978** (6), 286–8.

Godinot, M. (1981). Les Mammifères de Rians (Eocène Inférieur, Provence). *Palaeovertebrata*, **10** (2), 43–126.

Richter, G. (1981). Untersuchungen zur Ernährung von *Messelobunodon schaeferi* (Mammalia, Artiodactyla). *Senckenbergiana Lethaea*, **61** (3/6), 355–70.

Richter, G. (1987). Untersuchungen zur Ernährung eozäner Säuger aus der Fossilfundstätte Messel bei Darmstadt. In *Forschungsergebnisse zu Grabungen in der Grube Messel bei Darmstadt* (ed. S. Schaal), *Cour. Forsch-Inst. Senckenberg*, **91**, 1–33.

Rose, K. D. (1982). Skeleton of *Diacodexis*, oldest known artiodactyl. *Science*, **216**, 621–3.

Rose, K. D. (1985). Comparative osteology of North American dichobunid artiodactyls. *J. Paleontol.*, **59** (5), 1203–26.

Sudre, J. (1980). *Aumelasia gabineaudi* n.g. n. sp. nouveau Dichobunidae (Artiodactyla, Mammalia) du gisement d'Aumelas (Hérault) d'age Lutétien terminal. In *Palaeovertebrata, Mém. Jubil. R. Lavocat*, pp. 197–211. *Palaeovertebrata*, Montpellier.

Sudre, J., Russell, D. E., Louis, P., and Savage, D. E. (1983). Les Artiodactyles de l'Eocène inférieur d'Europe. *Bull. Mus. Nat. Hist. Natur., Paris (4)*, **5** (3–4), 281–333, 339–65.

Thewissen, J. G. M., Russell, D. E., Gingerich, P. D., and Hussain, S. T. (1983). A new dichobunid artiodactyl (Mammalia) from the Eocene of North-West Pakistan. Dentition and classification. *Proc. K. nederl. Akad. Wet.*, **B86** (2), 153–80.

Tobien, E. (1980). Ein anthracotherioider Paarhufer (Artiodactyla, Mammalia) aus dem Eozän von Messel bei Darmstadt (Hessen). *Geol. Jb. Hessen*, **108**, 11–22.

Tobien, E. (1985). Zur Osteologie von *Masillabune* (Mammalia, Artiodactyla, Haplobunodontidae) aus dem Mitteleozän der Fossilfundstätte Messel bei Darmstadt (S-Hessen, Bundesrepublik Deutschland). *Geol. Jb. Hessen*, **113**, 5–58.

23 Death and burial of the vertebrates

Franzen, J. L. (1987). Ein neuer Primate aus dem Mitteleozän der Grube Messel (Deutschland, S-Hessen). *Cour. Forsch-Inst. Senckenberg*, **91**, 151–87.

Franzen, J. L., Weber, J., and Wuttke, M. (1982). Senckenberg-Grabungen in der Grube Messel bei Darmstadt. 3. Ergebnisse 1979–81. *Cour. Forsch-Inst. Senckenberg*, **54**, 1–118.

Koenigswald, W. von (1979). Ein Lemurenrest aus dem eozänen Ölschiefer der Grube Messel bei Darmstadt. *Paläont. Z.*, **53**, 63–76.

Koenigswald, W. von and Wuttke, M. (1987). Zur Taphonomie eines unvollständigen Skelettes von *Leptictidium nasutum* aus dem Ölschiefer von Messel. *Geol. Jb. Hessen*, **115**, 65–79.

Richter, G. (1929). Gründung und Aufgaben der Forschungstelle für Meeresgeologie 'Senckenberg' in Wilhelmshaven. *Natur. u. Museum*, **59**, 1–30.

Schäfer, W. (1962). *Aktuo-Paläontologie nach Studien in der Nordsee*. Frankfurt-am-Main.

Weigelt, J. (1927). *Rezente Wirbeltierleichen und ihre paläobiologische Bedeutung*. Weg, Leipzig.

Wuttke, M. (1983). Aktuopaläontologische Studien über den Zerfall von Wirbeltieren. Teil I: Anura. *Senckenbergiana Lethaea*, **64** (5/6), 529–60.

Wuttke, M. (1988). Untersuchungen zur Morphologie, Paläobiologie und Biostratinomie der mitteleozänen Anuren von Messel. Mit einem Beitrag zur Aktuopaläontologie von Anuren und zur Weichteildiagenese der Wirbeltiere von Messel. Unpublished Ph.D. dissertation, University of Mainz.

24 Conservation—dissolution—transformation. On the behaviour of biogenic materials during fossilization

Franzen, J. L., Weber, J., and Wuttke, M. (1982). Senckenberg-Grabungen in der Grube Messel bei Darmstadt. 3. Ergebnisse 1979–81. *Cour. Forsch-Inst. Senckenberg*, **54**, 1–118.

Haupt, O. (1925). Die Paläohippiden der eocänen Süsswasserablagerungen von Messel bei Darmstadt. *Abh. hess. geol. L.-Amt*, **4**, 1–159.

Heil, R. (1964). Kieselschwamm-Nadeln im Ölschiefer der Grube Messel bei Darmstadt. *Notizbl. hess. L.-Amt Bodenforsch.*, **92**, 60–7.

Koenigswald, W. von and Storch, G. (1983). *Pholidocercus hassiacus*, ein Amphilemuride aus dem Eozän der 'Grube Messel' bei Darmstadt (Mammalia, Lipotyphla). *Senckenbergiana Lethaea*, **64**, 447–95.

Koenigswald, W. von and Wuttke, M. (1987).

Zur Taphonomie eines unvollständigen Skelettes von *Leptictidium nasutum* aus dem Ölschiefer von Messel. *Geol. Jb. Hessen*, **115**, 65–79

Koenigswald, W. von, Richter, G., and Storch, G. (1981). Nachweis von Hornschuppen bei *Eomanis waldi* aus der 'Grube Messel' bei Darmstadt (Mammalia, Pholidota). *Senckenbergiana Lethaea*, **61** (3/6), 291–8.

Kott, R. and Wuttke, M. (1987). Untersuchungen zur Morphologie, Paläokologie und Taphonomie von *Retifungus rudens* Rietschel 1970 aus dem Hunsrückschiefer (Bundesrepublik Deutschland). *Geol. Jb. Hessen*, **115**, 11–17.

Leo, R. F. and Barghorn, E. S. (1976). Silicification of wood. *Bot. Mus. Leaflets, Harvard University*, **25** (1), 1–47.

Martini, E. and Rietschel, S. (1978). Lösungserscheinungen an Schwammnadeln im Messeler Ölschiefer (Mittel-Eozän). *Erdöl-Erdgas Z.*, **94**, 94–7.

Müller, W. E. G., Zahn, R. K., and Maidhoff, A. (1982). *Spongilla gutenbergiana* n. sp., ein Süsswasserschwamm aus dem Mittel-Eozän von Messel. *Senckenbergiana Lethaea*, **63** (5/6), 465–72.

Pflug, H. D. (1984). *Die Spur des Lebens, Paläontologie chemisch betrachtet. Evolution, Katastrophen, Neubeginn.* Berlin.

Richter, G. (1985). Chitin abbauende Mikroorganismen aus dem Ölschiefer der Grube Messel bei Darmstadt. *Natur. u. Museum*, **115** (12), 390–3.

Richter, G. and Storch, G. (1980). Beiträge zur Ernährungsbiologie eozäner Fledermäuse aus der 'Grube Messel'. *Natur u. Museum*, **110** (12), 353–67.

Schäfer, W. (1962). *Aktuo-Paläontologie nach Studien in der Nordsee.* Frankfurt-am-Main.

Voigt, E. (1934). Die Fische aus der mitteleozänen Braunkohle des Geiseltales. *Nova Acta Leopoldina N.F.*, **2** (1–2), 1–146.

Voigt, E. (1935). Die Erhaltung von Epithelzellen mit Zellkernen, von Chromatophoren und Corium in fossiler Froschhaut aus der mitteleozänen Braunkohle des Geiseltales. *Nova Acta Leopoldina, N.F.*, **3**, 339–65.

Voigt, E. (1936). Weichteile an Säugetieren aus der eozänen Braunkohle des Geiseltales. *Nova Acta Leopoldina, N.F.*, **4**, 301–10.

Voigt, E. (1937). Weichteile an Fischen, Amphibien und Reptilien aus der eozänen Braunkohle des Geiseltales. *Nova Acta Leopoldina, N.F.*, **5**, 115–42.

Voigt, E. (1938). Fossile rote Blutkörperchen einer Eidechse aus der mitteleozänen Braunkohle des Geiseltales. *Forschungen und Fortschritte*, **14** (3), 32–3.

Voigt, E. (1988). Preservation of soft tissues in the Eocene lignite of the Geiseltal near Halle/S. In *Der eozäne Messelsee—Eocene Lake Messel* (ed. J. L. Franzen and W. Michaelis), Cour. Forsch-Inst. Senckenberg, **107**.

Waksman, S. A. and Iyer, K. R. N. (1933). Contribution to our knowledge of the chemical nature and origin of humus: IV. Fixation of proteins by lignins and formation of complexes resistant to microbial decomposition. *Soil Science*, **36**, 69–82.

Weber, J. and Zimmerle, W. (1985). Pyroclastic detritus in the lacustrine sediments of the Messel Formation. *Senckenbergiana Lethaea*, **66** (1/2), 171–6.

Wuttke, M. (1983). 'Weichteilerhaltung' durch lithifizierte Mikroorganismen bei mitteleozänen Vertebraten aus den Ölschiefern der 'Grube Messel' bei Darmstadt. *Senckenbergiana Lethaea*, **64** (5/6), 509–27.

25 From excavation to exhibition piece

Kühne, W. G. (1961). Präparation von flachen Wirbeltierfossilien auf künstlicher Matrix. *Paläont. Z.*, **35**, 251–2.

Lippmann, H. G. (1987). Präparation von Fossilien aus der Grube Messel. In *Fossilien der Messel-formation* (ed. R. Heil *et al.*), pp. 37–57. Hessisches Landesmuseum, Darmstadt.

Schaarschmidt, F. (1982). Präparation und Untersuchung der eozänen Pflanzenfossilien von Messel bei Darmstadt. *Cour. Forsch-Inst. Senckenberg*, **56**, 59–77.

Thiele-Pfeiffer, H. (1988). Die Mikroflora aus dem mitteleozänen Ölschiefer von Messel bei Darmstadt. *Palaeontographica*, **B211** (1–3), 1–86.

26 Fossilized gut contents: analysis and interpretation

Koenigswald, W. von, Richter, G., and Storch, G. (1981). Nachweis von Hornschuppen bei *Eomanis waldi* aus der 'Grube Messel' bei Darmstadt (Mammalia, Pholidota). *Senckenbergiana Lethaea*, **61** (3/6), 291–8.

Richter, G. (1987). Untersuchungen zur Ernährung eozäner Säuger aus der Fossilfundstätte Messel bei Darmstadt. *Cour. Forsch.-Inst. Senckenberg*, **91**, 1–33.

Richter, G. and Storch, G. (1980). Beiträge zur Ernährungsbiologie eozäner Fledermäuse aus der 'Grube Messel'. *Natur u. Museum*, **110** (12), 353–67.

Voigt, E. (1933). Die Übertragung fossiler Wirbeltierleichen auf Zellulose-Filme, eine neue Bergungsmethode für Wirbeltiere aus der Braunkohle. *Paläont. Z.*, **15** (1), 72–8.

Wilde, V. (1989). Untersuchungen zur Systematik der Blattreste aus dem Mitteleozän der Grube Messel bei Darmstadt (Hessen, Bundesrepublik Deutschland). *Cour. Forsch-Inst. Senckenberg*, **115**, 1–213.

Wuttke, M. (1983). 'Weichteilerhaltung' durch lithifizierte Mikroorganismen bei mitteleozänen Vertebraten aus den Ölschiefern der 'Grube Messel' bei Darmstadt. *Senckenbergiana Lethaea*, **64** (5/6), 509–27.

27 The Messel fauna and flora: a biogeographical puzzle

Antunes, M. T. and Russell, D. E. (1981). Le gisement de Silveirinha (Bas Mondego, Portugal): la plus ancienne faune de Vertébrés éocènes connue en Europe. *C.R. Acad. Sci. Paris, Ser. II*, **293**, 1099–102.

Antunes, M. T., Sigogneau-Russell, D., and Russell, D. E. (1986). Sur quelques dents de Mammifères du Crétacé supérieur de Taveiro Portugal (Note préliminaire). *C.R. Acad. Sci. Paris, Ser. II*, **303** (13), 1247–50.

Buffetaut, E. (1982). A ziphodont mesosuchian crocodile from the Eocene of Algeria and its implications for vertebrate dispersal. *Nature*, **30**, 176–8.

Collinson, M. E. (1986). Früchte und Samen

aus dem Messeler Ölschiefer. *Cour. Forsch-Inst. Senckenberg*, **85**, 217–20.

Collinson, M. E. (1987). The fruit and seed flora from Messel. In *International Messel Symposium Abstracts* 1. Frankfurt-am-Main.

Collinson, M. E. (1988). The special significance of the Middle Eocene fruit and seed flora from Messel, West Germany. *Cour. Forsch.-Inst. Senckenberg*, **107**, 187–97.

Dilcher, D. L. (1987). A comparative study of two similar lauraceous leaf species, one from Messel and one from North America. *International Messel Symposium Abstracts*. Frankfurt-am-Main.

Estravis, C. and Russell, D. E. (1989). Découverte d'un nouveau *Diacodexis* (Artiodactyla, Mammalia) dans l'Eocène inférieur de Silveirunha, Portugal. *Palaeovertebrata*, **19**, 29–44.

Godinot, M. (1978). Diagnoses de trois nouvelles espèces de mammifères du Sparnacien de Provence. *C.R. somm. Soc. géol. France*, **1978** (6), 286–8.

Godinot, M. (1981). Les mammifères de Rians (Eocène inférieur, Provence). *Palaeovertebrata*, **10** (2), 43–126.

Godinot, M. (1982). Aspects nouveaux des échanges entre les faunes mammaliennes d'Europe et d'Amérique du Nord à la base de l'Eocène. *Geobios, mem. spec.*, **6**, 403–12.

Hooker, J. J. (1980). The succession of *Hyracotherium* (Perissodactyla, Mammalia) in the English early Eocene. *Bull. Br. Mus. nat. Hist. (Geol.)*, **33** (2), 101–14.

Houde, P. (1986). Northern hemisphere origin of ostriches: an alternate hypothesis of ratite origins. *Nature*, **324**, 563–5.

Jähnichen, H., Mai, D. H., and Walther, H. (1977). Blätter und Früchte von *Engelhardia* Lesch ex Bl. (Juglandaceae) aus dem europäischen Tertiär. *Feddes Repertorium*, **88**, 323–63.

Koenigswald, W. von (1981). Paläogeographische Beziehungen der Wirbeltierfauna aus der alttertiären Fossillagerstätte Messel bei Darmstadt. *Geol. Jb. Hessen*, **109**, 85–102.

Mai, D. H. (1981). Entwicklung und klimatische Differenzierung der Laubwaldflora Mitteleuropas im Tertiär. *Flora*, **171**, 525–82.

McKenna, M. C. (1975). Fossil mammals and early Eocene North Atlantic land continuity. *Ann. Missouri Bot. Gard.*, **62** (2), 335–53.

McKenna, M. C. (1983). Cenozoic paleogeography of North Atlantic land bridges. In *Structure and development of the Greenland–Scotland ridge* (ed. Bott, Saxov, Talwani, and Thiede), Plenum, New York.

Peters, D. S. (1988). Ein vollstandiges Exemplar von *Paleotis weigelti* (Aves, Palaeognathae). *Cour. Forsch.-Inst. Senckenberg*, **107**, 223–33.

Russell, D. E. (1975). Paleoecology of the Paleocene–Eocene transition in Europe. In *Approaches to primate paleobiology* (ed. F. S. Szalay). Contr. Primat., **5**, 28–61.

Savage, D. E. and Russell, D. E. (1983). *Mammalian paleofaunas of the world*. Addison-Wesley, Reading, Massachusetts.

Sigé, B., Jaeger, J. J., Sudre, J., and Vianey-Liand, M. (1990). *Altiatlasius Koulchii* n.gen. et sp., primate omomyidé du Paléocène supérieur du Maroc, et les origines des Euprimates. *Palaeontographica*, **A214**, 31–56.

Storch, G. (1984). Die alttertiäre Säugetierfauna von Messel—ein paläobiogeographisches Puzzle. *Naturwiss.*, **71**, 227–33.

Storch, G. (1986). Die Säuger von Messel: Wurzeln auf vielen Kontinenten. *Spektrum der Wissenschaft*, **1986** (6), 48–65.

West, R. M. and Dawson, M. R. (1978). Vertebrate paleontology and the Cenozoic history of the North Atlantic region. *Polarforsch.*, **48**, 103–19.

Wilde, V. (1989). Untersuchungen zur Systematik der Blattreste aus dem Mitteleozän der Grube Messel bei Darmstadt (Hessen, Bundesrepublik Deutschland). *Cour. Forsch-Inst. Senckenberg*, **115**, 1–213.

Species Index

Page numbers in **bold** refer to illustrations.